Examining Neurocritical Patients

Eelco F. M. Wijdicks

Examining Neurocritical Patients

Springer

Eelco F. M. Wijdicks
Saint Marys Hospital
Mayo Clinic
Rochester, MN
USA

ISBN 978-3-030-69454-8 ISBN 978-3-030-69452-4 (eBook)
https://doi.org/10.1007/978-3-030-69452-4

This Springer imprint is published by the registered company Springer Nature Switzerland AG
The registered company address is: Gewerbestrasse 11, 6330 Cham, Switzerland

Preface

"Yes, but please describe the neurologic exam", is a question I ask often with a wary (and at times weary) voice. Unfortunatuly I may have to ask this question even more often now than in the recent past. Woe betides the hospitalist or neurointensivist who relies on ambiguous imaging and laboratory studies to the exclusion of key clinical pointers. We may not always rely wholly on a detailed enumeration of the patient's findings, but—it should be said—nobody should practice without incorporating the clinical examination.

Neurocritical care is a sui generis specialty. The neurology of neurocritical care obviously differs from the neurology practiced in the outpatient setting and on the ward—it focuses on changes as a result of rapid deterioration. Any patient with an acutely evolving neurologic disease requires multiple assessments to evaluate status and stability. Nothing can be taken for granted. Many disorders have specific neurologic patterns, but few have been adequately captured in commonly used scales and scores. Once amassed, the neurologic findings on examination require some deep thinking. The neurologic evaluation is not a scientific instrument. Neurology is not like that.

If you work in neurosciences intensive care unit—as I do—it seems that when you have seen one patient with an acute brain injury, well, you have seen one patient with an acute brain injury. Generalizations cannot be easily made, and this also applies to clinical examination. Seeing 100 patients with the same disorder may help, but this level of experience is only available to very few. Furthermore, technology is changing the way we examine patients, and there is growing concern that the neurologic examination is giving way to bedside technical tools and artificial intelligence. Imagine a medical world with computers providing a course of action and prognosis (a digital signal to let go?).

While the neurologic examination of a critically ill patient is complicated, it is rarely taught, audited, or debated. This book—a serious attempt to provide what can be called a master class— provides a detailed clinical assessment of the neurocritically ill patient. This systematic examination of the patient's acutely disturbed nervous function came into being due to the contributions of many "greats of modern neurology" and resulted in a precision never before reached. All of it was based on clinico-pathologic methodology, which must remain untouched and firmly engraved in our minds while evaluating these patients. Neurologic findings in acute brain injury have always seemed remarkable to me, and I hope that upon completion of this book, readers will share my fascination. Also, remember this important caveat: do not infer that I understand everything I have ever witnessed in an ICU. (I don't.) At a minimum, I tried to describe what can be seen at the bedside and challenged myself to create a picture in words.

In this work, I will explain why certain neurologic signs appear and provide the fundamentals of localization. Certain situations demand specific structured examinations, such as patients with permanent loss of consciousness and spinal cord compression. Some neurologic findings demand further differentiating neuroradiologic or electrophysiological studies and will be mentioned briefly, without detail. In the past, the distance between tests and clinical examination was a chasm, but now the two are more closely intertwined. Eventually, all have to be integrated logically.

The examination of the acutely ill neurologic patient has its demands, and there are more than a few reminders. We can expect the following issues. First, when pressed for time, a detailed history takes precedence over a detailed examination. We expect to miss important points in the history. Recounting circumstances immediately previous to or following the event falls mostly to laypersons. Answers may be evasive, even non-sequiturs. Overconfidence and framing bias lead to a bad start. Taking a history is critical to a reliable neurologic diagnosis. There are so many clues waiting to be discovered. Second, a number of clinical observations are frequently seen in the neurointensive care unit and often anticipated. Findings and deterioration are disease specific. Third, intensivists and other specialties often expect wonders from a neurologic consultation, but there is only so much we can do. An unresponsive patient for a neurointensivist is sometimes just that — unresponsive. The strength of a neurologist lies in thinking through what is offered, followed by careful organization.

Classification may not follow because neurologic findings are not always classifiable, and many syndromes with eponyms are incomplete or atypical. Moreover, while we may be frustrated by slowly deliberating, chin-stroking physicians not leaving the room until the diagnosis is established, this process must essentially remain in place, even when rapid-fire decisions are needed. Simplification (*reductio ad absurdum*) remains an ensconced basic concept, and this definitively applies to the complex, neurocritically ill patient with so many serious problems in the first place. We need to seek out what we need to see in the patients and not resort to extreme brevity and imprecise assessment. As we will discuss in this book, a hurried, sloppy approach may lead to ordering a series of inappropriate tests leading to even more inappropriate tests that never clarify the issue at hand. The neurointensivist extraordinaire is yet to be defined. However, a well-trained, sensible practitioner can expect to feel confident in the examination of the patient at times and insecure and skeptical at others.

This book contains very few treatment interventions; these can be found elsewhere. Instead, the narrative encourages the reader to concentrate on information gathering, examination, reexamination with change, and localization. However, because some clinical diagnoses may actually be a response to medication, I added some options. This book also contains no bedside technology such as transcranial Doppler, ICU sonography, or pupilometer. Their value for clinical diagnosis is less than we had hoped, and many new technologies require scientific scrutiny.

I also forego any suggestion how to notate examination findings—from + to +++ reflexes to ↑ ↓ for toe responses—the history of neurology notation is amusing, to say the least—and neither do I give in to the current ruthlessness of documentation demands by creating complex templates. You won't find recommendations on neurocritical care billing. In this book, I want to fully concentrate on the essentials of the neurology of neurocritical care.

Clinical examination at the bedside must start with a comprehensive history followed by examination (leading to tests) and communication of findings. Examination at the bedside is not much different than Osler's methods of inspection, palpation, auscultation, and contemplation. We have a number of stimuli that evoke reactions. Strength, sensation, reflexes, and movements (voluntary or involuntary) remain key observations. The cover summarizes what we mostly do—focussing on pupil size and recognizing a wide pupil, interpreting eye position, using pins to test sensation and finding sensory levels, and, of course, the emblematic reflex hammer. We should marvel at the simplicity of the tools and the complexity they can disclose.

A significant change is simulation as an adjunct to or even a replacement for traditional bedside teaching. Simulation centers have propelled education to a new level, and simulation of acute neurology and neurocritical care is achievable. Acute neurology can be simulated as long as scenarios concentrate on managing the disorder rather than portraying neurologic signs. However, this does not necessarily preclude testing and improving the analytical intelligence of the learner. Successful simulations can be accomplished even with complex scenarios such as traumatic brain injury, subarachnoid hemorrhage, stroke, status epilepticus, and neuromuscular respiratory failure. But the practice of systematically organizing findings, eliminating irrelevancies, and localizing lesions cannot be taught in simulation scenarios—at least not satisfactorily. The same arguments apply to telemedicine and, in particular, tele-neuro ICU. Neurointensivists are often asked questions about patients with neurocritical illness admitted to other centers in the USA and abroad. At best, we currently provide consultations imperfectly in an asynchronous form with evaluation of a clinical problem through e-consultation, emails, phone conversations, and occasionally short store-and-forward video snippets of some elements of a physical examination. However, this is generally insufficient. Present technology allows live synchronous, high-quality audio-video history acquisition, clinical examination, review of diagnostic studies, diagnosis, and discussion of the management of patients admitted to intensive care units (or emergency departments). Such a program will provide a wide range of neurologic expertise in acute and urgent settings. We expect there will be benefits not originally anticipated, cordial physician interactions better than hoped for, a measurable overall improvement of the critically ill neurologic patient, and pleased families.

I close the book with a chapter on communication and provide some guiding principles. We may know what we know, but does everyone get what we are saying? Doubt what you hear and verify yourself. Be a benign contrarian. The anatomy of miscommunication remains the most understudied topic in neurocritical care. It is a great feeling when teams are effective due to good communication, but there are serious challenges and, sometimes, brief moments of chaos.

Neurologic examination is poorly understood by many. Many "how to" examination books are wordy and bore the reader. In order to hold interest, I have written mostly in the first person. That is how I was taught and how I like to teach. I deliberately wrote the book in a quick and easy narrative style with controlled levity. I do not think such an approach minimizes nuance.

This book is primarily for budding neurointensivists, intensivists, neurosurgeons, neurosciences nursing staff, and learners in any of the intensive care-related specialties where the real work gets done. This work touches the core of my interest—rethinking the neurologic examination—but tailored, modified, and specialized for neurocritical illness.

Neurologists traditionally have put careful considerations of the examination findings before action and localization before tests. As a testimony, the book hopefully emphasizes the beauty of a clinical neurologic examination and how important it can be.

Rochester, MN, USA Eelco F. M. Wijdicks

Acknowledgments

This book has been expertly edited by Lea Dacy, who, by now, knows my grammatical quirks and missteps. I benefitted greatly by her nuanced editing and numerous suggestions. I cannot thank her enough.

I appreciate Editorial Director Richard Lansing's immediate enthusiasm some years ago when I proposed this book, which he guided to completion. I am thankful for the commitment of Sylvia Johnson and Sreebas Dutta of SPi Technologies India Private Ltd. I acknowledge a great debt to the Mayo Clinic Media Support Services, who were very helpful in illustrating key findings, and I especially appreciate the efforts of medical illustrators Paul Honermann, Joanna King, Seth Lambert, and Timothy Seelinger, who found new ways to express the common findings of neurologic examination. (Eadweard Muybridge was our inspiration with a number of still collages) Thanks again to Jim Rownd, who worked with me on many earlier books and, inspired by Picasso and Monet, made a perfect cover.

I enjoyed writing this book and I now better understand my own objectivity and subjectivity with taking a history, examination and communication of those terribly sick patients. But my problems are a bit yours too, and we all have to figure it out.

While writing this book, some material was published in a much-abridged form as articles in Springer Nature journals, and I appreciate their permission to re-use parts in the current work.

I hope I reach many health professionals with this work. This book greatly benefitted from the residents, fellows, and neurosciences ICU staff with whom I have worked and still work with. I could bounce off ideas and they nudged me to explain and explain. Because of them, the book wrote itself.

This book is dedicated to my family —Coen and Kathryn, Marilou and Rob, and my granddaughter Olivia—and with lots of love to my wife and companion nonpareil Barbara. Thanks to all her goodness and unwavering encouragement, I can do what I like to do.

Contents

1 Taking a History

What happened here? – is precisely what we will say (or think) while walking into a patient room and catching sight of a patient in neurologic distress. We might find ourselves in the emergency room with the intake nursing staff, at a rapid-response call with a resuscitating team, or in an intensive care unit (ICU) with the flight crew as the patient is transferred from a gurney. An established component of the neurocritical medical evaluation is, of course, taking a history (known also as anamnesis). This is not so easily done in an emergent or neurocritical setting. We will soon find out that others have asked the same question. Who will toil through the story? It is a formidable task, especially when patients are confused, aphasic, or, worse, sedated and intubated. The information-seeking healthcare provider relies on accompanying persons (ideally, close family members), but they often must travel separately, arriving in most instances after the patient enters the intensive care unit or the emergency room.

Frankly, these venues in no way resemble the outpatient setting in which the neurologist quietly greets the patient in the waiting room and leads him or her (an alert, fully aware individual) to a comfortable clinic office for a quiet, uninterrupted intake session. The ambience of the emergency room or the Neuro-ICU is often semi-chaotic, and the recounting of the circumstances of the event mostly falls to others. Furthermore, the narrative of the ictus and clinical trajectory may understandably be somewhat emotive. We will miss important points; they accumulate quickly in the heat of the moment.

Although a staple of our professional evaluation, it has been short changed in medical (and most certainly surgical texts). Few books address clinical history-taking in critically ill patients and none, I believe, at length. We rarely have the opportunity to have the patient describe the problem in his or her words. However, data-driven intensivists cannot rely on an incomplete history of chief complaints and its trajectory; they should make the effort to start from square one if information is muddled or contradictory.

Here's the thing. A clinical history consists of determining the onset of symptoms, which symptoms are absent, how severe they were, how they were perceived, and how the symptoms progressed. Taking a history is what we should do first according to educational dogma. This is only partly true. We see patients at arrival and get a first clinical impression, some snippets of what happened, and sometimes even a report from the flight crew. These pieces of information, after a patient is stabilized or resuscitated, are barely enough, and we will seek more from arriving family members or the patient. Clinical history thus never can be fully isolated as a separate task, and we will be influenced by what we see and hear.

This opening chapter addresses all aspects of the medical history: the obstacles, competencies, and recognition of a clinical trajectory from onset

E. F. M. Wijdicks, *Examining Neurocritical Patients*, https://doi.org/10.1007/978-3-030-69452-4_1

to presentation. Examples of the value of a good medical history are shown but also the familiar biases, prejudices, and potential for lost focus and digression when asking questions. The closing chapter is how we communicate all of this and how we prioritize information. We can decry the loss of clinical skills with some justification, but even the most skilled clinical neurointensivist must work from a solid history that includes major tenets such as chief complaint, history of present illness, past medical history, family history, and social and occupational history.

Obstacles and Competency

Obtaining a reliable history requires the collaboration of a bystander (ideally) or a close family member. Summarizing a medical history for the first time places a tremendous burden, often underappreciated by clinicians, on witnesses, who may be derided as "poor historians." This label is applied when families or patients provide vague descriptions, imprecise explanations ("doctor so-and-so said"), and mixed-up timelines. Moreover, neurologists should recognize that families may understand and define medical vocabulary differently. They may use words like "numbness" to describe weakness, that so-called "seizures" may actually be vasovagal collapse, and that "confusion" may be aphasia. More than in non-acute neurologic settings, families are overwhelmed with impaired observational skills. While this, in itself, often causes significant communication lapses, the situation may be further complicated by language and ethnic barriers and communication styles. Unfortunately, the hectic ICU environment may create an additional obstacle to rational explanations. The time needed for a carefully reconstructed timeline is infrequently available; essential information might only be recalled later. Particularly worrisome is the current "cut-and-paste" environment of electronic records and histories that are never questioned or confirmed.

A patient's history as presented in a handoff may be too easily accepted, even when generated in an environment of continuous interruptions. Moreover, miscommunication-related medical errors have been linked to handoffs with communication interruptions [1, 2]. Interruptions lead to poor recall and disagreement about the importance of communicated information. (See Chap. 12 for a more comprehensive analysis of communication problems.)

No scale in neurology takes the patient's history into account. In addition, there is little training in how to recognize cues with attentive listening, how best to communicate, how to avoid distractions in an urgent environment, and how to ask open-ended questions. Errors of commission, as these are known in legal circles, include trusting faulty memory, obtaining erroneous information, and failing to resolve contradictory statements in records or to review prior medical records. Urgent cases require quick histories, but thorough information-gathering (and, if needed, correctives) should come later, when the dust has settled. We tend to forget and move on to the next patient.

Experience is the best teacher on how to take a good history—knowing what to ask and how to redirect when needed. It also requires discernment about what can be ignored or minimized and what is crucially important. It is important to get a clear insight into what has happened, and some suggestions are shown in Table 1.1. Several elements of the history are critical, such as the onset and progress of symptoms. Clinical history can then be further deconstructed looking at more specifics (Table 1.2). Precipitating factors should be considered. For example, drug overdose must be considered in comatose patients with a normal computed tomography (CT) brain and for whom there are no neurologic signs other than coma. In addition to prescribed medications,

Table 1.1 Acute neurohistory in five numbers

#1: Do not make any decisions without hearing at least one version of what has happened.
#2: Remember a history needs verification by a second person.
#3: The necessary components of a clinical history are circumstance, ictus, and early course.
#4: Critical neurologic systems must include initial inquiries about cognition, cranial nerves, motor system, sensory system, and gait.
#5: Consider the possible later trajectories: recovered, stable deficit, improving, worsening, and stuttering.

Table 1.2 Deconstructing a clinical history

Elements	Details
Baseline function (and recent change)	Memory Productivity Mobility Responsibilities Safety
Prior hospitalizations or ED visits	Presumptive diagnosis
Drugs for infection	Antibiotics
Prior vices	Drinking habits Illicit drugs Over the counter
Prior psychiatry	Suicide attempts or considerations
Family history of predisposition	Aneurysms/AVM
Found down	Scene description Outside temperature Visible trauma Need for CPR Prior diabetes and insulin use Stroke clues (atrial fibrillation)

AVM arteriovenous malformation, *CPR* cardiopulmonary resuscitation, *ED* emergency department

Table 1.3 Urgent medication reconciliation

Drugs	Consequences
Anticoagulation (enoxaparin, NOACs)	Urgent reversal
Antibiotics (flagyl, cefepime)	Neurotoxicity
Withdrawal (baclofen; levodopa with carbidopa, opioids)	Treatment for rhabdomyolysis or seizures
Antiepileptics (levetiracetam, phenytoin)	Withdrawal or toxicity
Antidepressants (SSRI)	Serotonin syndrome

NOACs new oral anticoagulants, *SSRIs* selective serotonin reuptake inhibitors

neurointensivists must consider (and test for) other medications or drugs to which the patient might have had access. Patients with myasthenic crises virtually all had a precipitating infection. Guillain–Barré usually develops initially with symptoms that mimic an upper respiratory infection or a gastrointestinal complaint followed by (bloody) diarrhea. Neuromuscular respiratory failure often is worse at night because of a change to a supine position that precipitates diaphragmatic weakness.

High-pressure headaches are worse in the morning and have been attributed to a rise in intracranial pressure (ICP) during the night as a consequence of reclination, mild hypercarbia during sleep caused by respiratory depression, or a decreased cerebrospinal fluid (CSF) absorption. Low-pressure headache worsens with standing, coughing, sneezing, and exertion.

Biases and Pattern Recognition

There are situations in which the team looks at magnetic resonance imaging (MRI) findings and concludes that they "do not make sense." A clinical history may be incompatible with neurologic examination or test results. Examples of discrepancies abound. Red-flag symptoms missed in history-taking may include failure to inquire about fever, thunderclap-onset headache, intravenous drug abuse, empty medication bottles, or prolonged use of corticosteroids (Table 1.3). Essential to the history is documentation of medications the patient was taking or had recently (either deliberately or accidentally) discontinued. Some drugs are notoriously toxic, and others cause serious withdrawal symptoms; therefore, medication reconciliation is absolutely essential for a full history. Inaccurate history-taking leads to failure to order appropriate diagnostic imaging and missing the diagnosis.

And then there is a touchy issue. Although the young and less experienced are seldom bothered by the arrogance of their seniors, we are never absolved from making gross, regrettable errors through overconfidence. Arguably, errors should become less common as available knowledge increases—but not always. Even a seasoned, highly active, involved physician may be fooled by a slightly different presentation. Much of medical practice is pattern recognition, and knowing the patterns is critical to avoiding errors. However, clinical experience does not always reduce error rate; physicians can easily repeat the same mistake time and again. The most common bias is the so-called representativeness restraint. Physicians may miss important signs or try to force everything into a more recognizable scenario. History-taking is subject to confirmation bias, which reflects the tendency to discard contradictory data, to seek out data to confirm one's preconceived idea, and to jump quickly to a diagnosis [3]. Confirmation bias does not allow for consideration of other possibilities and may lead to "anchoring," when a history is cemented too early. Conditions that may seem acute may become chronic and vice versa after more information becomes available. While it is often useful to avoid getting distracted by incongruities, an important observation may occasionally require explanation. Errors are often in judgment and not exclusively procedural [4].

Hearing the Story: First Thoughts

We all think of something when we hear a story and often make provisional diagnoses after introductory comments. Thunderclap headache must be a ruptured aneurysm; ascending weakness after a period of tingling and upper respiratory infection must be Guillain–Barré syndrome; and headache, fever, and drowsiness must be a central nervous system (CNS) infection. Our thinking processes are "causal simplicity" and "unremarkable ordinary." [5] Since medical school, we have been familiar with the saying, "when you hear hoof beats, think horses, not zebras." Most physicians are taught to focus on the likeliest possibilities when making a diagnosis, not the unusual ones. Medicine is rife with "red herrings." But often the easy explanation just fits perfectly, and it feels good to be certain. Over time and after many years of experience, this satisfying feeling will disappear, and our minds will constantly play the Devil's advocate. We become less certain. We know that cookbook medicine does not exist. We know we do not like to be guided by the power of great anecdotes (the last-remembered case phenomenon). Be prepared that sometimes we will never know no matter how much we imagine. We must continue to doubt what we see and take nothing for granted.

When do we think there is an acute neuroemergency, and what are the early clues in the medical history? Here are my ten essential clues. These are (1) any new confusion, disorientation, or new, uncontrolled agitation; (2) acute decline in consciousness from sopor to stupor to coma; (3) any neuromuscular respiratory failure; (4) any acute new involuntary movement; (5) any acute new, persistent or progressing headache; (6) any obvious provoked or spontaneous asymmetry in movement; (7) any sensory sign over a large part of the body; (8) any motor sign that is progressing; (9) any acute inability to stand or walk; and (10) any new speech or language problem. Still, as we work diagnostically through known patterns and potential red flags (Table 1.4), we may decide the whole picture is out of context or aberrant. Often, unusual presentations of diseases require unusual inquiries, and these may come later when neuroimaging or laboratory tests are known. But causes often depend on the context. Coma in some emergency departments (EDs) is often attributable to a drug overdose.

Table 1.4 Red flags in history-taking

Red flags
Failure to recognice fever (meningitis, encephalitis, and epidural spinal abscess)
Failure to recognice immunosuppression and intravenous drug use (fungal or parasitic infections)
Failure to recognize comorbidity (medication errors due to failure of dose adjustment)
Failure to inquire about sphincter syndromes and loss of sensation (signs of cauda equina syndrome)
Failure to appreciate recent use of neuromuscular junction blockers (recently intubated comatose patients)
Failure to appreciate alcohol withdrawal or methanol intoxication (presenting intubated comatose patients with seizures and metabolic acidosis)

Construct a Clinical Trajectory

In the ICU, all is "acute" or "urgent," but the commonly used adjectives (e.g., acute, hyperacute, subacute, and rapidly progressive) likely have different meanings for different people. However, it is useful to view illness in definable events such as time of onset, time to nadir, time in nadir or time with stable deficit, and rate of recovery. Certain neurocritical disorders have predictable clinical trajectory patterns. Before the recently acquired luxury of detailed neuroimaging, neurologists predicted a diagnosis on the clinical trajectory and findings on examination. Many of us still try to practice a priori estimations, and neuroradiologists also use these when interpreting images and estimating a posteriori probabilities.

A history can be more specific when the diagnosis is quickly established. Key

Table 1.5 Key inquiries with major neurocritical illness

Traumatic brain injury	Fall from standing height or motor vehicle accident
	Assault
	Alcohol or drug use
	CPR at site
	Intubation at site
	Retrograde and anterograde amnesia
Middle cerebral artery occlusion	New hemiplegia
	Forced gaze
	Absent speech
	Acute neglect
	Prior face pain (dissection)
	Prior atrial fibrillation (embolus)
Cerebral hematoma	New hemiplegia
	New aphasia
	Hypertensive emergency
	Focal seizure (uncommon)
	Acute numbness (thalamus)
	Found comatose (catastrophic basal ganglia or pons)
	Confused only (caudate nucleus)
	Prior severe headache and OCP (cerebral venous thrombosis)
	Acute vertigo (cerebellum)
Aneurysmal subarachnoid hemorrhage	Split-second, extremely severe, never-before experienced headache or neck pain
	Vomiting
	syncope or brief loss of consciousness
	Seizure (uncommon)
	Acutely confused and nonsensical speech (ACA or MCA aneurysm)
	Visual loss (vitreous hemorrhage)
	Diplopia (third nerve palsy)
Basilar artery occlusion	Acutely comatose
	Alternating hemiparesis
	Diplopia
	Ataxia
Guillain–Barré syndrome	Tingling
	Facial diplegia
	Progressive leg weakness
	Recent dyspnea and orthopnea
Myasthenia gravis	Fluctuation of weakness
	Jaw and head drop
	Severe eyelid droop
	Unable to clear secretions
Encephalitis	Fever
	New aphasia or confusion
	Focal seizures (HSE)
	Arm flexion posturing (anti-NMDAR encephalitis)
	Stroke-like symptoms (VZV encephalitis)
Meningitis/abscess	Myalgias, ear pain, and sore throat
	Fever and vomiting
	Acute confusion and agitation
	Progressive decline in alertness
Acute spinal cord injury	Acute flaccid leg weakness
	Loss of sensation below a certain level
	Full bladder
	Acute constipation
	Vacillating weakness followed by rapid decline (necrotic myelopathy)

CPR cardiopulmonary resuscitation, *OCP* oral contraceptive pills, *ACA* anterior cerebral artery, *MCA* middle cerebral artery, *VZV* Varicella zoster virus, *HSE* herpes simplex encephalitis, *NMDAR* N-methyl D-aspartate receptor

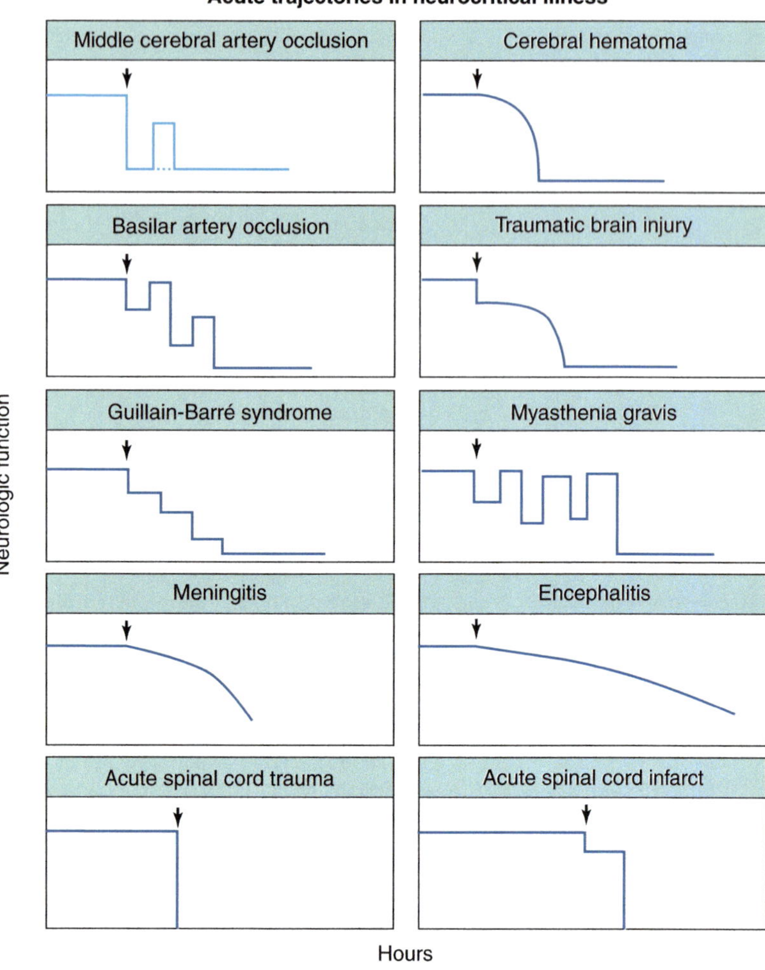

Fig. 1.1 Clinical trajectories for common neurocritical disorders in the first hours/days (not including recovery trajectories)

questions are shown in Table 1.5. Moreover, a good history can provide a clinical trajectory; several examples are shown in Fig. 1.1. Acute occlusions of the middle cerebral artery often suddenly present with a major deficit—but not always. These patients present with a flaccid hemiparesis, forced eye deviation, and muteness of severe dysarthria (left hemisphere), and hemi-body neglect (right hemisphere). A well-known clinical scenario is a patient found after being missing for days, markedly dehydrated, and having severe rhabdomyolysis from pressure-induced muscle necrosis, all unfortunately because the patient was helpless and his situation unknown. Some patients improve substantially only to worsen when the collateral circulation fails [6–8]. The dilemma of whether clot removal will result in marked improvement or cause a worsened deficit (if the clot should break) remains unresolved. Occluded vertebrobasilar arteries often fluctuate markedly in posterior-circulation signs such as ataxia, dysphagia, and dysarthria [9–11].

There are two categories of patients who present with acute basilar artery occlusion [12, 13]. One has a rapid onset with nearly all deficits at presentation and is often found comatose with terminal breathing and posturing. Rhythmical jerking movements may occur unexpectedly and usually last 5 seconds. A common mistake is to regard this presentation as a seizure. The patient may be misdiagnosed as a postictal state, certainly if the brainstem is not sufficiently examined. Embolus to the basilar artery may also present with a "locked-in syndrome" (an awake state with minimal or no ability to respond to the examiner).

The second category of basilar artery occlusion presents with early warnings often related to a clot in the vertebral artery causing a stroke in the cerebellum. These patients start with acute severe vertigo and vomiting; only a careful neurologic examination will bring on a nystagmus, dysmetria, and inability to stand and walk but no weakness. Involvement of the distal (tip) basilar artery, where the superior cerebellar artery and posterior cerebral arteries branch off, may cause ischemia of the mesencephalon, thalami, inferior temporal lobes, and occipital lobes, which may cause signs and symptoms of acute diplopia, unstable gait, reduced or fluctuating consciousness, with moments of vigilance alternating with stupor. These patients develop a rostral brainstem infarction, a syndrome characterized by visual field defects, disorders of vertical gaze, convergence, a skew deviation, and pupillary abnormalities mostly resulting in small and poorly reactive pupils. Behavior abnormalities, including hallucinosis, are common. Visual field defects, including visual perseverations and scintillations in an optic field, have also been described. Many occipital infarcts are bilateral, and cortical blindness may be noted. Although a clot can dissolve and fully resolve [14–16], it is better to retrieve it while there are residual symptoms; clinical history has taught us to anticipate worsening later. Major improvement of clinical signs may not be fully reassuring and becomes important in decisions on endovascular treatment in acute stroke.

Patients with an aneurysmal subarachnoid hemorrhage often appear stable initially, "looking (and feeling) great except for headache," but then acutely or gradually decline into a much worse neurological state. Rebleeding and acute hydrocephalus can pose major risks to the patient on the first day after aneurysmal rupture. Often, it is hard to overlook a rebleeding because the clinical changes are dramatic. Rebleeding may have occurred just before transport or even with transport. The seemingly well-looking patient may suddenly become stuporous or comatose, and new severe hypertension, tachypnea (or apnea), and tachycardia (or brief asystole) may accompany the altered consciousness. Motor responses change, and extensor posturing may occur [17–19]. The clinical trajectory of cerebral hemorrhage often involves more gradual symptoms than ischemic stroke. Small arterial bleeds lead to growing volume under pressure, damaging other arteries when they spread into the perivascular spaces, and tearing penetrating arteries or veins along the way. This domino effect progresses until platelet plugs appear surrounded by walls of red blood cells and fibrin – Fisher's fibrin globes [20]. (A recent hyperacute hemorrhage in the CT scanner fully supports this concept [21].) This pathophysiology has a clear clinical correlate. Although hemorrhages can be suddenly catastrophic (and lethal), neurologic deficits typically appear more gradually, with worsening weakness and new deficits over several hours. Some patients experience a further decline in consciousness and an abrupt change in breathing (from early mass effect and brain-tissue shift). This gradual course is more common in putaminal and thalamic hemorrhages than in cerebellar hemorrhages because the compartments are larger and more accepting of newly increased volume. Not all hemorrhages are arterial; some are venous, classifying more as hemorrhagic infarcts. The majority of patients with a cerebral venous sinus thrombosis present with a gradual onset of headache, a localizing deficit (weakness, aphasia, or visual field cut), focal seizures, or all of the above. We need to ask many questions including a personal or family history of thrombophilia, contraceptive use, recent childbirth, ongoing or recent infections (e.g., otitis media, sinusitis, or mastoiditis), systemic inflammatory disease, malignancy (acute leukemia), recent traumatic brain injury (often minor and trivial), or excessive dehydration caused by combined vomiting and diarrhea [22].

The vast majority of patients with severe traumatic brain injury (TBI) have clinical symptoms of diffuse axonal brain injury. Depending on the impact and the damage to brain parenchyma, patients may be alert and able to follow commands, or they may be combative, agitated, or comatose with abnormal motor responses [23, 24]. The clinical trajectory is determined by worsening hemorrhagic contusions, massive malignant cerebral edema, or further extension of an extra-axial hematoma. Extracranial blood, present either in the subdural or epidural compartment, can also determine the course in traumatic brain injury. The lucid interval after trauma may be indicative of epidural hemorrhages, but blossoming contusions may be more common than has been truly appreciated.

TBI commonly correlates with alcohol intoxication or drug use, thus easily confounding and masking the typical trajectory. High blood-alcohol levels correlate with an increased expansion rate [25]. Patients with TBI may have clinical deterioration from a systemic complication. The risk of early acute respiratory distress syndrome (ARDS) is high. Disseminated coagulopathy can be anticipated with penetrating gunshot wounds and change the presentation and trajectory dramatically.

The two most common acute neuromuscular disorders in the Neuro Intensive Care Unit are Guillain–Barré syndrome (GBS) and myasthenia gravis, but they have markedly different presentations. GBS proceeds in a stepwise fashion, using several days to reach a nadir. Myasthenia gravis, however, may fluctuate from a near-normal examination to marked weakness within the same day. Recognizing fatigable weakness often clinches the diagnosis. Myasthenia gravis patients often have abnormal eye movements, oropharyngeal weakness, and proximal limb weakness. Bulbar function must be specifically queried and examined, as it may determine the need for elective intubation. Myasthenia gravis-based muscle weakness occurs in the masseters, and jaw opening is typically stronger than jaw closure. Progressive neuromuscular transmission failure in the diaphragmatic muscle may occur in 30% of patients with myasthenia gravis at the time of diagnosis. One cause of in-hospital deterioration in myasthenia gravis is significant worsening within the first days of corticosteroid administration. Often, this clinical worsening is observed in patients exposed to corticosteroids for the first time, but it also can occur when the initial doses are high. When we see these patients for the first time, we can expect to see them visibly struggling to breathe or to sit up in bed, maintaining only marginal pulse-oximeter values, despite increasing oxygen requirements. When answering questions, patients appear unable to catch their breath or to resume normal breathing following the simplest exercise. Myasthenic crisis may also involve a serious overdose of cholinergic drugs with sweaty, salivating patients complaining of abdominal cramps and vomiting (in medical parlance, SLUDGE – salivation, lacrimation, urination, defecation, gastrointestinal upset, and emesis).

Traumatic spinal cord injury is acute and complete, although symptoms may improve when the spinal shock phase resolves [26]. Spinal cord infarct is also acute but may initially be gradual before a very sudden loss of motor function, which occurs up to 12 hours after initial onset [27].

CNS infections usually progress gradually, albeit more quickly in acute bacterial meningitis. Nevertheless, patients appearing mildly lethargic or obtunded on initial presentation may become deeply comatose within hours.

After the acute phase, a number of events determine the clinical course in patients with acute brain injury. First, obstructed CSF flow causes further deterioration unrelated to the primary event. Examples include aneurysmal subarachnoid hemorrhage, thalamic ventricle-trapping hemorrhages, cerebellar hematomas obliterating the fourth ventricle, or simply hemorrhage breakthrough into the ventricular system blocking outflow. Second, expansion due to cerebral edema and often driven by increased intracranial pressure can threaten the airway until patients need intubation. A neurosurgical intervention, such as evacuation of a hematoma or ventriculostomy placement, often interrupts a downward spiral in the clinical course. Later clinical trajectories are often determined by the aggressiveness of interventions and not by the natural history alone.

These specific clinical trajectories can often be obtained after a careful history and significantly add to a full description of the clinical picture. It is anticipatory information for attending neurointensivists – forewarned is forearmed.

More Reflections

Cabot and Adams wrote in 1938: "The history is the key to diagnosis .... More errors in diagnosis are traceable to lack of acumen in eliciting or interpreting symptoms than have ever been caused by a failure to hear a murmur, feel a mass, or take an electrocardiogram." [28] It has been known for centuries that a patient history is more than the patient's story and must also include the physician's interpretation [29]. The names patients give to their prior illnesses may be incorrect. Physicians are often required to correct self-diagnoses offered by the patient. The patient history is therefore a construct and not just an expression of the patient's narrative. The final physician history is a careful organization of true facts. The art of history-taking requires knowledge of disease and considerable experience to steer patients and families in the right direction. Test results are often emphasized by the person who provides the history, but physicians are concerned with symptoms. Physical examination cannot be only real path to diagnosis. Taking a clinical history takes time, which is not always available. Cell phones or pagers predictably interrupt the encounter and preclude careful listening. Physicians often redirect opening statements of patients and families but also direct questions toward a specific concern; we have no patience and redirect already after 20 seconds on average [30, 31]. Failures of information-gathering and integration are due to the provider feeling pressed for time. We must close out distractions and force the mind to slow down. Most errors are related to process breakdowns in the patient–practitioner clinical encounter. Diagnostic errors can be traced to cognitive errors by physicians, particularly in synthesizing the available information to identify the correct diagnosis. We may be "saved" by modern neuroimaging, which greatly increases the diagnosis of neurologic disorders, but diagnostic errors have far from disappeared. Better neuroimaging does not replace the need for systematic diagnostic reasoning, also because correct interpretation of brain imaging is very difficult and relies on both detailed clinical information and communication with the neuroradiologist.

Complacency is difficult to address, and I understand why. Physicians tend to generate a working diagnosis almost immediately upon hearing a patient's initial symptom presentation and want to seek a familiar pattern without exploration of other possibilities. *Mostly*, the diagnosis is correct, appropriate tests are ordered, and effective treatment begins. This pattern is common, difficult to change, and, counterintuitively, reassuring to families. However, exceptions must always be considered. According to Berner and Graber, physicians acknowledge the existence of diagnostic error but think the likelihood is less than it really is [32]. Vickrey advises deliberate pursuit of another angle: "Let's play devil's advocate" or "Let's re-review elements of the history." [3]

Deliberately considering a differential diagnosis prior to the final decision is essential to diagnostic reasoning. The most common breakdown points are test ordering and interpretation, performance of the medical history and physical examination, and initiation of consultations. While praising independence, we should also, once in a while, seek opinions from colleagues. Failure to order appropriate tests is the most frequent breakdown.

Getting the history right is a core principle. A phone call to family members or direct communication is essential to acquire a sense of the time course and urgency. Seemingly acute conditions may actually be chronic and vice versa as more information becomes available. Clinical presentations unique to neurointensive care include the comatose patient found down, rapidly progressive weakness, respiratory failure without obvious pulmonary or cardiac triggers, and, inevitably, the mysterious, progressive encephalopathy with abnormal CSF and hard-to-pinpoint MRI abnormalities. Often, with hindsight and a better history, the clinical diagnosis becomes obvious. None of us will forget the CSF we should have requested to diagnose meningitis had we but known of the looming infection and fever. If we had queried another patient about his unremitting back pain before he lapsed into unresponsive septic shock, we might have ordered an MRI of the spine to diagnose an epidural abscess. Painful, acute double vision indicates an urgent need for vascular studies and contrast-enhanced MRI to demonstrate a growing and unstable cerebral aneurysm. We expect cerebral vasospasm to follow a ruptured aneurysm, but it may come earlier than anticipated if we are unaware that the presenting headache was already a rebleed. Unexplained respiratory failure becomes clearly neurologic when the history reveals progressive dysphagia, muscle-mass loss and twitching (motor neuron disease), diplopia, and weakness increasing with exercise or repetitive use but with day-to-day variation, yet typically strong after a good night's rest (myasthenia gravis). I have seen a basilar artery embolus present as coma following a pulseless electrical activity arrest; only after the spouse arrived did a story of acute dysphagia, vertigo, and alternating weakness emerge to suggest that respiratory arrest preceded the cardiac arrest.

In the new era of intensive-care imaging with handheld and smartphone-connected devices, the clinical history may be sidelined even more. Who wants to know what really happened if we can see it already? Are examination and imaging not enough? Without diminishing these parts of the clinical assessment, we should approach them cautiously, with the intent to achieve accuracy. If pressed for time, it may be preferable to tailor the examination and spend more time taking a history. Information-gathering starts within the first moments of an encounter. Do not assume anything; verify everything. Generalization is not enough and we need to delve deeper if we can. We should warn against quick conclusions such as what follows a sign (*post hoc*) and therefore is caused by it (*ergo propter hoc*). Biases in taking history are prevalent, but hopefully we avoid them. (Table 1.6).

Table 1.6 Biases in taking a history

Anchoring	Stuck on features of the patient's presentation too early and failure to adjust when other information becomes available
Availability	The tendency to judge things as being more likely or frequently occurring, if they readily come to mind
Premature closure	The diagnosis is accepted before it has been fully verified.
Representativeness restraint	Pattern recognition and failure to think out of the box (atypical variants)
Unpacking principle	Not getting all information to establish a differential diagnosis
Errors of context	Too much information and losing track of the most important facts

Pointers and Takeaways

- Of course, taking a history and guiding the conversation remain the most critical skills of a neurointensivist, but don't assume they are acquired easily. Important information comes to those who keep asking.
- Urgency is not appreciated enough and should be established.
- Particularization of a clinical problem is needed for a rational decision, even in emergency settings. Be aware of the fallacy of generalization.
- Precipitating factors may provide important clues to unresolved cases.
- Clinical trajectories have been observed for most neurocritical illnesses and provide foresight into what may come next.

References

1. Gandhi TK. Fumbled handoffs: one dropped ball after another. Ann Intern Med. 2005;142:352–8.
2. Riesenberg LA, Leitzsch J, Massucci JL, et al. Residents' and attending physicians' handoffs: a systematic review of the literature. Acad Med. 2009;84:1775–87.
3. Vickrey BG, Samuels MA, Ropper AH. How neurologists think: a cognitive psychology perspective on missed diagnoses. Ann Neurol. 2010;67:425–33.
4. Rolston JD, Bernstein M. Errors in neurosurgery. Neurosurg Clin N Am. 2015;26:149–55. vii

5. Montgomery K. How doctors think: clinical judgment and the practice of medicine. Illustrated ed. New York: Oxford University Press; 2005.
6. Antunes Dias F, Castro-Afonso LH, Zanon Zotin MC, et al. Collateral scores and outcomes after endovascular treatment for basilar artery occlusion. Cerebrovasc Dis. 2019;47:285–90.
7. Hernandez-Perez M, Perez de la Ossa N, Aleu A, et al. Natural history of acute stroke due to occlusion of the middle cerebral artery and intracranial internal carotid artery. J Neuroimaging. 2014;24: 354–8.
8. Huttner HB, Schwab S. Malignant middle cerebral artery infarction: clinical characteristics, treatment strategies, and future perspectives. Lancet Neurol. 2009;8:949–58.
9. Conforto AB, de Freitas GR, Schonewille WJ, Kappelle LJ, Algra A, Group BS. Prodromal transient ischemic attack or minor stroke and outcome in basilar artery occlusion. J Stroke Cerebrovasc Dis. 2015;24:2117–21.
10. Greving JP, Schonewille WJ, Wijman CA, et al. Predicting outcome after acute basilar artery occlusion based on admission characteristics. Neurology. 2012;78:1058–63.
11. Schonewille WJ, Wijman CA, Michel P, et al. Treatment and outcomes of acute basilar artery occlusion in the Basilar Artery International Cooperation Study (BASICS): a prospective registry study. Lancet Neurol. 2009;8:724–30.
12. Kubik CS, Adams RD. Occlusion of the basilar artery; a clinical and pathological study. Brain. 1946;69:73–121.
13. Mattle HP, Arnold M, Lindsberg PJ, Schonewille WJ, Schroth G. Basilar artery occlusion. Lancet Neurol. 2011;10:1002–14.
14. Cornelius JR, Zubkov AY, Wijdicks EFM. Following the clot in spectacular shrinking deficit. Rev Neurol Dis. 2008;5:92–4.
15. Lee VH, John S, Mohammad Y, Prabhakaran S. Computed tomography perfusion imaging in spectacular shrinking deficit. J Stroke Cerebrovasc Dis. 2012;21:94–101.
16. Minematsu K, Yamaguchi T, Omae T. 'Spectacular shrinking deficit': rapid recovery from a major hemispheric syndrome by migration of an embolus. Neurology. 1992;42:157–62.
17. Kumar R, Friedman JA. Subarachnoid hemorrhage: the first 24 hours. A surgeon's perspective. Neurocrit Care. 2011;14:287–90.
18. Macdonald RL. Delayed neurological deterioration after subarachnoid haemorrhage. Nat Rev Neurol. 2014;10:44–58.
19. Rabinstein AA, Lanzino G, Wijdicks EFM. Multidisciplinary management and emerging therapeutic strategies in aneurysmal subarachnoid haemorrhage. Lancet Neurol. 2010;9:504–19.
20. Fisher CM. Pathological observations in hypertensive cerebral hemorrhage. J Neuropathol Exp Neurol. 1971;30:536–50.
21. Gunda B, Böjti P, Kozák LR. Hyperacute spontaneous intracerebral hemorrhage during computed tomography scanning. JAMA Neurol 2021;78: 365–6.
22. Bousser MG, Ferro JM. Cerebral venous thrombosis: an update. Lancet Neurol. 2007;6:162–70.
23. Carney N, Totten AM, O'Reilly C, et al. Guidelines for the management of severe traumatic brain injury. J Neurosurg. 2017;80:6–15.
24. Stocchetti N, Carbonara M, Citerio G, et al. Severe traumatic brain injury: targeted management in the intensive care unit. Lancet Neurol. 2017;16:452–64.
25. Carnevale JA, Segar DJ, Powers AY, et al. Blossoming contusions: identifying factors contributing to the expansion of traumatic intracerebral hemorrhage. J Neurosurg. 2018;129:1305–16.
26. Zalewski NL, Rabinstein AA, Krecke KN, et al. Characteristics of spontaneous spinal cord infarction and proposed diagnostic criteria. JAMA Neurol. 2019;76:56–63.
27. Atkinson PP, Atkinson JL. Spinal shock. Mayo Clin Proc. 1996;71:384–9.
28. Cabot R, Adams F. Physical diagnosis. London: Bailliere, Tindall & Cox; 1938.
29. Gillis J. The history of the patient history since 1850. Bull Hist Med. 2006;80:490–512.
30. Beckman HB, Frankel RM. The effect of physician behavior on the collection of data. Ann Intern Med. 1984;101:692–6.
31. Marvel MK, Epstein RM, Flowers K, Beckman HB. Soliciting the patient's agenda: have we improved? JAMA. 1999;281:283–7.
32. Berner ES, Graber ML. Overconfidence as a cause of diagnostic error in medicine. Am J Med. 2008;121:S2–23.

Scales and Scores

2

By their very nature, physicians have a tendency to classify, and so we find that Medicine is rife with scales and scores. There are clinical scores, imaging scores, and scores that predict outcome or the likelihood of a complication. Many scores combine clinical findings with radiologic findings, and some look at the results of intervention [1, 2]. Scales have been devised for specific disorder or grew out of a need to develop diagnostic criteria for a certain disease (e.g., El Escorial World Federation of Neurology diagnostic criteria for amyotrophic lateral sclerosis [3]), with some even adding levels of diagnostic certainty [4]. The number of scores and scales are increasing not only in the fields of medicine and intensive care but also in stroke, endovascular stroke intervention, and neurocritical care; redundancies are expected. Many are introduced, although few survive. Unsurprisingly, easy-to-use, simple scales are more likely to stay, which often implies that they may be too simplistic to be informative and have domains with little immediate clinical applicability. Moreover, with the use of numerical attributes, the sum may mean different things to different healthcare professionals.

Acute neurology and neurocritical care have their share of scales. Their use in practice is unknown, although some scales and scores are now mandated in stroke centers. Many scores and scales have not been validated repeatedly, and one (in-house) validation study is often the norm. Validation of these scales and scores may thus be marginal, although sometimes a scale gets scrutinized after years of usage and is found to have much less content validity. Intensive care health professionals need to be aware of some scales and scores and evaluate them, not in quality or quantity, but in kind. Although potentially useful in decision-making, scores and scales should be applied judiciously. Scales and scores do, however, assist the clinician to examine certain clinical (or neuroimaging) features more closely. The different components of the scale may impose a systematic methodology to an assessment, and that might be its greatest benefit. In this chapter, I will describe how scales can contain the essence of a clinical examination if well constructed. Many scores and scales are derivatives from older scales and are only mentioned if they are in common use. This chapter is mainly on clinical scales but must include neuroimaging scales when they can lead to a conclusive decision in clinical practice. The ones to know about are discussed and, for ease of reference, shown at the end of the chapter.

General Principles of Scales and Scores

Alvan R. Feinstein introduced the term *clinimetrics* in 1982 to coin a specialty that examined ratings to measure clinical symptoms and physical signs. The basic goals of clinimetry

E. F. M. Wijdicks, *Examining Neurocritical Patients*, https://doi.org/10.1007/978-3-030-69452-4_2

are to create scales with useful properties, but these properties vary with each scale. Some scales are evaluative instruments; others are used for prognostication or even decision-making. For example, National Institutes of Health Stroke Scale/Score (NIHSS) *evaluates deficits*, Alberta Stroke Program Early CT Scan (ASPECTS) score *determines treatment* such as endovascular intervention, and Intracerebral Hemorrhage (ICH) Score *assesses prognosis* at admission.

Scores measuring the severity of acute neurologic illness compare outcomes across different groups. Scoring systems have also been developed to assist in intensive care unit (ICU) discharge and admission decisions, but they are of limited usefulness, unless they can be accepted across a wide variety of diagnoses and patient populations.

Scores and scales can be critically evaluated. According to Feinstein [5, 6], scores and scales must meet the following five principles of clinimetry:

1. *A clear intended purpose:* what are its function, justification, and applicability?
2. *Comprehensibility:* Does the user understand what it is asking? Does the scale give enough clear directions for usage? Is the scale suitable?
3. *Face validity*: Does it accurately target the patient populations it wants to test?
4. *Content validity*: Does it contain all essential variables present?
5. *Ease of usage:* How much time and effort is required to obtain and organize the components of the scale?

Several terms have been associated with clinimetry. First, the term *consistency* depends on whether the same observer (clinician) or different observers rate the scale or score. This is indicated as the intra-observer variability or inter-observer variability. *Inter-observer reliability* is the agreement between observations made by two or more raters on the same patient or group of patients. *Intra-observer reliability* is the agreement between observations made by the same rater on two different occasions on the same patient or group of patients.

Clinimetry is also concerned with the quality of clinical measurements. Agreement between raters on categorical variables is often reported using *kappa* (K), which represents the agreement corrected for chance. Most scale evaluations assume that the two observers and the number of subjects are random samples from the tested populations and that the ratings of the two observers for each subject are independent (i.e., each observer is blinded to the rating given by the other observer) using K statistics. The problem of observer variability is widespread in any area of medicine. When interpreting K and the strength of agreement, the following values are useful: $K < 0.00$, poor; K 0.21–0.40, fair; K 0.61–0.80, substantial; and K 0.81–1.00, almost perfect.

Scales and Scores for Universal Use

Scales or scores have a substantially better chance of acceptance if they apply to a larger ICU population. These scales mostly quantify abnormal levels of consciousness, acute confusional states, or delirium. Outcome prediction applied indiscriminately in diverse populations also loses its specificity.

Coma Scales

Neurosurgeons in Glasgow recognized a need to improve characterization of transferred patients designated drowsy or somnolent. They were too often surprised at the condition of the arriving patient with traumatic brain injury and attributed the discrepancy to inadequate assessment and thus poor communication of the degree of decline in patient's responsiveness. They admirably constructed a scale that would summarize a number of responses not requiring detailed neurologic or neurosurgical knowledge. The Glasgow Coma Scale (GCS) evaluates eye opening and verbal (mostly orientation) and motor responses (Fig. 2.1). The lowest score is 1, and patients may already have a sum score of 3 essentially doing nothing. (Rumor has it that the number zero was not chosen because of practice in early computer programming to start counting

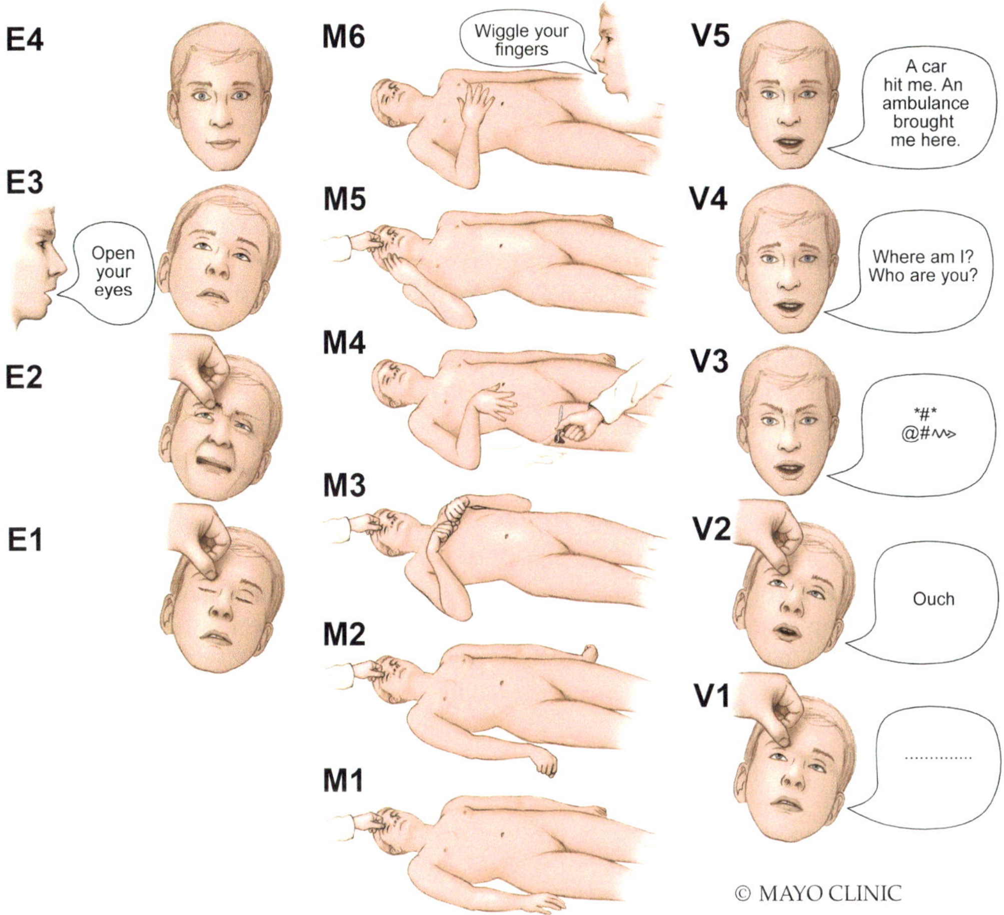

Fig. 2.1 The Glasgow Coma Score (From: Teasdale G, Jennett B. Assessment of coma and impaired consciousness. A practical scale. Lancet 1974 ; 2:81–84)

Eye opening	Verbal response	Motor response
4 = spontaneous	5 = orientated	6 = obeys commands
3 = to sound	4 = confused conversation	5 = localizes pain
2 = to pain	3 = inappropriate words	4 = flexion withdrawal
1 = none	2 = incomprehensible sounds	3 = abnormal flexion
	1 = none	2 = extension
		1 = none

at one). GCS easily led to a more widespread use, including inpatients already admitted to the unit. Nursing staff, who would otherwise describe gradually deteriorating patients as “sleepier,” would chart the GCS and were able to better gauge deterioration. GCS has stood the test of time remarkably well and has been used ubiquitously in emergency departments as well as in trauma practices throughout the world. GCS also has some value outside the hospital (some say perhaps it has the most value in the field), but it immediately loses its discriminatory value in intubated patients with traumatic facial swelling, in which both eye and verbal components become unreliable. GCS does not assess eye movements, brainstem reflexes, neurologic breathing patterns, or language in detail.

The Glasgow Coma Scale has been very useful in clinical outcome studies. However, a low GCS score does not portend poor outcome, and a high GCS score does not preclude deterioration. GCS became rapidly problematic in practice with its use of sum scores (e.g., GCS of 8). In the USA, few emergency physicians use the individual components in communication. When critically tested, practitioners have scored poorly in their knowledge of GCS, and mistakes are common. The originators of the scale warned against the use of sum scores, but alas it happened.

Concerns exist about appropriate use of the GCS, especially in assessing intubated patients, who represent approximately 30–40% of all ICU admissions. Sedation may cloud neurologic assessment and, most likely, affects the GCS. Although the designation "verbal" suggests that GCS addresses speech or language impairment, the verbal component actually measures orientation, and the motor response measures language comprehension, which is hampered if the patient is under sedation.

The Full Outline of UnResponsiveness (FOUR) score has been developed to address these major inadequacies in the GCS. Each category has four components as well as a maximal grade of 4, which makes it easy to remember and is reinforced by the acronym. These four components are *eye responses*, that is, eye opening and eye movements (tracking horizontally and vertically); *motor responses*, that is, following complex commands (squeezing hands can be reflexive) and response to pain stimuli (from localizing using multiple cortical areas to primitive brainstem responses involving rubrospinal and vestibulospinal tracts resulting in flexion or extension); *brainstem reflexes*, such as pupil, corneal, and cough reflexes; and *respiration*, including spontaneous respiratory rhythm or presence of respiratory drive on a mechanical ventilator (Fig. 2.2). Although several domains are tested, the FOUR score can be obtained in a few minutes. Because the FOUR score does not contain a verbal component, it can be measured with equal verity in intubated and nonintubated ICU patients. The three important pupil assessments in the FOUR score remain unaffected by any degree of sedation. One study suggested its combined use with pupilometer increased accuracy [7].

In addition to measuring eye opening, the FOUR score also assesses voluntary horizontal and vertical eye movements. It therefore detects a locked-in syndrome in a patient with the lowest possible GCS score of 3. It detects the presence of a vegetative state when the eyes open spontaneously but do not track the examiner's finger. The motor category includes the presence of myoclonus status epilepticus (persistent multisegmental, arrhythmic, jerk-like movements), an important prognostic sign after cardiac resuscitation. The motor component combines decorticate and withdrawal responses because the difference is often difficult to appreciate. The hand-position tests (thumbs-up, fist, and peace sign) further assess alertness and have proven validity. Three brainstem reflexes, which test mesencephalon, pons, and medulla oblongata functions, are used in different combinations. Breathing patterns are graded. The respiratory component measures rhythm [2–4] and drive (1–0), irrespective of a tracheostomy. Cheyne–Stokes respiration and irregular breathing can represent bi-hemispheric or lower brainstem dysfunction of respiratory control. In intubated patients, overbreathing of the mechanical ventilator or spontaneous ventilator-supported breathing represents functioning respiratory centers. The absence of a verbal-response test makes the FOUR score more useful in intensive care practices with a large number of intubated patients. It may also be more useful in young children. Some have rightfully argued that the hand-position test in the motor component already shows a similar orientation as is tested in the GCS verbal component. However, FOUR score assessment in our validation studies has been good to excellent with accurate outcome predictions in many patient populations (e.g., emergency departments, specialty and nonspecialty ICUs). The testing of each component can be easily mastered by physicians and interpreted satisfactorily by intensive care nurses.

Well, it appears now that the FOUR score—over 15 years since its publication—has been validated more than any score or scale in

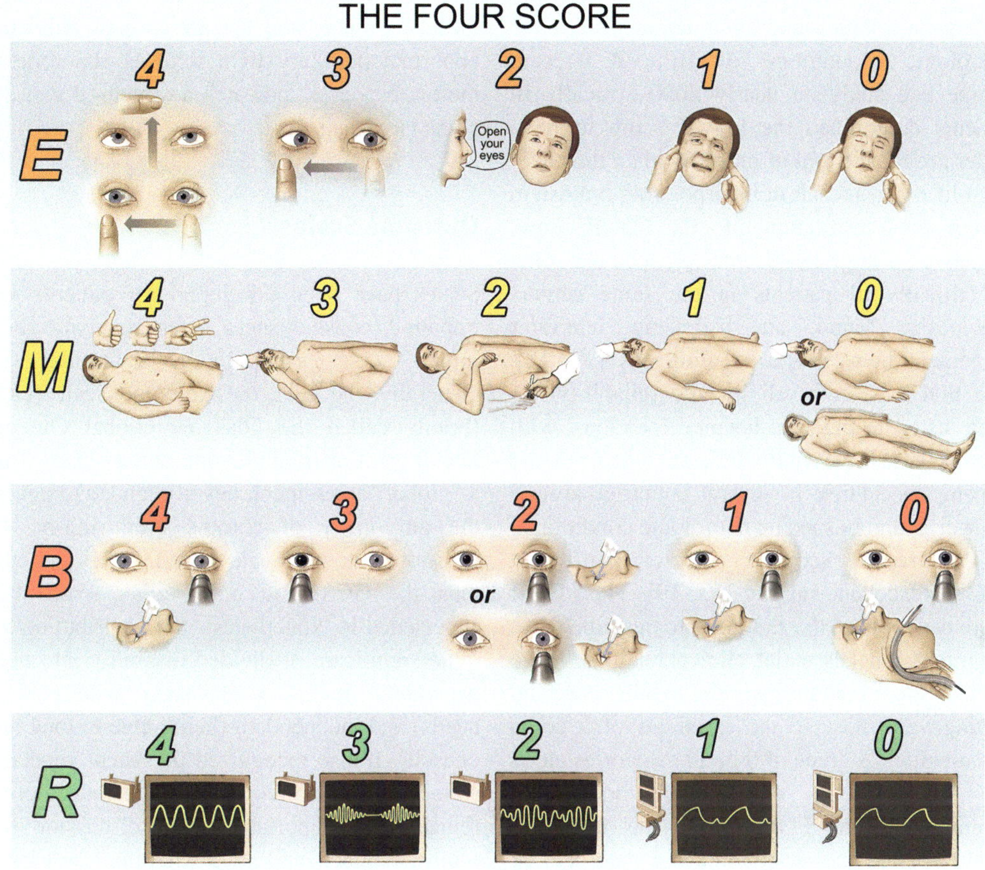

Fig. 2.2 The FOUR Score

Eye response (E)
- 4 = eyelids open, tracking, or blinking to command
- 3 = eyelids open but not tracking
- 2 = eyelids closed but open to loud voice
- 1 = eyelids closed but open to pain
- 0 = eyelids remain closed with pain

Motor response (M)
- 4 = thumbs-up, fist, or peace sign
- 3 = localizing to pain
- 2 = flexion response to pain
- 1 = extension response to pain
- 0 = no response to pain or myoclonus status

Brainstem reflexes (B)
- 4 = pupil and corneal reflexes present
- 3 = one pupil wide and fixed
- 2 = pupil or corneal reflexes absent
- 1 = pupil and corneal reflexes absent
- 0 = absent pupil, corneal, and cough reflex

Respiration (R)
- 4 = not intubated, regular breathing pattern
- 3 = not intubated, Cheyne-Stokes breathing pattern
- 2 = not intubated, irregular breathing
- 1 = breathes above ventilator rate
- 0 = breathes at ventilator rate or apnea

Wijdicks EFM, Bamlet WR, Maramattom BV, et al. Validation of a new coma scale: The FOUR score. Ann Neurol. 2005;58:585–593.

neurology or neurosurgery. It has also been successfully validated in the emergency department, and this suggests that it can be used by emergency physicians at any level of training and by the nursing staff. The FOUR score validity has also been tested in several countries around the world and with different physicians and nursing specialties and in a number of acute neurologic conditions [8–16]. A recent prospective study on nearly 2000 critically ill patients determined the FOUR score to be a better prognostic tool of mortality than the GCS, most likely as a result of incorporating brainstem reflexes and respiration into the FOUR score. Moreover, a large multicenter prospective study of critically ill patients in the same centers previously found an excellent interrater agreement between paired clinicians [17, 18]. The FOUR score's validity and reliability has been tested in multiple hospital locations, with telemedicine [12], within different physician and nursing specialties, in several countries around the world, and in specific neurologic conditions.

Two recent scoping reviews confirmed good-to-excellent validity [13, 19]. The FOUR score necessitates the examiner to describe these important, if not essential, clinical features. There is much more granularity in the FOUR score and getting very close to the essentials of a coma examination. A grade of 0 in all categories alerts the examiner to consider a brain-death examination. The FOUR score has not been tested outside the realm of acute disorders of consciousness but may also be useful in more chronic patients in a minimally conscious state or in a persistent vegetative state, the other valid coma-recovery scales notwithstanding. Due to its greater neurologic detail and improved prognostic indicators, the FOUR score is more helpful than the GCS in predicting outcome in the intensive care unit. However, the GCS remains entrenched in clinical assessments, despite its shortcomings. Deadoption of old systems with newer, better ones is slow in medicine.

Of course, coma scales are more useful for initial assessment but should never replace a full neurologic assessment of the comatose patient. Therefore, initial neurologic assessment should involve the assessment of motor examination and motor response. It should include the assessment of tone—rigid, flaccid, or in between; the presence of spontaneous (nystagmus, bobbing, and dipping) or induced eye movements (oculocephalic and oculovestibular testing) and so many other features seen with coma. The brainstem reflexes have a localizing value, and the presence of anisocoria, abnormal pupillary light reflexes and reflexive motor responses indicates a structural cause of coma (see Chap. 7).

Outcome Scales

Scales have been developed for patients with impaired consciousness after traumatic brain injury or cardiopulmonary resuscitation. Generally, the Glasgow Outcome Scale (GOS) (briefly called the Glasgow Global Outcome Scale) was one of the first outcome scales. It was a "global" assessment, never intended to get into the nitty-gritty of chronic brain injury. The Glasgow Outcome Scale (GOS) was less accepted than the GCS, and many other scales have superseded it. Nonetheless, the contribution was quite significant; it divided disability into being dependent (requiring assistance with personal needs) and independent (being able to look after oneself). It also recognized persistent vegetative state as a separate condition. These distinctions remain very important in any discussion with family members and often what they are looking for when discussing prognosis. Much later, an extended form was used providing more details (Table 2.1). Teasdale (and Jennett earlier) recognized that the word "independence" can have many interpretations, and "going home" is not a valid outcome criterion. Neither is returning to work a valid outcome criterion; some patients will not be able to hold a job for long, will be reassigned to less challenging responsibilities, or may require closer supervision. Other factors such as economic circumstances may play a much larger role [20].

The *Cerebral Performance Categories* in later studies have been used more often, although they are basically modifications of the GOS. Functional outcome is usually dichotomized as good (Cerebral Performance Category [CPC] 1–2, indicating low-to-moderate disability) or poor (CPC 3–5, indicating severe disability, coma, or death). The CPC was introduced in the 1986 thiopental-loading study after cardiac arrest and has been used ever since in cardiac arrest studies [21]. The CPC is the GOS in reverse. The CPC score is the most commonly used tool for both research and audit purposes. Studies define a good outcome as a CPC score of 1 or 2 and a poor outcome (severe neurological disability, persistent vegetative state, or death) as a CPC score of 3, 4, or 5. The scale has not been subjected to intra- and inter-observer agreement assessment, and CPC 2 and 3 have overlapping characteristics (which would certainly show problems in a rigorous validation study). Death includes both "brain death and death by traditional criteria," and it failed to distinguish between the two (Table 2.2) [22].

Rating patient improvement in disorders of consciousness is needed not only to measure the range of behaviors more effectively but also to assist in prognostication in patients with severe brain injury. Probabilistic models improve accuracy when compared to clinical prediction, but uncertainty continues to exist about the most meaningful and interpretable score [23–25]. Current scales that are used in practice and research studies are Disability Rating Scale, Glasgow Outcome Scale (and the Cerebral Performance Scale), Glasgow Outcome Scale-Extended (GOSE), Neurobehavior Rating Scale-Revised and, more recently, the Neurologic Outcome Scale for Traumatic Brain Injury [26]. Outcome measures in rehabilitation are most commonly considered in relation to the International Classification of Functioning, Disability and Health, realms of impairment (how the examination differs from normal), activity limitations (the so-called activities of daily living (ADL) and instrumental activities of daily living (IADL)), and restrictions to participation (e.g., personal, family, vocational, community roles). The NIH toolbox, Patient-Reported Outcomes Measurement Information System (PROMIS) measures, Activity Measure for Post Acute Care (AM-PAC), and Traumatic Brain Injury-Quality of Life (TBI-QOL) are examples of current measures. Multiple functional scales have been developed to measure performance at different levels of improvement (e.g., SMART) [27].

A validated scale is the JFK Coma Recovery Scale (CRS) (Table 2.3), which is skewed toward motor responses. A total JFK Coma Recovery Scale-Revised (CRS-R) score of 10 or higher provides a strong evidence of conscious awareness but has resulted in a false-negative diagnostic error in 22% of patients who demonstrated conscious awareness based on the CRS-R diagnostic criteria. A cutoff score of 8 provides the best balance between sensitivity and specificity, accurately classifying 93% of cases.

The functional independence measure (FIM) (Table 2.4) has been validated and adequately evaluates abilities in self-care, mobility, sphincter control, and social cognition among others. Clinical training of the rater is less important; even family members can use it to rate the performance ability of the patient. It assesses the burden of care or the amount of assistance required by the patient and is a straightforward measure of activity limitations. Not unexpectedly, FIM is correlated to socioeconomic status [28].

Although the FIM is most widely known, the Disability Rating Scale has been used in research studies and requires eight items. The total scores vary from 0 to 30 (vegetative state ≥22). The first three items are the Glasgow Coma Score (the lowest number here is 0 rather than 1), and the other scores are for self-care activities and level of functioning (physically, mentally, emotionally, and socially). The interrater reliability is good, as is the comparison of ratings by family members and rehabilitation professionals.

It remains to be seen if better (more detailed) outcome scales are useful because independence is the most important outcome for many patients and families. Moreover, "dependence on others" as an outcome measure in clinical trials should be better explained and detailed. It is far more complex than simply getting some outside help or relying on a close-knit family.

Scales and Scores for Specific Disorders

The most common acute neurologic disorders seen in the intensive care units have all been subjected to scrutiny of their clinical presentation, leading to assessment of features with statistical significance in outcome determination, numerical assignment (higher number with higher level of significance), and eventually a score. Patients now arrive in the unit with an admission note that contains several scores. Moreover, hospital and training accreditation agencies may require this notation to provide some standardization of assessment. A few commonly used scales and scores are presented here.

Traumatic Brain Injury Scales

Large databases, such as the Corticosteroid Randomization After Significant Head Injury (CRASH) and the International Mission for Prognosis and Analysis of Clinical Trials in Traumatic Brain Injury (IMPACT) databanks, provide estimates of mortality and unfavorable outcome. There are multiple prediction models in traumatic brain injury (TBI). The largest database of TBI (International Mission for Prognosis and Analysis of Clinical Trials in Traumatic Brain Injury [IMPACT]) uses admission characteristics to calculate an estimate of prognosis. Variables include age, motor response, pupil responses, presence of hypoxia and hypotension, computed tomography (CT) categorization of lesion severity and mass effect, presence of traumatic subarachnoid hemorrhage (tSAH) and epidural mass; an extended model adds serum glucose and hemoglobin levels. The second largest database (Corticosteroid Randomization After Significant Head Injury [CRASH]) incorporates a major extracranial injury into the equation and the country of the patient to account for different resources in developing countries [29]. (Calculators are freely available online.) The IMPACT databank is often more optimistic. It would be a grave mistake to clinically rely on these scales (see Chap. 11 for more details and discussion).

Clearly, the most significant variables on outcome in traumatic brain injury are the age of the patient and severity of the injury, as evidenced by the presence or absence of pupillary reactivity, motor response, computed tomography (CT) findings, and associated systemic insults [29, 30].

Penetrating TBI has also been investigated for outcome prediction. One study in a large, 2-center US cohort investigated important clinical and radiologic predictors (Social Phobia Inventory [SPIN] score) associated with survival after penetrating TBI. The score was, however, derived from a small number of patients and a large number of situations. The individual elements are numerous, potentially leading to overfitting of the model. They include motor GCS (mGCS), pupil reactivity, self-inflicted or not, patient transferred or not, gender, injury severity score, and international normalized ratio (INR). Higher mGCS and pupillary reactivity on admission were, by far, the strongest independent predictors of survival; all other independent predictors added very little to the model. The SPIN score has not been validated and does not contain radiologic factors. Therefore, it is currently a very preliminary tool that cannot provide guidance for physicians and families in their direction-of-care decision-making in patients with penetrating TBI [31].

Stroke Scales and Scores

The history of stroke scales is short, and most are developed now. A number of scales performed very well (e.g., the Canadian Neurological Scale and Middle Cerebral Artery Neurological Score). Eventually, the dust settled, and one scale prevailed [32]. Assessment of stroke has become more formalized with the universal adoption of the NIH Stroke Scale (Table 2.5) [22]. The scale varies from 0 to 43, and the official description is as follows:

> The NIHSS is a 15-item neurologic examination stroke scale used to evaluate the effect of acute cerebral infarction on the levels of consciousness, language, neglect, visual-field loss, extraocular movement, motor strength, ataxia, dysarthria, and sensory loss. A trained observer rates the patient's ability to answer questions and perform activities. Ratings for each item are scored with 3 to 5 grades with 0 as normal, and there is an allowance for untestable items. The single patient assessment requires less than 10 minutes completing. The evaluation of stroke severity depends upon the ability of the observer to accurately and consistently assess the patient [33].

Reliability of the NIHSS among examiners has been previously evaluated [34]. Interrater agreement was lowest in the assessment of ataxia, facial motor function, extraocular movements, and dysarthria when each component was evaluated individually. The NIHSS fails miserably by not examining standing and walking in detail; in patients with low scores, a disabling leg weakness (not weight-bearing but antigravity) can be easily missed. Distal, disabling hand and finger weakness from a stroke in the hand knob area (precentral gyrus) is not rated. Moreover, signs exclusive to the posterior circulation are omitted, creating another opportunity to overlook critical signs. High NHISS can be recorded in posterior circulation strokes and are a summation of neurologic signs (e.g., unconscious with quadriplegia starting with 20 points). Low NIHSS stroke scales, however, could mask significant brainstem injury. Both major shortcomings emphasize the importance of a thorough neurologic examination, particularly of the cranial nerves, eye position, and movement, further to evaluate the posterior circulation in patients where the clinical syndrome is uncertain [22, 34, 35]. Prediction of a large vessel occlusion is usually determined by the severity of the stroke and whether there are cortical findings such as significant aphasia or agnosia. The RACE Scale includes 5 items (facial, arm, leg, gaze-eye deviation, aphasia-agnosia) and a score more than 5 predicts 40% or more chance of large vessel oclcusson. RACE perform only marginally better than a high NIHSS [36].

Numerous scores exist for estimating postthrombolysis risk including the Stroke-Thrombolytic Predictive Instrument (Stroke-TPI), iSCORE, DRAGON, Stroke Prognostication using Age and NIH Stroke Scale-100 (SPAN-100), Acute STroke Registry and Analysis of Lausanne (ASTRAL), Postthrombolysis Risk Score (PRS), Hemorrhage After Thrombolysis (HAT), SEDAN, and Safe Implementation of Treatments in Stroke Symptomatic Intracerebral Hemorrhage (SITS-ICH); most of these are cumbersome. TURN has been considered easier to use but only includes prestroke MRs Score and baseline NIHSS score. (Platelet count was later dropped [37]).

In cerebral hematoma, a number of scores have been developed, often to assess a baseline and, more recently, to predict enlargement [38]. The ICH Score devised by Hemphill and colleagues (Table 2.6) [39], which incorporates multiple clinical and imaging characteristics, is a commonly used assessment tool to assist clinicians in such situations, providing an estimated mortality risk at 30 days posthemorrhage. Several scores (Table 2.7) have been developed to predict the risk of expansion within 24 hours, namely the Bleeding Assessment Tool (BAT) score [40], PREDICT (Predicting Hematoma Growth and Outcome in Intracerebral Hemorrhage Using Contrast Bolus Computed Tomography) score [41], and the 9-point score; the BRAIN score was developed due to lack of ready access to CT angiography (CTA) [42]. They have been recently validated and were comparable in prediction [42]. Another is the HIT Expert Probability (HEP) score by Yao [43], which has been externally validated.

The Intracerebral Hemorrhage Outcomes Project (ICHOP) scores take into account factors used by other ICH scoring systems (GCS, NIHSS, and hematoma volume for a 3-month outcome), while also incorporating novel factors such as premorbid functionality and Acute Physiology And Chronic Health Evaluation II (APACHE II) score to develop a more extensive model for prognosticating functional outcome [44]. All these ICH scales involved spontaneous hemorrhages.

Grading for outcome in ruptured arteriovenous malformation (AVM) was recently developed. The Arteriovenous Malformation-Related Intracerebral Hemorrhage (AVICH) score (Table 2.8), a grading system to predict clinical outcome in arteriovenous malformation-related intracerebral hemorrhage, predicts outcome of patients with ruptured AVM and associated ICH better than the ICH score, the Spetzler–Martin, or the supplemented Spetzler–Martin grading system [45, 46].

Aneurysmal subarachnoid hemorrhage has been graded with the World Federation of Neurological Surgeons (WFNS) or Hunt and Hess system, but both are far from adequate summaries of the patient's condition and, by any standard, insufficient (Table 2.9) [47]. The Hunt and Hess scale has the lowest inter-observer agreement [48]. In the WFNS, level of consciousness is the major determinant, but there is little differentiation other than the presence of "focal signs." Most studies have divided subarachnoid hemorrhage of WFNS grades I to III from poor-grade SAH (WFNS grade IV or V), but this assumes that the poor clinical grade is caused by the initial impact alone. We found the timing of grading to be important, since a considerable number of patients improved after initial neurocritical care management [49]. The WFNS score does very little in describing aneurysmal subarachnoid hemorrhage.

Neuromuscular Diseases and Scales

A basic, tested way of evaluating weakness is through the Medical Research Council (MRC) scale (Table 2.10) [47]. It defines weakness for each tested muscle and grades from 5 to 0. It has been validated in inter- and intra-observer studies, but several limitations have been noted. First, the scale does not consider clinically relevant changes in grades 3 and 4, which are particularly important in the recovery phase. Second, within grades 3 and 4, the MRC scale does not include the range of motion for which a movement (e.g., 10° or 60°) can be performed; this is important to assess muscles innervated by a single peripheral nerve.

For intensivists, three disorders are common — Guillain–Barré syndrome (GBS), myasthenia gravis, and amyotrophic lateral sclerosis. Scales have been developed to categorize the degree of weakness and disability, but none addresses the acuteness of the worsening. There are a number of questionnaires that address later functionality, and some are specific for neuropathies (such as the GBS disability scale and chemotherapy-induced neuropathy) [50].

There is no shortage of scoring systems for myasthenia gravis. Many are time consuming but provide the opportunity to look carefully at individual, crucially important muscles. Muscular weakness in myasthenia gravis (MG) is commonly assessed using Quantitative Myasthenia Gravis (QMG) Score. The QMG is a 13-item (3 ocular, 2 bulbar, 1 respiratory, 1 neck, and 6 limbs) scale that measures muscle strength and endurance [51]. The manual muscle testing (MMT) measures the strength or function of 18 muscle groups (3 ocular, 3 bulbar, 2 neck, and 10 limbs, scored bilaterally) [51–53]. The MG-ADL is an 8-item questionnaire (2 ocular, 3 bulbar, 1 respiratory, and 2 limbs) assessing common MG symptoms and dysfunction. All three instruments have been validated. The MG Composite has 3 ocular, 3 bulbar, 1 respiratory, 1 neck, and 2 limb items (Table 2.11) [54]. It differs from the QMG and MMT, which have a greater representation of upper- and lower-extremity strength items [55]. The Oculobulbar Facial Respiratory (OBFR) score objectively measures the bulbar function in myasthenia gravis [56]. The OBFR score consists of facial muscle-strength assessment, including five facial muscles, assessment of palatal contractility, tongue appearance, swallow time for 100 ml of water, and respiratory assessment applying the forced vital capacity (FVC), which is a simple bedside measure of respiratory muscle function. The facial muscle score (FMS) involves the assessment of five facial muscles including orbicularis oculi (oculi, the muscle that closes eyelids), frontalis (raises eyebrows), corrugator supercilii (the frowning muscle), orbicularis oris (responsible for closing mouth and puckering lips), and buccinator (the muscle that blows cheeks out). The palatal-contractility test looks specifically for asymmetry. It assesses swallow time for 100 ml of water (preferably at room temperature) including the ranges for normal swallow time and mild, moderate, or severe prolongation. Tongue appearance, with specific attention to the lateral margins and central portions, assesses muscle bulk of the tongue. Forced vital capacity (FVC) values are compared with predicted values based on an individual's gender, age, and height. The cutoff ranges for mild, moderate, or severe

FVC reduction were based on the QMG scoring system.

Contrary to MG, the number of useful scales and scores in GBS is much smaller. For many years, the Hughes severity scale has been used, particularly to decide on plasma exchange or intravenous gamma globulin (Table 2.12). Other GBS scales of severity have not been developed, except for the Erasmus GBS Respiratory Insufficiency Score (EGRIS) (Table 2.13) [57]. Vital capacity and electrophysiological measurements were not used, and no other assessment of respiratory function (breathing effort) exists. However, the score emphasizes that rapidity and degree of weakness culminating in oropharyngeal involvement increase the chance of ventilator support—a well-known clinical observation that does not need inclusion in a scoring system.

Neuroimaging Scales and Scores

A number of CT-scan grading systems have been developed. Virtually all systems operate on the basis of semiquantitative assessment. One of the first disorders to receive a careful CT study was aneurysmal subarachnoid hemorrhage. CT scans of patients with aneurysmal subarachnoid hemorrhage have been scrutinized for years, which led to a number of grading systems starting with Fisher. The Fisher scale [58], although deeply ingrained in neurological practice, remains a gross estimate of the amount of subarachnoid blood and has significant inter-observer variability. It has been modified to include intraventricular hemorrhage (IVH) (Table 2.14). Some scales have been proposed to provide early identification of patients at risk of developing delayed cerebral ischemia (DCI) [59–62]. One study asked four independent clinicians to assess five CT scan-based grading systems—Fisher, modified Fisher, Claassen, Hijdra, and the Barrow Neurological Institute (BNI) [63]. The Hijdra grading system (Fig. 2.3) had the best inter-observer agreement and was a better independent early predictor of 6-month clinical outcome than the other systems. A Hijdra score $\geq$ 22 was associated with poor outcome [63]. A number of other comparative studies suggest that grading per cistern and summation of the amount of blood have more predictive value than estimating "thickness" of blood layers [64–66].

Due to a revolutionary change in ischemic-stroke treatment spearheaded by intravenous thrombolysis success and followed by major improvement after endovascular-clot retrieval in selected patients, additional scales and scores were needed to guide decisions and to assess success of the intervention. The Alberta Stroke Program Early CT Scan (ASPECTS) score (Fig. 2.4) has prevailed and is now used to determine endovascular eligibility. The ASPECTS score—starting at 10—subtracts 1 point for each area of hypo-attenuation or loss of gray-white differentiation. An ASPECTS score of <7 corresponds with a >70–100 mL core and, thus, may be a cutoff of poor outcome. However, the interrater reliability of the ASPECTS score using this cutoff point is poor (weighted K of 0.53), and a small caudate, lentiform nucleus and insular infarct may already result in an ASPECTS score close to 7. Not everyone involved with stroke care feels confident with this CT rating scale when using it for crucial treatment decisions. Moreover, neuroradiological input increases the reliability of the CT reading. It would be problematic to choose endovascular approaches based solely on this score, but ASPECTS grading does force physicians to look at CT scans in more detail. In one study, two experienced neuroradiologists assessed per-region agreement for ASPECTS on CT images from 375 patients with acute ischemic stroke: the greatest agreement was found in the caudate, lentiform, and M5 regions, with the M3 and internal-capsule regions having the lowest inter-observer agreement. This limitation is relevant for interventional treatment decisions based on the ASPECTS [67].

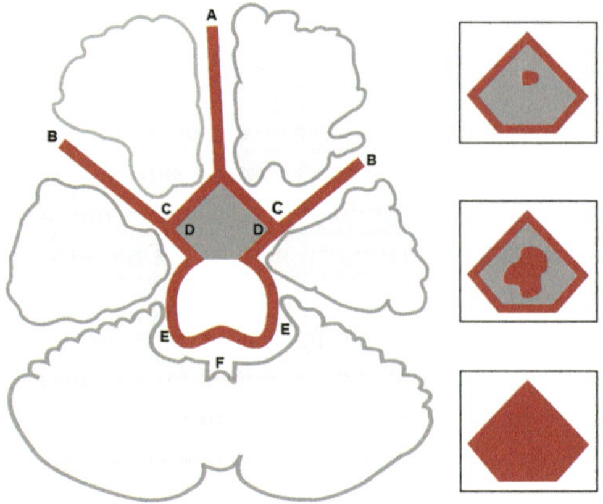

Fig. 2.3 The Hijdra score: A method of grading subarachnoid hemorrhage identifies 10 basal cisterns and fissures: (*A*) frontal interhemispheric fissure; (*B*) sylvian fissure, lateral parts; (*C*) sylvian fissure, basal parts; (*D*) suprasellar cistern; (*E*) ambient cisterns; and (*F*) quadrigeminal cistern. The amount of blood in each cistern and fissure is graded 0, no blood; 1, small amount of blood; 2, moderately filled with blood; and 3, completely filled with blood. (see inserts) The score ranges from 0 to 30 points

With the increased use of endovascular treatment scales, scores will increasingly document the success rate of recanalization. The thrombolysis in cerebral infarction (TICI) score (Table 2.15) evaluates cerebral perfusion before and after endovascular treatment of stroke. Modified TICI 2b and modified TICI 3 assess the technical success of endovascular treatment by reviewing the effectiveness and safety of mechanical thrombectomy. Patients with TICI 3 reperfusion achieve better clinical outcomes and less hemorrhagic transformation than patients achieving TICI 2b reperfusion [68]. Others have disagreed and found that patients with TICI 2c grade are distinguished from those with 2b, because 2c is clinically equivalent to 3 and has a better outcome than 2b. Therefore, achieving 2c or 3 is likely to be closer to the successful aim of endovascular thrombectomy in acute ischemic stroke than achieving 2b [69]. However, a recent meta-

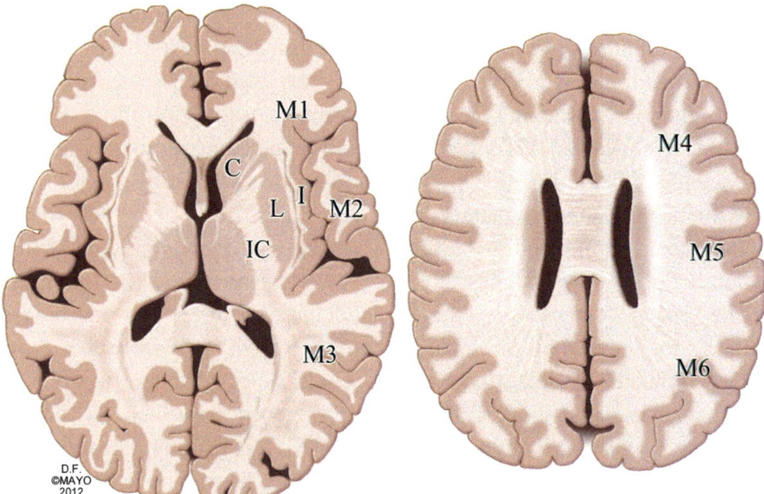

Fig. 2.4 ASPECT score. For ASPECTS, the territory of the middle cerebral artery (MCA) is allotted 10 points. 1 point is subtracted for an area of early ischemic change, such as focal swelling, or parenchymal hypoattenuation, for each of the defined regions. A normal CT scan has an ASPECTS value of 10 points. A score of 0 indicates diffuse ischemia throughout the territory of the middle cerebral artery. ASPECTS A anterior circulation, P posterior circulation, C caudate, L lentiform, IC internal capsule, I insular ribbon, MCA middle cerebral artery, MI anterior MCA cortex, M2 MCA cortex lateral to insular ribbon, M3 posterior MCA cortex, M4, M5, and M6 are anterior, lateral, and posterior MCA territories immediately superior to M1, M2, and M3, rostral to basal ganglia. Subcortical structures are allotted three points (C, L, and 1C). MCA cortex is allotted seven points (insular cortex, M1, M2, M3, M4, M5, and M6)

analysis urged the consideration of successful revascularization as a TICI 3 score and complete revascularization as a goal. This also correlated with lower rates of ICH among TICI-3 patients, thus countering the concern that complete revascularization might cause a greater risk of reperfusion hemorrhage [70].

TBI must include a careful assessment of the CT scan findings, which not only can be communicated but can be used in research studies. TBI scales have been validated [71] to evaluate the performance of three head CT classification systems (e.g., Marshall CT Classification [Table 2.16], Rotterdam CT Score [Table 2.17], and Helsinki CT Score [Table 2.18]) in predicting 6-month mortality and 6-month functional outcome independently and together with known TBI-outcome predictors. These CT-classification systems demonstrated equally accurate predictions of 6-month unfavorable outcome, with no significant difference between the individual CT scores. Only for 6-month mortality predictions did the Helsinki CT Score show slightly better accuracy than the other CT scores. Although existing head CT-classification systems demonstrate mostly good-to-excellent statistical performance in outcome prediction, they do not significantly improve the performance of a simple model based on age, motor response, and pupil responsiveness.

More Reflections

Scales and scores are a mixed (overstuffed) bag. If used in practice, notes may become loaded with acronyms and unexplained component ratings. The uninitiated would certainly be overwhelmed to read

> A 57 M with NIHSS 20, ASPECT4, TICI 3, mRANKIN3 or 60 F with SAH, GCS 4, FOUR E3M2B4R1, WFNS 3, and mFISHER 3.

In large clinical trials with heterogeneous populations, the usefulness of scores and scales is established; they allow a metric where none previously existed. However, as scales become increasingly complex, they are more likely to be quickly forgotten. For scale and score designers, the challenge is to find the right amount of necessary information and avoid filling the scale with irrelevant components. Scores and scales must remain valid despite treatment and interventions and cannot lose any of their components if patients are subjected to critical care interventions (e.g., GCS verbal score invalidated with endotracheal intubation). Neuroimaging has become an integral part of decision-making, and scale "reading" of these images is useful for comparison. It is surprising that both the leading stroke scales (NIHSS and ASPECT) do not evaluate for posterior circulation disease, which would be missed if we relied exclusively on these two scales. The TICI score is a useful metric to establish the degree of revascularization in patients with ischemic stroke. Unfortunately, many scales and scores have made it into required documentation for accreditation of medical centers. Such a demand may lead to compliance and some thoroughness but not necessarily accuracy. The practice may easily change into "throwing some numbers around." Arbitrary numerical cutoffs for therapeutic intervention may become the norm. Numerical changes may not mean improvement or worsening. Commonly, the physician communicating a number may have difficulty remembering where it came from, and repeat counting rarely leads to the same number. Still, as previously noted, scales and scores are approximations of what can be assessed neurologically or radiologically, and clinicians must accept this major limitation. Although many are here to stay (and nearly sacred), practice would be better without many of the unvalidated scores and scales also because they can never be the most appropriate description of a patient's condition. We should be better than that. When they deal with outcome, low sum scores may suggest an irreversible condition and may discourage a neurosurgical or other intervention, while in fact these interventions may improve the score.

Pointers and Takeaways

- Scales and scores must be understood and appreciated.
- Scales and scores must be used judiciously.
- Scales and scores help to achieve a more systematic evaluation.
- Scales and scores must not replace a focused neurologic examination.
- Scales and scores may show greatly varying specificity and sensitivity.
- Scales and scores are generally insufficient for accurate disability determination.

Appendix: Additional Scales and Scores

Table 2.1 Glasgow outcome scale (original and revised)

1 = Dead (*Dead*[a])
2 = Vegetative state (*Vegetative State*)
3 = Severe disability (3 = *lower severe disability needs full assistance; 4 = upper severe disability needs partial assistance*)
4 = Moderate disability (5 = *lower moderate independent but not back to work; 6 = upper moderate disability partly to resume work*)
5 = Good recovery (7 = *lower good recovery minor physical or cognitive deficits; 8 = upper good recovery fully recovered or minor symptoms not affecting daily life*)

[a]The Glasgow Outcome Scale Extended (GOSE) is given in italics with adjusted numbering

Table 2.2 Cerebral performance category (CPC)

CPC 1	Good cerebral performance (normal life but may retain nondisabling deficits)
CPC 2	Moderate cerebral performance (disabled but independent)
CPC 3	Severe cerebral disability (conscious, dependent on others for support)
CPC 4	Coma
CPC 5	Dead (brain death or death by traditional criteria)

CPC cerebral performance category

Table 2.3 JFK Coma recovery scale

Patient: **Diagnosis** **Etiology:**

Date of onset **Date of admission:**

Date

Week	ADM	2	3	4	5	6	7	8	9	10	11	12	13	14	15	16
Auditory function scale																
4-Consistent movement to command[a]																
3-Reproducible movement to command[a]																
2-Localization to sound																
1-Auditory startle																
0-None																
Visual function scale																
5-Object recognition																
4-Object localization: Reaching[a]																
3-Visual pursuit[a]																
2-Fixation[a]																
1-Visual startle																
0-None																
Motor function scale																
6-Functional object use[b]																
5-Automatic motor response[a]																
4-Object manipulation																
3-Localization to noxious stimulation[a]																
2-Flexion withdrawal																
1-Abnormal posturing																
0-None, flaccid																
Oromotor/verbal function scale																
3-Intelligible verbalization[a]																
2-Vocalization/oral movement																
1-Oral reflexive movement																
0-None																
Communication scale																
2-Functional: Accurate[b]																
1-Nonfunctional; Intentional																
0-None																
Arousal scale																
3-Attention																
2-Eye opening without stimulation																
1-Eye opening with stimulation																
0-Unarousable																
Total score																

[a]Denotes minimally conscious state (MCS)
[b]Denotes emergence from MCS

Table 2.4 Functional independence measure

(FIM™) Instrument Score	
Self-Care	1–7
A. Eating	
B. Grooming	
C. Bathing	
D. Dressing—upper body	
E. Dressing—lower body	
F. Toileting	
Sphincter control	1–7
G. Bladder management	
H. Bowel management	
Transfers	1–7
I. Bed, chair, wheelchair	
J. Toilet	
K. Tub, shower	
Locomotion	1–7
L. Walk/wheelchair	
M. Stairs	
Motor Subtotal Score	
Communication	1–7
N. Comprehension	
O. Expression	
Social cognition	1–7
P. Social interaction	
Q. Problem solving	
R. Memory	
Cognitive Subtotal Score	
Total FIM Score	18–126

Scores range from 1 to 7 (1 = worst, 7 = best). Grade: 7—Complete Independence (Timely, Safely), 6—Modified Independence (Device), 5—Supervision (Subject = 100%+), 4—Minimal Assist (Subject = 75%+), 3—Moderate Assist (Subject = 50%+), 2—Maximal Assist (Subject = 25%+), 1—Total Assist (Subject = less than 25%). As a rough estimate, patients with FIM scores less than 40 are most commonly discharged to a skilled nursing facility

FIM functional independence measure

Table 2.5 Modified National Institutes of Health Stroke Scale

Item name	Score
Level of consciousness questions	0 = answers both correctly 1 = answers one correctly 2 = answers neither correctly
Level of consciousness commands	0 = performs both tasks correctly 1 = performs one task correctly 2 = performs neither task
Gaze	0 = normal 1 = partial gaze palsy 2 = total gaze palsy
Visual fields	0 = no visual loss 1 = partial hemianopsia 2 = complete hemianopsia
Left arm	0 = no drift 1 = drift before 10 seconds 2 = falls before 10 seconds 3 = no effort against gravity 4 = no movement
Right arm	0 = no drift 1 = drift before 10 seconds 2 = falls before 10 seconds 3 = no effort against gravity 4 = no movement
Left leg	0 = no drift 1 = drift before 5 seconds 2 = falls before 5 seconds 3 = no effort against gravity 4 = no movement
Right leg	0 = no drift 1 = drift before 5 seconds 2 = falls before 5 seconds 3 = no effort against gravity 4 = no movement
Sensory	0 = normal 1 = abnormal
Language	0 = normal 1 = mild aphasia 2 = severe aphasia 3 = mute or global aphasia
Neglect	0 = normal 1 = mild 2 = severe

From Lyden et al. [22]. With permission of the American Heart Association

Table 2.6 Intracranial hemorrhage (ICH) score

Component	ICH score points
Glasgow coma scale (GCS)	
3–4	2
5–12	1
13–15	0
ICH volume (mL)	
≥30	1
<30	0
Intraventricular hemorrhage (IVH)	
Yes	1
No	0
Infratentorial origin of ICH	
Yes	1
No	0
Age (years)	
≥80	1
<80	0
Total ICH score	0–6

ICH intracerebral hemorrhage, *GCS* GCS score on initial presentation (or after resuscitation)
ICH volume on initial CT calculated using the ABC12 method
IVH indicates the presence of any intraventricular hemorrhage on initial computed tomography

Table 2.7 ICH Expansion scores

BAT Score (range 0–5)	Points
Blend sign	
Present	1
Absent	0
Any hypodensity	
Present	2
Absent	0
Time from onset to NCCT	
<2.5 hours	2
≥2.5 hours or unknown	0
Baseline ICH volume	
≤10 ml	0
10–20 ml	5
>20 ml	7
Recurrent ICH	
No	0
Yes	4
Anticoagulation with warfarin at onset	
No	0
Yes	6
Intraventricular extension	
No	0
Yes	2
Number of hours from onset to baseline CT	
>1	5
1–2	4
2–3	3
3–4	2
4–5	1
>5	0
HEP score (range 0–18)	
Time from ICH onset to baseline CT <3 hours	
No	0
Yes	3
Diagnosis of dementia	
No	0
Yes	4
Current smoking	
No	0
Yes	3
Antiplatelet drug use	
No	0
Yes	3
GCS Score at presentation	
3–5	3
6–8	2
9–11	1
12–15	0
SAH on baseline CT	
No	0
Yes	2

BAT bleeding assessment tool, *NCCT* noncontrast head CT, *ICH* intracerebral hemorrhage, *CT* computed tomography, *HEP* HIT expert probability, *GCS*= Glasgow Coma Scale, *SAH* subarachnoid hemorrhage

Table 2.8 Arteriovenous malformation-related intracerebral hemorrhage (AVICH) score

Parameter	Definition	Points
Arteriovenous malformation size (cm)[a]	<3	1
	3–6	2
	>6	3
Deep venous drainage[a]	No	0
	Yes	1
Eloquence[a]	No	0
	Yes	1
Age	<20 years	1
	20–40 years	2
	>40 years	3
Diffuse nidus[b]	No	0
	Yes	1
Glasgow Coma Scale score[c]	13–15	0
	5–12	1
	3–4	2
Intracerebral hemorrhage volume[c]	<30 cm^3	0
	>30 cm^3	1
Intraventricular hemorrhage[c]	No	0
	Yes	1

[a]Derived from the Spetzler–Martin grading system
[b]Derived from the Lawton–Young grading system
[c]Derived from the intracerebral hemorrhage score: maximum, 13 points; minimum, 2 points

Table 2.9 Grading System proposed by the World Federation of Neurological Surgeons (WFNS) for the Classification of subarachnoid hemorrhage

WFNS Grade	Glasgow coma scale score	Motor deficit
I	15	Absent
II	14–13	Absent
III	14–13	Present
IV	12–7	Present or absent
V	6–3	Present or absent

WFNS World Federation of Neurological Surgeons

Table 2.10 Medical Research Council muscle power scale

Grade 0 No contraction
Grade 1 Flicker of contraction
Grade 2 Active movement, with gravity eliminated
Grade 3 Active movement against gravity
Grade 4 Active movement against gravity and resistance
Grade 5 Normal strength (full resistance)

Table 2.11 Myasthenia gravis composite score

Ptosis upward gaze	
>60 seconds	0
11–60 seconds	1
1–10 seconds	2
Spontaneous	3
Double vision on lateral gaze	
Left or right >60 seconds	0
11–60 seconds	1
1–10 seconds	3
Spontaneous	4
Eye closure	
Normal	0
Mild weakness	0
Moderate weakness	1
Severe weakness	2
Talking	
Normal	0
Intermittent slurring or nasal speech	2
Constant slurring or nasal speech but can be understood	4
Difficult to understand speech	6
Chewing	
Normal	0
Fatigue with solid food	2
Fatigue with soft food	4
Gastric tube	6
Swallowing	
Normal	0
Rare episode of choking	2
Frequent choking necessitating dietary changes	5
Gastric tube	6
Breathing	
Normal	0
Shortness of breath with exertion	2
Shortness of breath	
At rest	4
Ventilator dependence	9
Neck flexion	
Normal	0
Mild weakness	1
Moderate weakness	3
Severe weakness	4
Shoulder abduction	
Normal	0
Mild weakness	2
Moderate weakness	4
Severe weakness	5
Hip flexion	
Normal	0
Mild weakness	2
Moderate weakness	4
Severe weakness	5
Total range	*0–50*

Table 2.12 Hughes functional grading scale for Guillain–Barré syndrome

Score	Description
0	Healthy
1	Minor symptoms or signs, able to run
2	Able to walk 5 m independently
3	Able to walk 5 m with a walker or support
4	Bed or chair-bound
5	Requiring assisted ventilation
6	Death

Table 2.13 EGRIS

Measure	Categories	Score
Days between the onset of weakness and hospital admission	>7 days	0
	4–7 days	1
	≤3 days	2
Facial and/or bulbar weakness at hospital admission	Absence	0
	Presence	1
MRC sum score at hospital admission	60–51	0
	50–41	1
	40–31	2
	30–21	3
	≤20	4
EGRIS		0–7

EGRIS Erasmus GBS Respiratory Insufficiency Score, *MRC* Medical Research Council

Table 2.14 Modified Fisher scale

Grade	Finding
1	Focal or diffuse thin SAH without IVH
2	Focal or diffuse thin SAH with IVH
3	Thick SAH present without IVH
4	Thick SAH present with IVH

SAH subarachnoid hemorrhage, *IVH* intraventricular hemorrhage

Table 2.15 TICI Reperfusion scale

0	No perfusion
1	Perfusion past the initial obstruction, but little or slow distal perfusion with limited branch filling
2a	Partial perfusion of less than half of the vascular distribution of the occluded artery
2b	Partial perfusion of more than half of the vascular distribution of the occluded artery
3	Full perfusion with filling of all distal branches

Table 2.16 The Marshall computerized tomography classification

Marshall CT classification	Definition
DI I	No visible intracranial pathology seen on CT
DI II	Cisterns present with midline shift 0–5 mm and/or lesion densities present; no high- or mixed-density lesion >25 cm^3; may include bone fragments and foreign bodies
DI III	Cisterns compressed or absent with midline shift of −5 mm; no high- or mixed-density lesion
DI IV	Midline shift >5 mm; no high- or mixed-density > 25 cm^3
Evacuated mass lesion	Any lesion surgically evacuated
Nonevacuated mass lesion	High- or mixed-density lesion cm^3; not surgically evacuated

CT computerized tomography, *DI* diffuse injury

Table 2.17 The Rotterdam computerized tomography score

Rotterdam CT score	Score
Basal cisterns	
Normal	0
Compressed	1
Absent	2
Midline shift	
No shift or ≤ 5 mm	0
Shift > 5 mm	1
Epidural mass lesion	
Present	0
Absent	1
IVH or tSAH	
Absent	0
Present	1
Sum score	+1

CT computerized tomography, *IVH* intraventricular hemorrhage, *tSAH* traumatic subarachnoid hemorrhage

Table 2.18 The Helsinki computerized tomography score chart

Variable	Score
Mass lesion type(s)	
Subdural hematoma	2
Intracerebral hematoma	2
Epidural hematoma	-3
Mass lesion size >25 cm^3	2
Intraventricular hemorrhage	3
Suprasellar cisterns	
Normal	0
Compressed	1
Obliterated	5
Sum score	-3 to 14

References

1. Sembill JA, Gerner ST, Volbers B, et al. Severity assessment in maximally treated ICH patients – the max-ICH score. J Neurol Sci. 2017;381:68–9.
2. Ziai WC, Siddiqui AA, Ullman N, et al. Early therapy intensity level (TIL) predicts mortality in spontaneous intracerebral hemorrhage. Neurocrit Care. 2015;23:188–97.
3. Brooks BR. El Escorial World Federation of Neurology criteria for the diagnosis of amyotrophic lateral sclerosis. Subcommittee on Motor Neuron Diseases/Amyotrophic Lateral Sclerosis of the World Federation of Neurology Research Group on Neuromuscular Diseases and the El Escorial "Clinical limits of amyotrophic lateral sclerosis" workshop contributors. J Neurol Sci. 1994;124:96–107.
4. Fokke C, van den Berg B, Drenthen J, Walgaard C, van Doorn PA, Jacobs BC. Diagnosis of Guillain-Barre syndrome and validation of Brighton criteria. Brain. 2014;137:33–43.
5. Feinstein AR. An additional basic science for clinical medicine: IV. The development of clinimetrics. Ann Intern Med. 1983;99:843–8.
6. Feinstein AR. The intellectual crisis in clinical science: medaled models and muddled mettle. Perspect Biol Med. 1987;30:215–30.
7. Olsen MH, Jensen HR, Ebdrup SR, et al. Automated pupillometry and the FOUR score – what is the diagnostic benefit in neurointensive care? Acta Neurochir. 2020;162:1639–45.
8. Bruno MA, Ledoux D, Lambermont B, et al. Comparison of the Full Outline of UnResponsiveness and Glasgow Liege Scale/Glasgow Coma Scale in an intensive care unit population. Neurocrit Care. 2011;15:447–53.
9. Fischer M, Ruegg S, Czaplinski A, et al. Inter-rater reliability of the Full Outline of UnResponsiveness score and the Glasgow Coma Scale in critically ill patients: a prospective observational study. Crit Care. 2010;14:R64.
10. Idrovo L, Fuentes B, Medina J, et al. Validation of the FOUR Score (Spanish Version) in acute stroke: an interobserver variability study. Eur Neurol. 2010;63:364–9.
11. Marcati E, Ricci S, Casalena A, Toni D, Carolei A, Sacco S. Validation of the Italian version of a new coma scale: the FOUR score. Intern Emerg Med. 2012;7:145–52.
12. Adcock AK, Kosiorek H, Parich P, Chauncey A, Wu Q, Demaerschalk BM. Reliability of robotic telemedicine for assessing critically ill patients with the Full Outline of UnResponsiveness Score and Glasgow Coma Scale. Telemed E-Health. 2017;23:555–60.
13. Almojuela A, Hasen M, Zeiler FA. The Full Outline of UnResponsiveness (FOUR) score and its use in outcome prediction: a scoping review of the pediatric literature. J Child Neurol. 2019;883073818822359
14. Weiss N, Venot M, Verdonk F, et al. Daily FOUR score assessment provides accurate prognosis of long-term outcome in out-of-hospital cardiac arrest. Rev Neurol-France. 2015;171:437–44.
15. van Ettekoven CN, Brouwer MC, Bijlsma MW, Wijdicks EFM, van de Beek D. The FOUR score as predictor of outcome in adults with bacterial meningitis. Neurology. 2019;92:E2522–6.
16. Zeiler FA, Lo BWY, Akoth E, et al. Predicting outcome in subarachnoid hemorrhage (SAH) utilizing the Full Outline of UnResponsiveness (FOUR) Score. Neurocrit Care. 2017;27:381–91.
17. Kramer AA, Wijdicks EFM, Snavely VL, et al. A multicenter prospective study of interobserver agreement using the Full Outline of Unresponsiveness score coma scale in the intensive care unit. Crit Care Med. 2012;40:2671–6.
18. Wijdicks EFM, Kramer AA, Rohs T Jr, et al. Comparison of the Full Outline of UnResponsiveness score and the Glasgow Coma Scale in predicting mortality in critically ill patients. Crit Care Med. 2015;43:439–44.
19. Almojuela A, Hasen M, Zeiler FA. The Full Outline of UnResponsiveness (FOUR) Score and its use in outcome prediction: a scoping systematic review of the adult literature. Neurocrit Care. 2019;31:162–75.
20. Jennett B, Teasdale G. Management of Head Injuries, vol. 20. Philadelphia: F. A. Davis Company; 1981.
21. Brain Resuscitation Clinical Trial ISG. Randomized clinical study of thiopental loading in comatose survivors of cardiac arrest. N Engl J Med. 1986;314:397–403.
22. Lyden PD, Lu M, Levine SR, Brott TG, Broderick J, Group NrSS. A modified National Institutes of Health Stroke Scale for use in stroke clinical trials: preliminary reliability and validity. Stroke. 2001;32:1310–7.
23. Seel RT, Dijkers MP, Johnston MV. Developing and using evidence to improve rehabilitation practice. Arch Phys Med Rehabil. 2012;93:S97–100.
24. Seel RT, Macciocchi S, Velozo CA, et al. The safety assessment measure for persons with traumatic brain injury: item pool development and content validity. NeuroRehabilitation. 2016;39:371–87.

25. Seel RT, Steyerberg EW, Malec JF, Sherer M, Macciocchi SN. Developing and evaluating prediction models in rehabilitation populations. Arch Phys Med Rehabil. 2012;93:S138–53.
26. McCauley SR, Wilde EA, Moretti P, et al. Neurological outcome scale for traumatic brain injury: III. Criterion-related validity and sensitivity to change in the NABIS hypothermia-II clinical trial. J Neurotrauma. 2013;30:1506–11.
27. American Congress of Rehabilitation Medicine BI-ISIGDoCTF, Seel RT, Sherer M, et al. Assessment scales for disorders of consciousness: evidence-based recommendations for clinical practice and research. Arch Phys Med Rehabil. 2010;91:1795–813.
28. Garcia-Rudolph A, Cegarra B, Opisso E, Tormos JM, Bernabeu M, Sauri J. Predicting length of stay in patients admitted to stroke rehabilitation with severe and moderate levels of functional impairments. Medicine (Baltimore). 2020;99:e22423.
29. Collaborators MCT, Perel P, Arango M, et al. Predicting outcome after traumatic brain injury: practical prognostic models based on large cohort of international patients. BMJ. 2008;336:425–9.
30. Maas AI, Marmarou A, Murray GD, Teasdale SG, Steyerberg EW. Prognosis and clinical trial design in traumatic brain injury: the IMPACT study. J Neurotrauma. 2007;24:232–8.
31. Mikati AG, Flahive J, Khan MW, et al. Multicenter validation of the survival after acute civilian penetrating brain injuries (SPIN) score. Neurosurgery. 2019;85:E872–9.
32. Muir KW, Weir CJ, Murray GD, Povey C, Lees KR. Comparison of neurological scales and scoring systems for acute stroke prognosis. Stroke. 1996;27:1817–20.
33. NIH Stroke Scale International Home. http://www.nihstrokescale.org/2017/index.html. Accessed 15 Sept 2020.
34. Lyden PD, Lau GT. A critical appraisal of stroke evaluation and rating scales. Stroke. 1991;22:1345–52.
35. Goldstein LB, Bertels C, Davis JN. Interrater reliability of the NIH stroke scale. Arch Neurol. 1989;46:660–2.
36. Duvekot MHC, Venema E, Rozeman AD et al. Comparison of eight prehospital stroke scales to detect intracranial large-vessel occlusion in suspected stroke (PRESTO): a prospective observational study. Lancet Neurol. 2021;20:213–21.
37. Asuzu D, Nystrom K, Schindler J, et al. TURN score predicts 90-day outcome in acute ischemic stroke patients after IV thrombolysis. Neurocrit Care. 2015;23:172–8.
38. Gregorio T, Pipa S, Cavaleiro P, et al. Assessment and comparison of the four most extensively validated prognostic scales for intracerebral hemorrhage: systematic review with meta-analysis. Neurocrit Care. 2019;30:449–66.
39. Hemphill JC 3rd, Bonovich DC, Besmertis L, Manley GT, Johnston SC. The ICH score: a simple, reliable grading scale for intracerebral hemorrhage. Stroke. 2001;32:891–7.
40. Morotti A, Dowlatshahi D, Boulouis G, et al. Predicting intracerebral hemorrhage expansion with noncontrast computed tomography the BAT score. Stroke. 2018;49:1163–9.
41. Huynh TJ, Aviv RI, Dowlatshahi D, et al. Validation of the 9-point and 24-point hematoma expansion prediction scores and derivation of the PREDICT A/B scores. Stroke. 2015;46:3105–10.
42. Wang X, Arima H, Al-Shahi Salman R, et al. Clinical prediction algorithm (BRAIN) to determine risk of hematoma growth in acute intracerebral hemorrhage. Stroke. 2015;46:376–81.
43. Yao X, Xu Y, Siwila-Sackman E, Wu B, Selim M. The HEP score: a nomogram-derived hematoma expansion prediction scale. Neurocrit Care. 2015;23:179–87.
44. Gupta R, Arora VK. Performance evaluation of APACHE II score for an Indian patient with respiratory problems. Indian J Med Res. 2004;119:273–82.
45. Neidert MC, Lawton MT, Kim LJ, et al. International multicentre validation of the arteriovenous malformation-related intracerebral haemorrhage (AVICH) score. J Neurol Neurosurg Psychiatry. 2018;89:1163–6.
46. Neidert MC, Lawton MT, Mader M, et al. The AVICH score: a novel grading system to predict clinical outcome in arteriovenous malformation-related intracerebral hemorrhage. World Neurosurg. 2016;92:292–7.
47. Aids to examination of the peripheral nervous system. Vol Memorandum no. 45. London: Her Majesty's Stationery Office; 1976.
48. Degen LAR, Mees SMD, Algra A, Rinkel GJE. Interobserver variability of grading scales for aneurysmal subarachnoid hemorrhage. Stroke. 2011;42:1546–9.
49. Giraldo EA, Mandrekar JN, Rubin MN, et al. Timing of clinical grade assessment and poor outcome in patients with aneurysmal subarachnoid hemorrhage clinical article. J Neurosurg. 2012;117:15–9.
50. Curcio KR. Instruments for assessing chemotherapy-induced peripheral neuropathy: a review of the literature. Clin J Oncol Nurs. 2016;20:144–51.
51. Sanders DB, Tucker-Lipscomb B, Massey JM. A simple manual muscle test for myasthenia gravis: validation and comparison with the QMG score. Ann N Y Acad Sci. 2003;998:440–4.
52. Barnett C, Herbelin L, Dimachkie MM, Barohn RJ. Measuring clinical treatment response in myasthenia gravis. Neurol Clin. 2018;36:339–53.
53. Barohn RJ, McIntire D, Herbelin L, Wolfe GI, Nations S, Bryan WW. Reliability testing of the quantitative myasthenia gravis score. Ann N Y Acad Sci. 1998;841:769–72.
54. Hughes RA, Newsom-Davis JM, Perkin GD, Pierce JM. Controlled trial prednisolone in acute polyneuropathy. Lancet. 1978;2:750–3.

55. Burns TM, Conaway MR, Cutter GR, Sanders DB, Muscle SG. Less is more, or almost as much: a 15-item quality-of-life instrument for myasthenia gravis. Muscle Nerve. 2008;38:957–63.
56. Farrugia ME, Harle HD, Carmichael C, Burns TM. The oculobulbar facial respiratory score is a tool to assess bulbar function in myasthenia gravis patients. Muscle Nerve. 2011;43:329–34.
57. Walgaard C, Lingsma HF, Ruts L, et al. Prediction of respiratory insufficiency in Guillain-Barre syndrome. Ann Neurol. 2010;67:781–7.
58. Fisher CM, Roberson GH, Ojemann RG. Cerebral vasospasm after ruptured aneurysm – clinico-radiologic correlation. Stroke. 1977;8:11–11..
59. Claassen J, Bernardini GL, Kreiter K, et al. Effect of cisternal and ventricular blood on risk of delayed cerebral ischemia after subarachnoid hemorrhage: the Fisher scale revisited. Stroke. 2001;32:2012–20.
60. Dengler NF, Sommerfeld J, Diesing D, Vajkoczy P, Wolf S. Prediction of cerebral infarction and patient outcome in aneurysmal subarachnoid hemorrhage: comparison of new and established radiographic, clinical and combined scores. Eur J Neurol. 2018;25:111–9.
61. Hijdra A, Brouwers PJ, Vermeulen M, van Gijn J. Grading the amount of blood on computed tomograms after subarachnoid hemorrhage. Stroke. 1990;21:1156–61.
62. van der Jagt M, Hasan D, Bijvoet HW, Pieterman H, Koudstaal PJ, Avezaat CJ. Interobserver variability of cisternal blood on CT after aneurysmal subarachnoid hemorrhage. Neurology. 2000;54:2156–8.
63. Woo PYM, Tse TPK, Chan RSK, et al. Computed tomography interobserver agreement in the assessment of aneurysmal subarachnoid hemorrhage and predictors for clinical outcome. J Neurointerv Surg. 2017;9:1118–24.
64. Dupont SA, Wijdicks EFM, Manno EM, Lanzino G, Brown RD Jr, Rabinstein AA. Timing of computed tomography and prediction of vasospasm after aneurysmal subarachnoid hemorrhage. Neurocrit Care. 2009;11:71–5.
65. Dupont SA, Wijdicks EFM, Manno EM, Lanzino G, Rabinstein AA. Prediction of angiographic vasospasm after aneurysmal subarachnoid hemorrhage: value of the Hijdra sum scoring system. Neurocrit Care. 2009;11:172–6.
66. van Norden AG, van Dijk GW, van Huizen MD, Algra A, Rinkel GJ. Interobserver agreement and predictive value for outcome of two rating scales for the amount of extravasated blood after aneurysmal subarachnoid haemorrhage. J Neurol. 2006;253:1217–20.
67. Nicholson P, Hilditch CA, Neuhaus A, et al. Per-region interobserver agreement of Alberta Stroke Program Early CT Scores (ASPECTS). J Neurointerv Surg. 2020;12:1069–1071.
68. Dargazanli C, Consoli A, Barral M, et al. Impact of modified TICI 3 versus modified TICI 2b reperfusion score to predict good outcome following endovascular therapy. AJNR Am J Neuroradiol. 2017;38:90–6.
69. Jang KM, Nam TK, Ko MJ, et al. Thrombolysis in cerebral infarction grade 2C or 3 represents a better outcome than 2B for endovascular thrombectomy in acute ischemic stroke: a network meta-analysis. World Neurosurg. 2020;136:e419–39.
70. Rizvi A, Seyedsaadat SM, Murad MH, et al. Redefining 'success': a systematic review and meta-analysis comparing outcomes between incomplete and complete revascularization. J Neurointerv Surg. 2019;11:9–13.
71. Lindfors M, Vehvilainen J, Siironen J, Kivisaari R, Skrifvars MB, Raj R. Temporal changes in outcome following intensive care unit treatment after traumatic brain injury: a 17-year experience in a large academic neurosurgical centre. Acta Neurochir. 2018;160:2107–15.

3 Patterns: Interpreting Focal Findings

Neurology is, by definition, a localizing specialty. Therefore, once specific findings appear, we must establish location (i.e., hemispheres, brainstem, cerebellum, spine, or the motor unit) and then get more granular detail (i.e., cortex, thalamus, pons, and vermis), sometimes even up to the tracts (i.e., posterior columns, pyramidal tract). Findings not localized to specific structures may indicate global insults to the brain, often secondary to revived vital signs or abnormal physiology. Moreover, not all regions represent eloquent parts of the brain; significant injury may cause minor findings and vice versa.

How does this work when one is standing beside a critical care bed and looking at the patient? [1] The neurologic examination has a few guiding principles. First, the main aim of a neurologic examination is discovery and exposition of a response and often requires a look at subtleties. Good observation involves knowing where to look and what to test. Simple handheld tools suffice. Next, find a possible pattern. Simplicity usually overrides complexity; many "neurologic manifestations" must be discarded as irrelevant. Finally, expect to examine the patient more than once; as the old saying goes (paraphrased), "a repeat neurologic examination is the best diagnostic test."

The reader will recognize that the neurologic examination in critically ill patients is different. With acute brain lesions, we look at alertness (or lack thereof) and judge the level of consciousness, presence of eye movements, motor responses upon provocation by a noxious stimulus, a series of brainstem reflexes, major (and, less often, subtle) asymmetries, and abnormal movements. In acute lesions of the spinal cord, we identify levels within the cord suggested by patterns of weakness, sensory loss, and bladder function. In acute injury of the peripheral nervous system or muscle, we look at progression of weakness, the mechanics of respiration, handling of secretions, swallowing, vocal tone, and pauses in speech. Initially, we are much less concerned with chronic neurologic disease manifestations.

Let us start with the main question. How do we process the symptoms presented to us? In the previous chapter, we learned that scales summarizing neurologic examination are inherently insufficient; few scales measure features needed for urgent decisions. Scales are approximations that should be viewed as the basics of a neurologic examination—not the whole thing. We have to be seriously worried that we will be buried under scores with cut-off points for severity.

Let us focus on focality. This is the opposite of general decline in function; rather, some parts work well and others do not. Focal signs may be obvious to everyone (e.g., hemiparesis) or subtle and easily overlooked by a rushed clinician (e.g., hemispatial neglect or anomic aphasia). Laureno said "neurologists see asymmetries daily" [2] and "functional asymmetries correlate poorly with

E. F. M. Wijdicks, *Examining Neurocritical Patients*, https://doi.org/10.1007/978-3-030-69452-4_3

anatomical asymmetries and functional asymmetries do not correlate with each other." Indeed, we do not believe too much in symmetries and try to accept normal asymmetries when we think it is normal. The neurologic examination should separate the wheat from the chaff—what matters, what counts, and what to disregard. Narrowing it down is what we do. Moreover, we need to guard against the post hoc fallacy of attributing one finding to another without a causal relationship. We must not fixate on a certain scenario to the exclusion of inconvenient, contradictory findings.

Modern intensive-care practice often involves sedation, which complicates assessment. A not-so-unusual scenario is an unresponsive patient arriving at the emergency department in respiratory distress, which requires intubation. Anesthetic drugs and paralytic agents make the neurologic examination a game of guesswork, and the prior immediate history becomes an important guide. The attending physician may order a series of neuroimaging tests (hopefully, the right ones) only to find out later that none were contributory or necessary. Sedated patients can render good judgment impossible. Patients with prior IV infusions and organ dysfunction need time to wash these out. Physicians seldom fully account for these effects. (For a more detailed discussion, see Chap. 5.)

Deficiencies in neurologic skills have been laid bare, and part of it may be "neurophobia" [3]. It leads to near-total reliance on the neurologist. Recent Telestroke programs have also revealed something worrisome. Telestroke patients may present with some facial asymmetry or deviation of tongue protrusion, which often means little. Providers may struggle to assess visual fields, eye movements, and degree of weakness; they may overlook the crucial findings (i.e., forget to examine for neglect, use double stimulation).

Many works have described the neurologic examination, and there is a wide choice of educational books [4–18]. However, most textbooks are deficient in describing how to localize abnormalities in acutely ill neurologic patients. Understanding the meaning of the physical findings comes with experience, which requires seeing many patients at presentation, following the clinical course, and learning the signs of clinical deterioration. This requires years and years of neurointensive care experience and practice. Taking only a few services a year (because there are many intensivists in your medical center) won't cut it.

This chapter strives to achieve the following: (1) explain the affected anatomy and examination techniques; (2) discuss how to collate the findings; and (3) discover how to localize the lesion in acutely ill neurologic patients. The findings of the neurologic examination of the neurocritically ill patient often have major consequences that lead to medical or surgical interventions; the decision to intervene may even be irrespective of neuroimaging results.

The Neurologic Examination—Method and Function

Neurologic examination in acutely ill patients is limited due to most patients' inability to cooperate with testing and disorders that produce clinical pictures not seen in ambulant patients. The approach therefore differs from a clinic-based examination but is also partly the same. First, the brainstem reflexes are of utmost importance, and abnormalities become immediately significant. Second, the neurologic findings can predict a certain clinical course and deterioration. Third, syndromic neurology in deteriorating patients exists and can be recognized. Fourth, neurologists use tools unique to neurology, as well as tools used by other specialties. These include safety pins, reflex hammers, tuning forks (used in ENT), and ophthalmoscopes (used in ophthalmology) [19, 20]. Examining for acute injury to the central and peripheral nervous system requires a specific method to avoid missing crucial symptoms. Each part of the nervous system, including the cerebral hemispheres, brainstem, spinal cord, peripheral nerves, and muscles, has its own method. We will always somehow need a "top-down" neurologic examination, which includes language, executive function, brainstem reflexes, tone, strength, coordination, sensation, and reflexes (tendon, cutaneous, grasp, plantar, and coordinated, nocifensive motor responses).

Any examination starts with inspection, which is often done insufficiently or perfunctorily. Note the overall demeanor. When approaching the bedside, we may see something seriously amiss. Patients stare into space seeking contact, may not be "present," mumble speech, may not move certain parts, or move parts of the body "too much." Facial expressions may be reduced. Unresponsive patients may be more responsive than they appear, and not all unresponsive patients are comatose. Inspection can be quite subjective, and terms such as "frailty," "older than biological age," or, at the end of the spectrum, "moribund" have different meanings for different observers. The inspection should take note of body hygiene, odors, and dishevelment. Some are very specific (e.g., uremia and urine, hyperammonia and rotten eggs) but not always; remember, for example, that vodka-intoxicated patients do not smell like alcohol. Bruises, cuts, tattoos, and needle marks may have meaning (and may indicate abuse), and we look for them specifically after we attain a good sense of the cause of injury. A mastoid bruise (ecchymosis) indicates a petrous temporal bone fracture and possible CSF leak. Nails and fingertips inspection can be diagnostic (e.g., splinter hemorrhages or Osler's nodes point to endocarditis). A tongue bite is not routinely sought by others, but we do, and is clinically relevant (better proof of a seizure than EEG). Pressure areas or sores may indicate the patient has been lying on the ground for long, often days. When spotted, these indicate the potential of rhabdomyolysis. Some of these consequential observations in patients with acute brain injury are shown in Fig. 3.1. General physical findings are much less common in patients with acute neuromuscular diseases, but the highly diagnostic skin rash in dermatomyositis will have to be recognized (Fig. 3.2).

Next, be aware of warning signs such as episodic bradycardia, reduced breathing frequency, apneic periods, fever, or hypothermia. Each may require correction to more normal values before proceeding to the neurologic examination. How seriously abnormal vital signs affect neurologic examination (and by what degree) has never been studied. Systolic hypertension surprisingly is tolerated by many, until it suddenly causes a decline in responsiveness (approximately at 60 mmHg of systolic pressure).

After inspection (and resuscitation/correction of abnormal vital signs), there are five tiers to the neurologic examination.

1. Once the examination is complete, list the main abnormal findings.
2. Normal findings are mentioned if they do not fit the emerging clinical picture.
3. Find the affected structure (i.e., localization).
4. Confirm the affected structure with additional neuroimaging.
5. Predict nature of the injury.

This sequence must be followed each time and will allow some standardization of the examinations. Gradually working from a general clinical picture to a more specific picture is a main concept of neurologic assessment. Moreover, less is more. Once an injury is established, a cause must be considered. (The Bradford Hill criteria [21], which look at aspects of an association and when a factor is a cause or contributing factor, offer some guidance. These criteria examine (1) strength, consistency, and coherence of the association; (2) specificity of the association; (3) temporal relationships; (4) biological plausibility; and (5) analogy to a known phenomenon and, if relevant, dose-response association and experimental confirmation.)

Finally, management decisions (and occasionally triage) follow. The examination will have to be repeated after an intervention is performed. This approach has efficiency and coherence and could lead to improved practice quality.

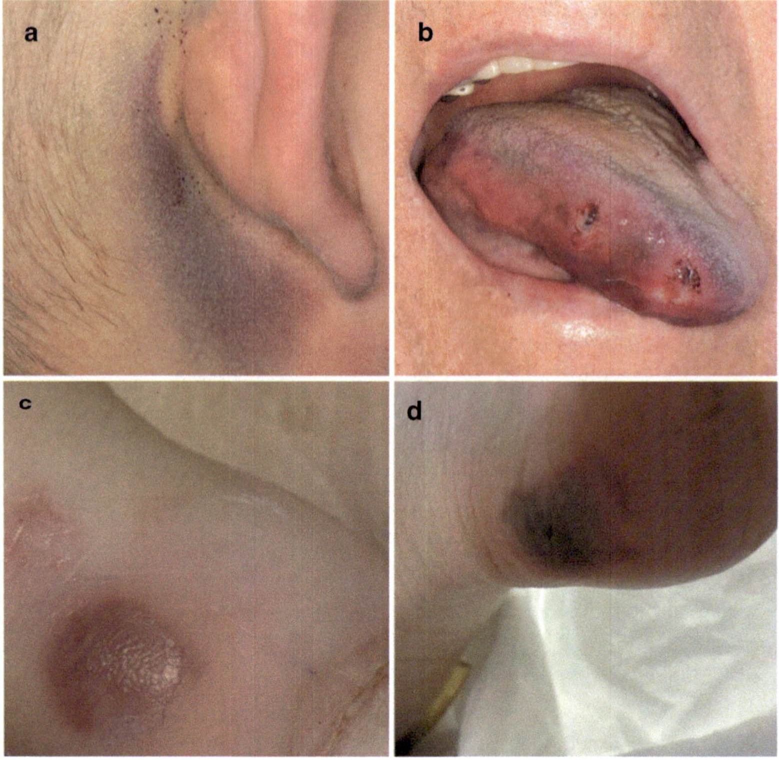

Fig. 3.1 Physical findings with consequences (**a**: Mastoid ecchymosis or Battle sign, **b**: tongue bite from seizure, **c**: Osler node on a finger, **d**: Heel pressure ulcer)

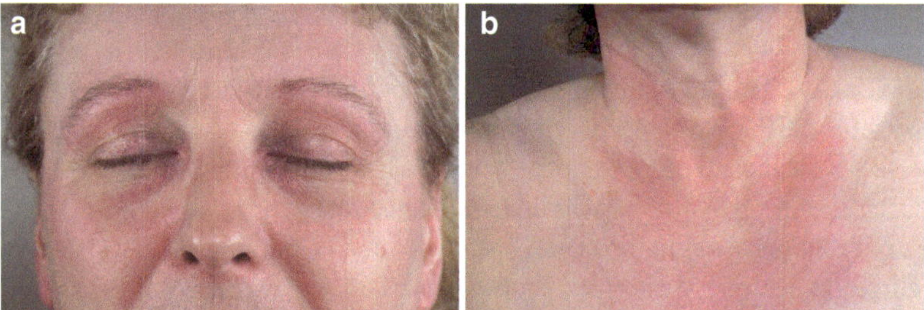

Fig. 3.2 (**a, b**) Dermatomyositis rash

Let's work our way through these above mentioned components and let's start with how to localize the neurologic findings—in other words, which part of the brain is predominantly dysfunctional and producing these signs.

Neurologic Examination: Localizing Findings

Each domain of the neurologic examination of the acutely ill patient can be further explored and documented. How does one go about it? In this stage of the examination, the area in the brain from which the findings most likely originate becomes important. It has always been useful to "go by systems," just as internists work by organ systems to ensure completeness. We can skip over a few things if the diagnosis is clear, but then we can focus in detail on what is important for decision-making.

Orienting Macro Localization

It helps organizationally to divide the brain into the hemispheres with separate lobes, the brainstem, and the cerebellum. Each lesion produces a combination of symptoms explained by the involved structures. These are shown in Fig. 3.3. Usually, small lesions in the brainstem produce major findings due to compactness of structures. In the cerebral hemispheres, large lesions can still be clinically silent, particularly in the right frontal lobe (actually prefrontal). Only some enlargement or mass effect impinging on other structures could result in noticeable symptomatology. Cerebral hemispheric lesions usually present themselves with disturbances in multitasking and switching, reasoning, abnormality of perception, memory disturbances, as well as weakness and, eventually, abnormal levels of consciousness. Moreover, a visual field defect with hemiparesis places the lesion in the hemisphere. Lesions in the brainstem characteristically produce a cranial nerve deficit, with or without hemiparesis.

Because stroke is such a common occurrence in the intensive care unit, localization using the cerebral circulatory system helps to explain acute neurologic findings. Circulation to the brain has been traditionally been divided into the anterior and posterior circulation (Fig. 3.4). In acute ischemic stroke, theoretically, each acute brain-structure dysfunction (and often combinations) can be linked to a certain arterial territory. However, it is actually more complicated and dependent on arterial collaterals; the circle of Willis varies considerably, with frequent asymmetry and rarely an ideal configuration, as well as the distal collateral artery connections. These factors may determine if someone is immediately facing a large ischemic territorial stroke and whether removal of the clot still makes sense. Interhemispheric blood flow across the anterior communicating artery and reversal of flow in the proximal anterior cerebral artery provide collateral support in the anterior portion of the circle of Willis. The posterior communicating arteries may supply collateral blood flow in either direction between the anterior and posterior circulations. Distal branches of the major cerebellar arteries similarly provide collateral links across the vertebral and basilar segments of the posterior circulation [22–24]. Additionally, the anterior and posterior spinal arteries communicate with branches of the proximal intracranial arteries supplying the medulla and the pons. Thus, an acute occlusion of the largest cerebral artery can be silent when perfusion is fully compensated (about 30% of cases), while a single perforating artery to the brainstem can cause major disabling findings.

There are a number of other caveats. The most important is that the classic syndromes rarely present as described; more commonly, they are incomplete. The figure shows these commonly affected territories and their key findings. Note that large arteries are divided into numbered segments from stem to branch (e.g., M1 stem and M2 first branch). This information is useful in a patient with a large-vessel occlusion going into an endovascular retrieval procedure because we can predict which clinical signs remain if certain branches are not engaged or cannot be opened.

Anterior circulation syndromes, such as a carotid artery syndrome or middle cerebral artery syndrome, present with motor weakness affecting the opposite side involving face, arm, and leg, with hemi-sensory loss, conjugate gaze or preference of eye position, and head movement directed toward the side of the lesion. Broca's aphasia can occur with an infarct in the dominant hemisphere, and a more global aphasia results when the infarct is more extensive. An infarction in the nondominant hemisphere can, with careful

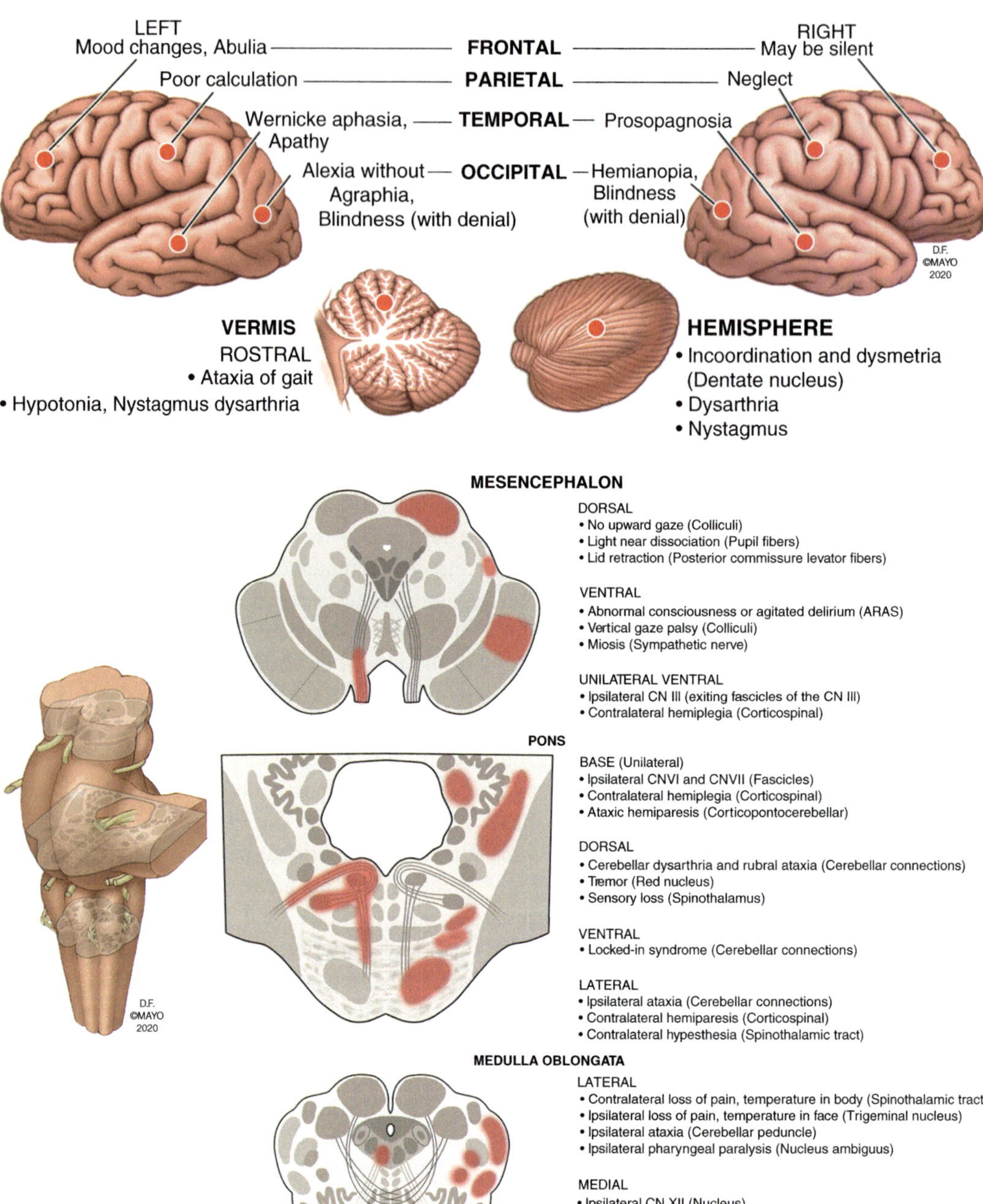

Fig. 3.3 Localization templates

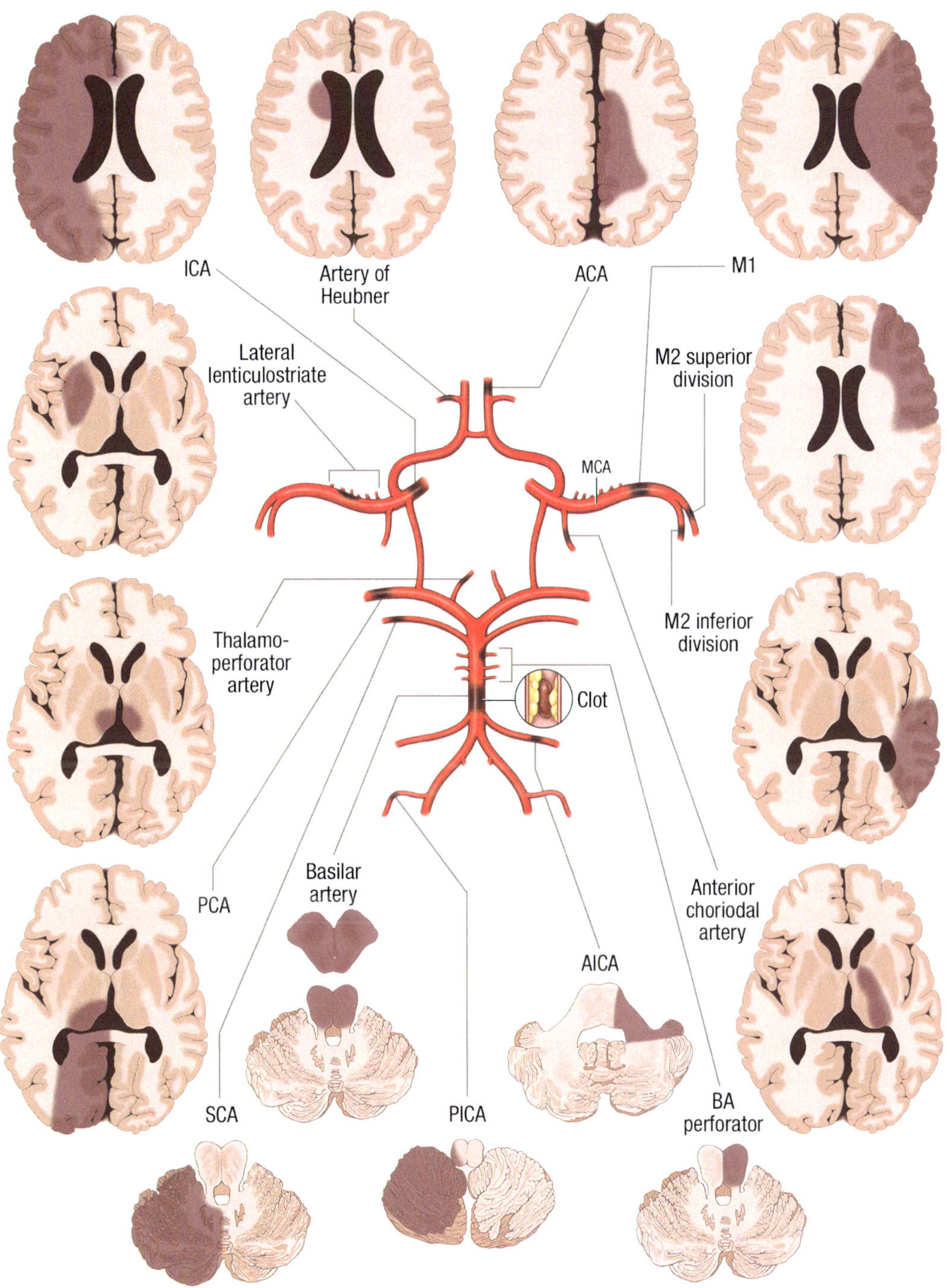

ICA: Fluctuating hemiplegia, hemianesthesia, stupor, silent Heubner: Faciobrachial weakness ACA: Hemiparesis (proximal shoulder, distal leg) Lenticulostriate: Abulia or disinhibited Thalamoperforator: Apathy, amnesia, clumsiness PCA: Visual field cut, limb numbness, alexia (left) SCA: Dysmetria, contralateral pain loss, CN IV palsy, Horner's syndrome PICA: Nystagmus, ataxia, dysarthria Basilar: Anisocoria, skew, tetraparesis, coma, locked-in AICA: Dysarthria, ataxia, hemiparesis Basilar perforator: Ataxic hemiparesis, dysarthria, clumsy hand Anterior choroidal: Hemiplegia without aphasia or gaze preference MCA M1: Eye deviation, hemiplegia, global aphasia or neglect MCA M2 superior: Aphasia, hemineglect, hemiparesis MCA M2 inferior: Visual field cut, Wernicke aphasia, fingeragnosia, alexia

Fig. 3.4 Vascular territories (key findings mentioned; see text for more clinical details)

examination, reveal attention deficits, neglect, denial, apraxia, or impaired prosody.

Posterior circulation syndromes are far more complex and frequently due to a basilar artery occlusion. Because this artery provides blood to the brainstem, its nuclei and tracts are involved and result in acute ophthalmologic findings, usually involving the third or sixth nerve but which may also result in acute ophthalmoplegia and pupillary abnormalities. If the lesion is localized exclusively to the posterior cerebral artery, an acute hemianopia is present. With bilateral occipital lobe involvement, the patient may deny or be unaware of blindness. Bilateral occipital or para-occipital infarctions may be associated with optic ataxia (inability to group objects), disturbance of visual attention, and simultanagnosia (i.e., Balint syndrome, the inability to see two objects at the same time, such as pen and glasses).

Thalamic infarctions due to involvement of the thalamo perforators originating from the most distal segment of basilar artery and posterior communicating arteries are notoriously difficult to recognize. These infarcts may involve different regions of the thalami, and thus, most of them present with an abnormal level of consciousness, vertical gaze abnormality, abnormal sleep, and anomic aphasia.

Praxis and Behavior

The mental faculties and their examination, as established in neurology clinics, are far more difficult to examine in critical illness. Many patients have perceptional and attentional difficulties, cannot retain memories, and are unable to perform a recall test. Patients who appear cognitively impaired may improve when organ failure recovers or when sedative drugs and opioids have washed out. However, some tests may still be possible in the "altered," "confused," and "meandering" patient. A simple test of attention (a forerunner of the FOUR score) involves asking the patient to raise a hand when hearing words that include the letter A (e.g., "Schools and Highways cost money; we all pay for them through taxes") and to show a fist, peace sign, or thumbs-up sign [25]. We can test perception by asking the patient to name an object (e.g., keys). Orientation can be tested by asking for the place and the time. Failure to identify oneself is only seen with advanced (chronic) dementia, functional disorder, or (acute) psychosis. Memory for remote events might involve asking the patient to name his/her wedding date. Ability to reason might include asking patients to differentiate between water and ice or to explain a proverb such as "the early bird catches the worm."

Although it is less common, patients may occasionally display clinical signs of delusions defined by erroneous beliefs and often centered on a single thought. Hallucinations are typically visual and not aural.

Frontal lobe syndromes appear in many guises, such as loss of vitality and notably slow thinking, and may be manifested by weird behavior, sexual harassment, cynically inappropriate remarks in an attempt to be humorous, or intense irritability. Any executive function requiring advanced planning or some degree of organization is disturbed but may be covered up by euphoria, platitudes in speech, or "robot-like" behavior; many will display conserved social graces. Temporal lobe masses may also generate behavioral changes and changes in dominance (i.e., left in formerly right-handed persons).

Temporal masses may change a normal personality to depressive and/or apathetic. More posterior localization in the dominant temporal lobe may produce Wernicke's aphasia. This classic type of aphasia is recognized by continuously "empty" speech, often with syllables, words, or phrases at the end of sentences and, characteristically, with incomprehensible content (words that cannot be found in any dictionary, also known as neologisms). Involvement of the nondominant temporal lobe may be manifested only by an upper-quadrant hemianopia, if found, and nonverbal auditory agnosia (inability to recognize daily familiar sounds, such as a loud clap or tearing of paper).

Parietal lobe masses also produce localization-dependent effects. Nondominant right parietal lesions usually cause neglect of the paralyzed right limb up to entire unawareness but also cause marked inertia and aloofness. A dominant parietal lobe impairs normal arithmetical skills, recognition of fingers, and right–left orientation. A nonfluent aphasia may occur as well.

Occipital lobe masses produce hemianopia. When only the inferior occipital cortex is involved, achromatopsia (loss of color vision in a hemianopic field) or abnormal color naming ("What is the color of the sky, an apple, and a tomato?") may result. Extension into the subcortical area from edema might produce alexia without agraphia but all in a left-dominant occipital lesion. An alert patient (or one with some minor drowsiness) will allow us to look for higher function. Neurologists often use Latin and Greek combined and love the Greek A prefix (Table 3.1).

Table 3.1 The A (Greek for without) prefix in neurology

Diagnosis	Testing
Aphasia	Agrammatism
	Paraphasia
	Impaired naming
	Impaired writing
	Abnormal prosody
	Overuse of connectors (*if, and, but*)
	Neologisms
Apraxia	No weakness but no motor performance
	Inability to copy or pantomime
Abulia	Grasp, snout reflex
	Motor impersistence
	Paratonia (gegenhalten)
Agnosia	Neglect left side
	Cortical blindness
	Prosopagnosia

Assessing Speech and Language

For a simple organization of the common aphasias, see Fig. 3.5. Albeit a key component of assessment, examination of speech and language is likely shortchanged in the neurosciences intensive care unit. There are three reasons: (1) patients are intubated, and many later undergo tracheostomy; (2) patients are agitated and confused, which confounds routine assessment of language domains, fluency, and clarity; (3) lingering sedation may make speech slow and slurred; it is anyone's guess whether more was present. The diagnosis of acute dysarthria and aphasia has major implications and should not be treated in a cavalier manner. I have observed several instances where a patient was asked how he or she felt and answered with just a single syllable, even "fine" or "OK." Instead of moving along, as some are tempted to do, physicians should ask more questions to reveal marked difficulty with expression or repetition. I have observed instances in which marked dysarthria was attributed to absent dentures and dry mouth

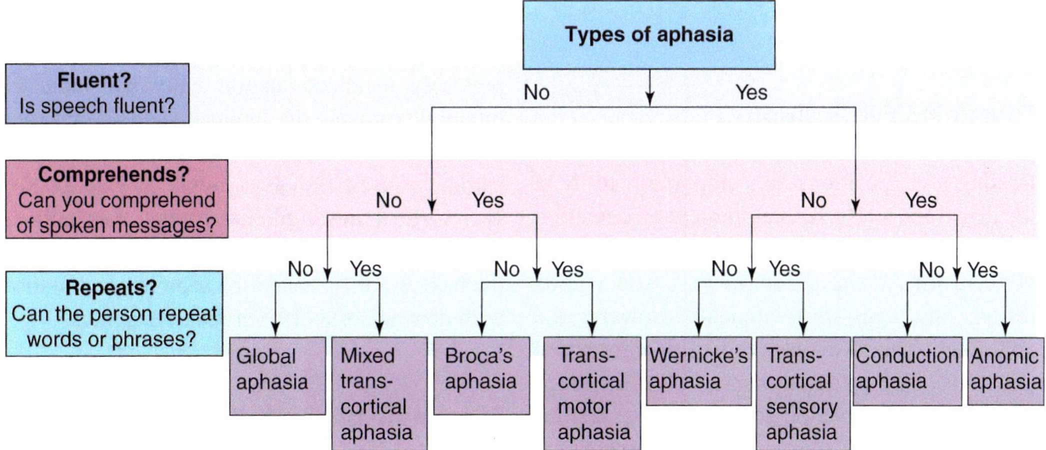

Fig. 3.5 Aphasia classification

and thus dismissed. Cerebellar dysarthria is perhaps most often overlooked in patients with an acute infarction of the posterior–inferior cerebellar artery. These patients are notoriously difficult to diagnose; essential clues, such as dysarthric speech and marked instability while standing and walking, may be overlooked in a patient arriving on a gurney.

In general, dysarthria is incoordination of speech with normal language formulation. Aphasia has been traditionally classified into a number of abnormalities of syntax (grammar structure), lexicon (meaning of words), and fluency (flow of words) [26, 27]. Abnormalities thus result in ungrammatical, unintelligible monotone and slow, labored speech. Fluent and melodic aphasia often has neologisms (e.g., *costamontala* for *wallet*) and word substitutions (e.g., *stuff* or *thing).* The worst form is global aphasia, where words are used rarely (the most oft-used are expletives), but global aphasia is rarely permanent and often changes to an expressive aphasia. Aphasia is then subdivided into classic forms with differences in comprehension and repeats. The main categories localize to certain areas in the brain, typically left, the dominant hemisphere of right-handed persons. An unusual aphasia in admissions to the neurosciences ICU is thalamic aphasia with attentional and memory deficits acutely, fluently, and confabulating (due to its connection with the mammary bodies). Word repetition, however, is strikingly intact [28].

Examining Eye Position

Resting eye positions are important to help localize injury. Horizontal-conjugate deviation indicates a hemispheric lesion. The gaze is directed toward the lesion. With a frontal lobe lesion, the eyes turn tonically toward the abnormality because one eye field is unopposed (Fig. 3.6). The oculomotor nuclei are localized between mesencephalon and pons; thus, lesions in the pons below this level of crossing fibers may produce tonic deviation away from the lesion. Persistent horizontal gaze preference indicates a substantial hemispheric lesion.

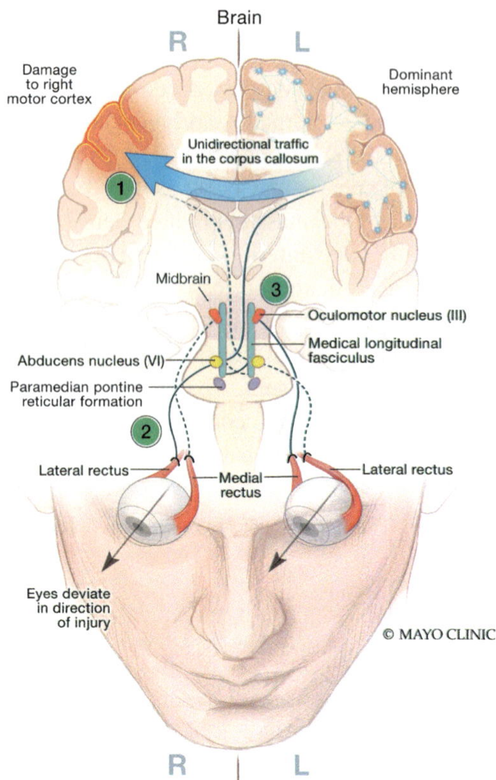

Fig. 3.6 The frontal eye fields and eye deviation with a cortical lesion on the right

Midbrain lesions at the tegmentum interrupt vertical eye movements and may cause forced downward movement. Downward eye deviation localizes either in the thalamus or in the dorsal midbrain, often from a massive thalamic hemorrhage extending into the mesencephalon. Upward deviation, on the other hand, is less precisely localized anatomically but indicates bilateral hemispheric damage, such as that seen with extensive hypoxic–ischemic insult after cardiac resuscitation or asphyxia. Skew deviation is a vertical misalignment of the eyes and, in comatose patients, diagnostic of an embolus to the basilar artery. It is often seen in combination with new anisoria or a vertical nystagmus.

Examining the Brainstem Reflexes

There are a number of tests and a number of ways to approach a comatose patient. Not infrequently we speak loudly to them followed by testing of a blink reflex. The blink-reflex arc is complex and depends on the stimulus (Fig. 3.7a, b). This stimulus is usually sound (with the acoustic nerve as the afferent part) or a rapid-hand approach (with the optic nerve as the afferent part). The efferent parts are the orbicularis oculi muscles innervated by the facial nerve. The acoustic arc is through the brainstem, and blinking does not need cortical input after a loud sound. An intact visual system, including a functioning visual cortex, is needed to cause blinking after a visual stimulus. This difference is useful when examining patients with severe hemispheric injury and intact brainstem function such as in a persistent vegetative state. Visual tracking (Fig. 3.7c) requires cortical input from the primary visual cortex and frontal eye fields. Horizontal visual tracking is a coordinated response using the lateral gaze center or pontine paramedian reticular formation. It activates the oculomotor nucleus in the mesencephalon and the abducens nucleus in the pons. The ascending pathway from the lateral-gaze center to the oculomotor center is known as the medial longitudinal fasciculus. Vertical visual tracking is coordinated through the vertical gaze center present in the periaqueductal gray matter of the mesencephalon. It projects to the oculomotor and trochlear nuclei.

Before examining the brainstem reflexes further, it is important to visualize the optic disk and retina despite being limited in dilating the pupil. Swelling of the disk is ideally measured in diopters; this is done by finding a clear disk with the strongest glass in the ophthalmoscope and then comparing it with the strongest glass where the retina is clear. Blurring of the disk is noted. Retinal hemorrhages and vitreous hemorrhages are noted in patients with aneurysmal subarachnoid hemorrhage. We then proceed with a methodological examination of brainstem reflexes.

The examination of brainstem reflexes is underutilized by nonneurologists; this omission is unfortunate. These reflexes are extraordinarily important because changes can signal acute brainstem injury before a CT scan is done. Ideally, they can even prompt the addition of a CT angiogram that locates a proximal basilar artery embolus.

The first cranial nerve, the olfactory nerve, is not tested until later if the patient complains about anosmia (loss of smell and taste). This test can be done using commercially available test strips. The second cranial nerve is the optic nerve, and two components are tested. Abnormal visual fields involve a larger path and are from interruptions beyond the chiasm or at the chiasm (i.e., pituitary tumor). The pupillary light reflex involves the optic tract (cranial nerve II), and the signal enters the midbrain at the pretectal area, synapsing to Edinger Westphal nucleus and further on to fibers in the oculomotor nerve (cranial nerve III) connecting with constrictor pupillae (Fig. 3.7d). A lesion in the tectum or pretectum with involvement of the posterior commissure fixes the pupil to light when light is swung in front of the pupil. Stretching or compressing the oculomotor nerve or compressing the midbrain oculomotor complex results in a dilated pupil due to an intact, unopposed, sympathetic pathway. When the descending sympathetic tracts are involved, the pupil fixes in mid-position. Conversely, pinpoint pupils are seen in pontine lesions from damage to descending sympathetic fibers, but light reflex is intact. Several neurosciences intensive care units make use of pupilometers [4, 29–31]. Pupilometers provide an objective measurement and may show pupil contraction when the observer's eye fails (very rarely, the opposite). They may also show the appearance of "sluggishness," which is a poorly defined term, often used by nursing staff if they feel uncertain about the response. Frequently, inexperienced nursing staff will characterize pupils as sluggish while they are effectively fixed. Mathematical detection of sluggishness may not have a clinical meaning, but some studies suggest a link to increased intracranial pressure. Generally, pupilometers can confirm clinical examination and may be useful for nursing staff with variable experience. They do not describe the pupil configuration. It remains unclear if a pupilometer is inferior to or better than a magnifying glass.

Pupil size is estimated and documented as pinpoint (1–2 mm), mid-position (3–6 mm), or dilated (6–8 mm). The pupil can be round, pear-shaped, or oval. Anisocoria can exceed 2 or

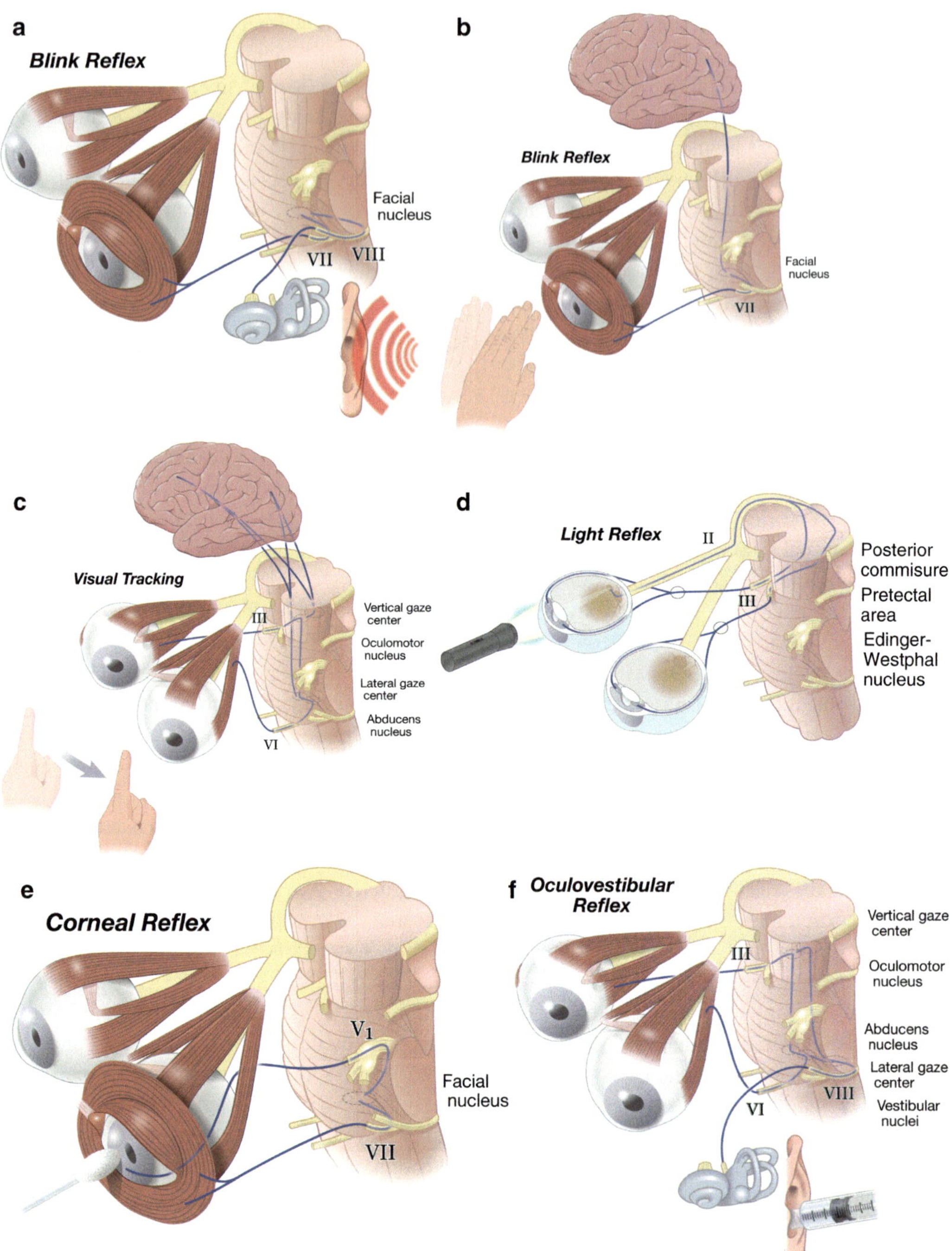

Fig. 3.7 Blink to sound and threat (**a**, **b**), visual tracking (**c**), pupil light reflex (**d**), corneal reflex (**e**), oculovestibular responses (**f**)

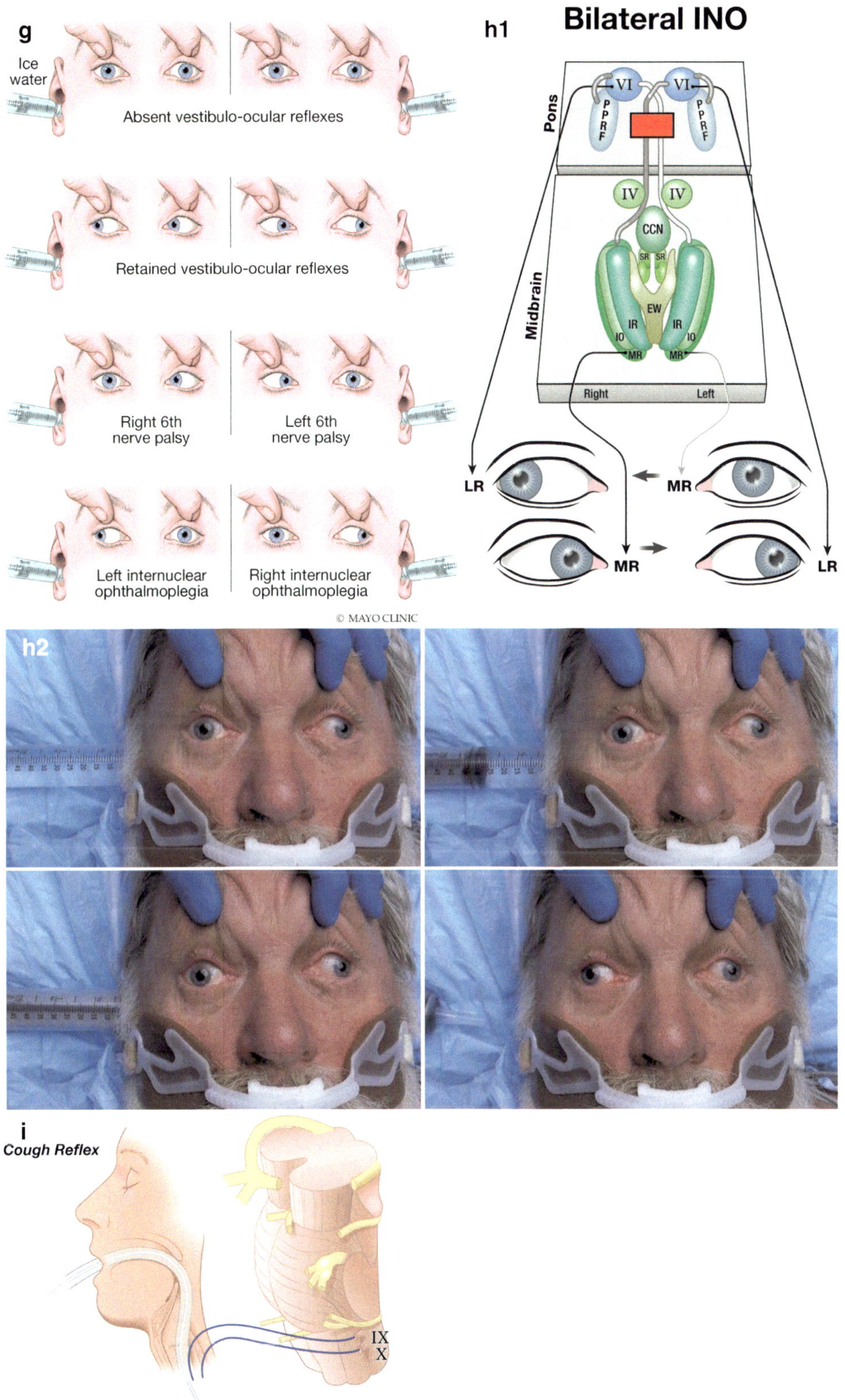

Fig. 3.7 (continued) Internuclear ophthalmoplegia (anatomy and eye movements) (**g**), reflex circuit (**h1**), and patient example (**h2**), cough reflex (**i**)

3 mm. Using a magnifying glass for assessment is very helpful. Moreover, slowly or barely responding pupils can indicate intoxication due to alcohol or high-dose anesthetics. In pinpoint pupils, the light reflex cannot be reliably assessed. Such an afferent pupillary defect (Marcus-Gunn pupil) is traditionally examined by the swinging-flashlight test. The patient is asked to fixate on a distant target to eliminate the meiotic effect of accommodation. A bright light is moved from one eye to the other. The response may vary from minimal asymmetry to pupils failing to constrict or dilate when the penlight moves to the affected eye. In its most pronounced form, pupils dilate immediately when the light shines into the diseased eye. Afferent pupillary defect is linked to optic neuropathy, but a retinal lesion or massive intravitreous hemorrhage (Terson syndrome) may produce similar findings. Fortunately (and unfortunately), the most often-examined brainstem reflex is the pupillary reflex; for many, that is where the examination stops. Anisocoria is often irrelevant, and pupillary reflexes are often insignificant. However, five classic pupillary conditions carry major implications: large and small pupil (acute basilar artery occlusion); unilateral wide, fixed pupils (acute brain shift); middle-position, fixed pupils (major upper-brainstem injury); or intoxication pupils, which widen with cocaine and constrict with heroin.

The corneal reflex is elicited by touching the cornea. The ophthalmic (I) division of the trigeminal nerve (nasociliary branch) synapses with the motor division of the facial nerve that contracts the orbicularis oculi muscles [32]. (Fig. 3.7e).

An oculovestibular reflex can be elicited in comatose patients (Fig. 3.7f). It requires introduction of cold (iced) water into the ear canal, detection of cold water by the semicircular canals, signaling through the eighth cranial nerve to the vestibular nuclei, and projecting to the third, fourth, and sixth nuclei via the gaze centers. In comatose patients, this induces a gradual movement of the eyes toward the cold stimulus. The responses are shown in Fig. 3.7g. Internuclear ophthalmoplegia involves a lesion of the medial longitudinal fasciculus in the upper pons and can be documented with caloric irrigation testing. Its presence reveals a pontine lesion. For which tracts are injured see Fig. 3.7h1; for a patient example, see Fig. 3.7h2. Oropharyngeal function involves hypoglossal testing (tongue protrusion deviates to the affected side); the median is the space between the incisors. (Tongue deviation is often mistakenly diagnosed in facial palsy.) The cough reflex (Fig. 3.7i) is usually best elicited with tracheal suctioning. It consists of a pathway from the sensory laryngeal nerve to the efferent vagal nerve [33].

Motor Responses

Motor responses are important in acutely ill neurologic patients because they often change first. The examination consists of six assessments: (1) the tone and position; (2) spontaneous movements; (3) evoked movements; (4) strength; (5) symmetry, and (6) a motor response. At its highest level, a motor response involves responding to a simple request (e.g., "point to the ceiling") or a more complex one (e.g., "touch your left ear with your right arm and then close your eyes"). With more complex requests, attention plays a bigger role rather than inability to perform a task (apraxia). Praxis can be tested by a simple calculation or by carrying out a simple test such as making interlocking rings with thumb and an index finger.

In wakeful patients, individual muscles are tested for strength (see Chap. 10). Patterns become quickly apparent. This involves comparing proximal with distal weakness (myopathy vs neuropathy) and looking for a pyramidal distribution (extensors more involved than flexors).

In patients with an acute brain injury leading to stupor, motor responses evoked by noxious stimuli are graded from localization to no response. Flexion or extension at the arms can be observed and may indicate decorticate responses (stereotyped slow flexion in elbow, wrist, and

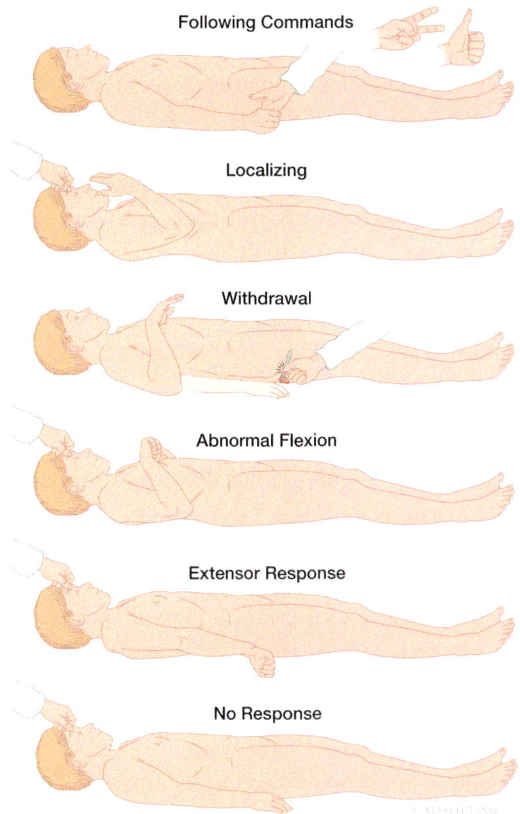

Fig. 3.8 Motor responses

finger grasping) or decerebrate responses (adduction and internal rotation of shoulder, arm extension, and wrist pronation with fist formation); there may be more-difficult-to-classify arm excursions with alternating formes frustes of pathological flexion and extension (Fig. 3.8).

Generalized myoclonus can be seen when anoxic-ischemic injury follows cardiopulmonary resuscitation, lithium intoxication, penicillin and cephalosporin intoxication, and pesticides. Consisting of brief, rapid jerks in all extremities and the face or eyelids, it can be quite forceful and may involve the abdominal muscles, causing difficulty in patient-ventilator synchrony. Shivering is common after awakening from anesthesia, but in patients admitted to the emergency department, it can also indicate hypothermia or early sepsis. Fine shivering may also occur after upper-brainstem destruction and may be difficult to distinguish from spontaneous clonus. Shivering without sweating or piloerection is therefore due to reticulospinal injury. Other movement abnormalities are discussed in detail in Chap. 4.

Coordination

Cerebellar function testing includes finger-to-nose testing, demonstrating the classic decomposition of movement. The normal synchronicity is absent, resulting not only in loss of fluidity but also overshoot of movement. Many variants of this exam exist, such as heel-to-knee testing and tapping the foot to the hand of the examiner when examined in bed. Rapid hand movements (such as "do as if you turn a lightbulb in a socket") are fragmented (known as dysdiadochokinesis). During examination in a standing position, we ask patients to bend their head and trunk backwards while standing with eyes closed, which often brings out dysbalance.

Sensation

This is a difficult part of the examination because it requires focused attention. Superficial sensation is best tested with a piece of paper or wisp of cotton wool. Pain is tested with pinpricks and has to be slightly uncomfortable to distinguish from superficial sensation. Temperature sense is tested with a cold piece of metal or hot water in a test tube. Discrimination is tested with two stimuli and followed by joint-position sense and vibratory sense. Vibration perception with a tuning fork placed on bony surfaces (ankle, phalanx) stimulates mechanoreceptors conveyed both in posterior columns and corticospinal pathways in the cord. Astereognosis is failure to recognize simple objects such as a coin or key. Visuospatial function is impaired with neglect and an important indicator of cortical involvement. It may be present in the absence of significant left-sided weakness and is often underappreciated as a very disabling neurologic deficit. Landmarks in sensory levels are provided in Chap. 9.

Stance and Gait

Patients on a gurney or in bed in the emergency department are rarely asked to walk. When patients with a "normal neurologic examination" or "minimal findings" do try to walk; however, a major new disability may be revealed. Not only may it be impossible for them to stand unassisted, but patients with acute cerebellar lesions may not even manage to put one foot forward. Others may drag a leg or veer to one side. Neglect may not be apparent until the patient starts to walk.

A normal walk can be broken down in a fluent movement of legs with heel striking the ground first and ankles passing very close (millimeters) to each other. Arms swing accordingly. None of that is seen in abnormal gait, which can be cerebellar ataxic (wide based and irregular cadence), sensory ataxic from spinal cord or sensory ganglia disease (stamping feet while watching the ground for stability), steppage gait with polyneuropathy (i.e., foot drop with increased flexion in hip), toppling gait associated with medulla oblongata or vestibular lesions (consistently falling to one side), and spasticity (stiffness in leg and flexion in arm). Gait abnormalities are also associated with movement disorders such as Parkinsonism (hesitation, shuffling, absent arm swing). A hysterical gait with suddenly buckling knees presents an etiological puzzle.

Static posture is tested by the Romberg test. A standing position with feet close together and eyes closed can be maintained for 30 seconds in normal individuals younger than 70 years. Crossing the arms against the chest adds complexity to the test and may induce more subtle swaying. Tandem gait walking (10 steps) evaluates vestibular function with eyes closed and cerebellar function with eyes opened. Imbalance due to an acute vermian or cerebellar hemisphere lesion can be dramatic because patients are unable to sit steadily upright or to stand unassisted.

Reflexes

The reflex hammer, too often and unfortunately substituted with the head of the stethoscope, identifies the neurologist—as does the safety pin—but other specialists make use of it [19]. The tendon stretch reflexes are far less important in neurocritical illness than admitted. Asymmetries may indicate a new lesion, but there should be obvious differences (normal vs. clonus). Absent tendon reflexes are diagnostic in a patient with GBS complaining of tingling. Clonus is diagnostic in a spinal cord lesion. However, presence of any tendon stretch reflex also indicates that the patient is not on large doses of neuromuscular-blocking agents or toxins. Abdominal reflexes are underutilized but help to determine a sensory level in patients without a potbelly. The test that nearly defines neurology for many is the Babinski sign. This is a plantar reflex and can be seen after firmly stroking the sole of the patient's foot with a wooden stick or pointed handle of the reflex hammer. The technique requires starting at the beginning at the lateral heel, following the margin of the foot, and ending across the ball of the foot curving to the base of the big toe. The essential feature is extension (dorsiflexion) of the hallux, produced by active contraction of extensor hallucis longus. Flexion of the big toe (downward) at the metatarsophalangeal joint is the normal response; extension (upward) occurs with upper motor neuron lesions. (The often-heard term, "up-going Babinski's," would make Joseph Babinski roll over in his grave.) Babinski's sign is often also associated with toe abduction (fanning). Babinski's sign is a clinically significant observation but seen in any condition that depresses function of the forebrain or its projections. So expect to see it in acute brain injury but also general anesthesia, coma from organ failure, during a seizure and thereafter. An interesting observation is its possible appearance in the opposite non-paralyzed leg when the brainstem shifts to the opposite tentorium damaging the pyramidal tracts

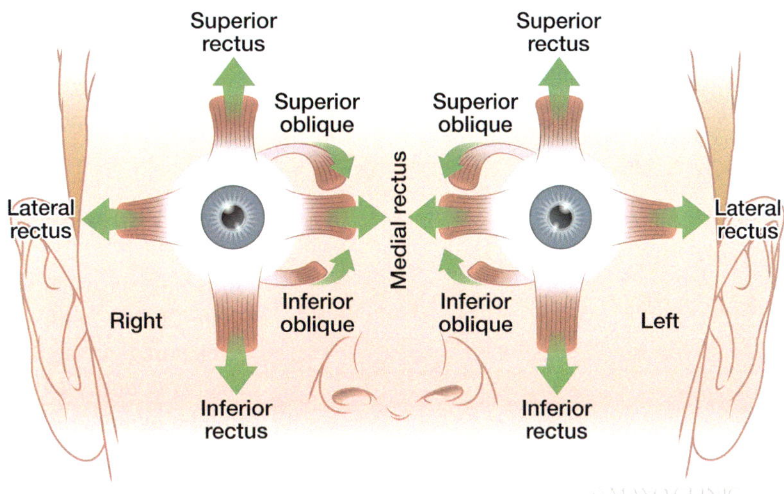

Fig. 3.9 Eye muscles and function

(Chap. 5). As a monitoring sign for brain shift, it all sounds too good to be true, and it has not been systemically studied.

Primitive reflexes also appear with chronic and acute injury. These are snout, suck, and root reflexes (lightly tapping the upper lip or the side of the mouth); the palm mental reflex (stroking the palm causing contraction of the mentalis muscle at the same side); and grasp reflex (sliding two fingers in the palm of the patient to and fro results in involuntary grasping). These inhibited reflexes often surface in patients with acute frontal syndromes, previously undiagnosed neurodegenerative disease (such as progressive supranuclear palsy), or later characteristically in a patient in a permanent vegetative state due to severe structural injury.

Specific Conditions and Syndromes

Certain syndromes need to be part of the vocabulary. This section addresses the ones that are clinically pertinent in acute neurology.

Diplopia

Before understanding the pathophysiology, we need to remind ourselves of the pairs of eye muscles, which move the eyes in a certain direction. Their individual function is shown in Fig. 3.9.

Acute diplopia (monocular or binocular) is very complex to analyze, and the underlying lesion or dysfunction may remain ambiguous. (We have all been taught. When imaging studies are nondiagnostic, to consider myasthenia gravis or some strabismus decompensation.) Neurologic causes of diplopia localize in several structures and consider these: cortex, brainstem/subcortex, cavernous sinus, and the oculomotor cranial nerves. The eye position could be noted and labeled. The commonly used terms are orthotropia (in balance or straight), esotropia (inward), exotropia (outward) hypertropia (upward), and hypotropia (downward).

Monocular diplopia, almost always due to abnormalities in the refractive media, precludes further neurologic work-up. Binocular diplopia is difficult to assess because multiple cranial nerve involvement is present in some patients. Questions to clarify the chief complaint in acute diplopia should include mode of onset, diplopia disappearing after one eye is closed, whether vertically or horizontally oriented, whether always present or fluctuating, and whether more pronounced in a certain gaze. A most commonly observed cranial nerve deficit next to an abducens paresis (CN VI), which also is often a false localizing sign (Chap. 5), is the oculomotor palsy with its typical down-and-out position (Fig. 3.10). Horizontal movement of the involved eye is limited inward (adduction), and there are limited excursions with vertical movement.

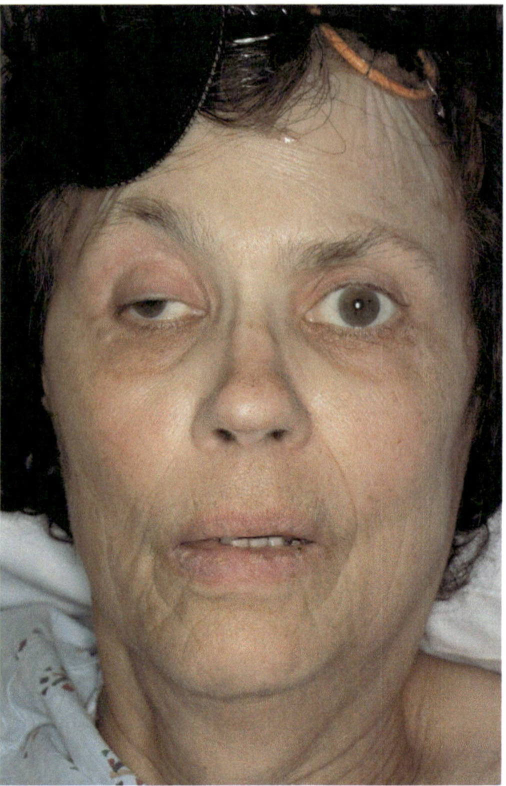

Fig. 3.10 Three major characteristics of oculomotor palsy: ptosis and down and out eye position

Skew deviation may be associated with diplopia and indicates an internuclear lesion. It is a result of abnormalities in vertically ascending fibers from vestibular nuclei within the medial longitudinal fasciculus and, not infrequently, is due to a pontine stroke in elderly patients and multiple sclerosis in younger patients. The cause of acute diplopia, however, may also include other factors, such as difficulty moving the globe due to mass effect in the orbit (thyrotoxicosis) [34], diplopia caused by an acute manifestation of myasthenia gravis, and chronic progressive external ophthalmoplegia, particularly if ptosis is bilateral. A cavernous sinus lesion should be considered when an abducens lesion is associated with Horner syndrome. Acute oculomotor palsy with preceding retro-orbital pain may be a sign of an unruptured posterior communicating aneurysm [35], and two-thirds may be smaller than 6 mm; it may herald rupture and indicate rapid aneurysm growth [36]. Development of pupil involvement, albeit uncommon, may be particularly worrisome for pending rupture. Pupil-sparing may indicate a recent ischemic stroke in the mesencephalon. The differential diagnosis also includes multiple sclerosis.

Acute Visual Loss

Blindness is usually defined as vision of less than 20/200 with correction or a field not subtending an angle greater than 20 degrees. *Acute blindness* may involve eyes and, excluding ophthalmologic disorders, points to bilateral involvement of the occipital lobes. Differential diagnosis involves acute basilar artery embolus, sagittal sinus thrombosis, posterior reversible encephalopathy syndrome, and many drug-induced encephalopathies including vincristine, methotrexate, cyclosporine, and tacrolimus.

Monocular blindness is more common than acute loss of entire vision. In addition, transient monocular visual loss is more common than persistent monocular defect. Transient monocular visual loss often includes embolization due to lesions of the aortic arch, heart valves, or carotid artery but may also include abnormalities associated with increased viscosity or hypercoagulability. Many patients need admission to further evaluate its mechanism.

Monocular vision loss often indicates an ophthalmologic disorder. A neurologic cause for monocular visual loss is most likely optic neuropathy. It typically manifests with markedly reduced visual acuity (20/200), inability to recognize color or brightness (particularly red), and, often, no obvious findings on neurologic examination except an afferent pupil defect. The optic disk may take time to become abnormal but may show pallor or elevation. Optic neuritis is associated with periocular pain and pain on eye movement in 90% of cases. The causes of optic neuritis are manifold and typically can be divided into inflammatory causes and the first manifestations of multiple sclerosis (5-year probability of 30%).

Bilateral visual loss may result from lesions at any topographic location of the afferent visual system ending in the occipital poles, but it is now often recognized in a considerable number of patients presenting with posterior reversible encephalopathy syndrome (PRES).

Acute Vertigo and Nystagmus

When acute spinning indicates vertigo, the history may provide additional clues. Autonomic symptoms such as vomiting, nausea, pallor, and sweating are less pronounced in central lesions, but these same symptoms are so common and can have so many degrees of severity that they cannot be used as major discriminating factors. Vertigo due to positional change, coughing, sneezing, fluctuating hearing loss, nonpulsatile tinnitus, and hearing loss is more typical of peripheral (vestibular) disease.

Classification of nystagmus is simple (i.e., jerk to one direction, slow return to original position) or complex (dancing, bobbing, dipping, or rotational). The movement of nystagmus is best judged by looking at surface conjunctival blood vessels. Caloric tests will set up an endolymph current and cause a nystagmus in wakeful persons and a slow eye deviation toward the cold stimulus. Controlled only by subcortical circuits and the brainstem, eyes may "bop," roll, "see-saw," and retract [37]. Diffuse cortical injury is likely despite findings not specific or sensitive enough for prognostication. Dissociation between paired movements, due to failure of one eye to abduct or adduct (using ice-water tests), indicates brainstem involvement.

Oscillopsia, in which patients feel images are moving or bouncing, is another important symptom. When combined with spontaneous nystagmus and transient vertigo, a peripheral source is likely. Central causes may be strongly considered when head movement induces oscillopsia. Cerebellar lesions produce a spinning sensation but more often severe scanning speech; impaired finger-to-nose and heel-knee-shin testing predominate in the clinical picture. Ipsilateral hearing loss points to an occlusion of the anterior inferior cerebellar artery [38, 39]. Ipsifacial numbness, hypophonia, and Horner syndrome localize in the lateral medulla oblongata.

The characterization of nystagmus into central (brainstem—cerebellum) or peripheral (vestibular) causes is an important determinant. *Central nystagmus* has characteristic direction dependence. When present, gaze turned to the beat of the fast component of nystagmus increases the frequency and amplitude in any type of nystagmus. In central causes of nystagmus, gaze turned away from the direction of the fast component will achieve the opposite effect and may abolish it or, in extremes of gaze, reverse the direction of nystagmus. A vertical nystagmus (up- or downbeat) is almost always central but can be drug-induced (particularly, by opioids). Gaze-evoked nystagmus with similar amplitudes in both directions is due to medication but has been reported in myasthenia gravis, multiple sclerosis, and cerebellar atrophy. Periodic alternating nystagmus (nystagmus changing direction) has typically been considered in diseases of the craniocervical junction but can also be due to phenytoin or lithium overdose.

Examination should focus on three components, summarized as follows. First, the type of nystagmus is noted and is a good indication of the source of the lesion (horizontal, rotational, or vertical). Second, the direction of the nystagmus is considered. In vestibular lesions, unidirectional nystagmus with fast-phase beating does not change in either direction. In central lesions, the nystagmus becomes more intense in the fast-beating direction and less intense in the slow-beating direction. Third, the vestibular ocular reflex is tested. This reflex is abnormal in a peripheral vestibular lesion. To test this reflex, patients are asked to fix their gaze on the examiner's nose. (This requires the patient's cooperation and attention, which may be impaired in brainstem lesions.) A fast 15- to 20-degree head tilt to one side should maintain this gaze fixation with a central lesion, but fixation is temporarily lost with a peripheral lesion.

Peripheral vestibular and congenital nystagmus can be markedly muted when the patient fixes his gaze [40, 41]. Nystagmus can be observed with eye closure, but it is easier to examine the eye with an ophthalmoscope covering the opposite eye. This maneuver eliminates fixation and brings on the alternating drift and correcting jerks of the retina that are a manifestation of nystagmus.

Positional nystagmus can be documented by performing the Dix-Hallpike maneuver, in which the patient changes position rapidly, from sitting on the examination table, to hanging the head, turned sideways, over the edge of the table. In most cases, a torsional and vertical nystagmus appears after 10 seconds' delay and produces a vertiginous sensation that fades away with repeated testing. Both delay and fatigability are characteristic of positional nystagmus and help to localize a vestibular lesion, mostly due to canalithiasis. Absence of delay and fatigability may suggest a central cause.

Disorders of the vestibulospinal reflexes (i.e., neuronal connections from labyrinths and vestibular neurons to anterior horn cells) are tested by past-pointing, the Romberg test, and tandem walking. Past-pointing has the patient touch the hand of the examiner with an extended arm, close his or her eyes, point up, and try to touch the examiner's hand again in a repeated to-and-fro movement but with eyes closed. Past-pointing occurs toward the damaged side. Abnormal past-pointing tests may not be replicated by the more traditional finger-to-nose test because joint and muscle proprioception during this coordinated movement may compensate. Another technique is vertical writing with eyes closed. This test identifies unilateral vestibular dysfunction but may be due to both peripheral and central causes [42].

More Reflections

Localization involves identification of the anatomical structure that explains all findings, which is difficult in the hemisphere because we still do not understand exactly which structures correspond to certain areas. For example, only recently was the posterior (cuneus) cortex function determined to play a role in awakening from coma. Understanding the organization of the brainstem is helpful and, in particular, junctions where one tract meets another tract or nucleus. Single cranial nerve deficits are an example of simple localization and found by tracing the nerve to its origin nucleus in the brainstem. When a single cranial nerve is involved and other signs are present, we look for areas to match the rest of it. It can be simple in Weber syndrome with a third-nerve deficit and crossed pyramidal lesion in the anterior midbrain where they cross. Similarly, with an equally small lesion in the lateral medulla oblongata, where many structures are close to each other. For example, Wallenberg syndrome, with a large number of clinical signs, localizes to a small area in the lateral medulla oblongata where nuclei and tracts are abundant (Table 3.2). Eponyms remind us of constellations of findings and, incidentally, to honor the neurologists who figured them out for us, usually without benefit of the diagnostic tools available today. Localization eponyms (Table 3.3) were known long before any imaging was available, and it is a testament to what neurologists can deduce. "Eponymization" reduces an

Table 3.2 Wallenberg syndrome

Vertigo, nausea and vomiting, and symptoms like skew deviation, diplopia, and severe gait ataxia could be due to the pathology of the vestibular nuclei or vestibular–cerebellar connections
Spinothalamic tract—Contralateral impairment of pain and temperature in trunk and limbs
Trigeminal nerve (V)—Ipsilateral loss of pain and temperature on the face
Vestibular nucleus—Ipsilateral nystagmus, vertigo, nausea, and vomiting
Nucleus ambiguus—Ipsilateral dysphagia, dysarthria, and dysphonia
Descending sympathetic—Ipsilateral Horner syndrome
Cerebellum—Ipsilateral gait ataxia
Glossopharyngeal (IX)—Ipsilateral absent gag reflex and hoarseness
Vagus nerve (X)—Ipsilateral deficit of reflex cough test
Trigeminal nerve (V)—Trismus due to masseter and temporalis hypercontraction

Table 3.3 The more common eponyms of diseases

Brown-Sequard syndrome
Guillain–Barré syndrome
Fisher's variant
Wernicke–Korsakoff syndrome
Parinaud syndrome
Wallenberg syndrome

academic practitioner's entire work to a single observation.

Caplan stated in his text [43] that the most important and most frequently missed signs of brain function involve abnormalities of higher cortical function, level of alertness, the visual and oculomotor systems, and gait and concluded that "these are the parts of the examination most overlooked by non-neurologists, which provide key clues to anatomical localization." (I should add brainstem reflexes, eye position, and spontaneous eye movements.) One thing is certain. Neurologic localization is not some quaint thing neurologists do. Furthermore, neurologists cannot be the old default position; knowledge of the specialty must be shared widely.

Pointers and Takeaways

- In each patient, the examiner should identify the proximate topography of the lesion.
- MRI may help in localization but is not "the last word."
- Accept that textbook syndromes are rarely complete and variants are more common.
- Remember neurologic assessment is often about dismissal of irrelevant signs.
- Take some time to examine the patient; you may find something you would have otherwise missed.
- Document what you see and reason what you think.

References

1. Nicholl DJ, Appleton JP. Clinical neurology: why this still matters in the 21st century. J Neurol Neurosurg Psychiatry. 2015;86:229–33.
2. Laureno R. Foundations for clinical neurology. New York: Oxford University Press; 2017.
3. Jozefowicz RF. Neurophobia: the fear of neurology among medical students. Arch Neurol. 1994;51:328–9.
4. Larson MD, Muhiudeen I. Pupillometric analysis of the 'absent light reflex'. Arch Neurol. 1995;52:369–72.
5. Monrad-Krohn GH, Refsum S. The clinical examination of the nervous system. 12th ed. London: H. K. Lewis & Co.; 1964.
6. Donaghy M. Brain's diseases of the nervous system. 12th ed. New York: Oxford University Press; 2009.
7. Aird RB. Foundations of modern neurology: a century of progress. New York: Lippincott Williams & Wilkins; 1994.
8. Steinberg DA. Scientific neurology and the history of the clinical examination of selected motor cranial nerves. Semin Neurol. 2002;22:349–56.
9. Denny-Brown D. Handbook of neurological examination and case recording. Cambridge, MA: Harvard University Press; 1946.
10. Campbell WW, editor. DeJong's the neurologic examination. 7th ed. Philadelphia: Lippincott Williams & Wilkins; 2012.
11. Alpers BJ, Mancall EL. Clinical neurology. Philadelphia: F. A. Davis; 1971.
12. Steegman AT. Examination of the nervous system: a student's guide. 3rd ed. Chicago: Year Book Medical Publishers; 1970.
13. Mayo Clinic Sections of Neurology and Section of Physiology. Clinical examinations in neurology. Philadelphia: W. B. Saunders Company; 1956.
14. Schwartzman RJ. Neurologic examination. 1st ed. Malden: Blackwell Publishing Inc.; 2006.
15. Fine EJ, Ziad DM. History of the development of the neurological examination. Handb Clin Neurol. 2010;95:213–33.
16. Biller J, Gruener G, Brazis P. DeMyer's the neurologic examination: a programmed text. 6th ed. New York: McGraw-Hill Education; 2011.
17. Fuller G. Neurological examination made easy. 5th ed. Edinburgh: Churchill Livingstone; 2013.
18. Lewis SL. Field guide to the neurologic examination. Philadelphia: Lippincott Williams and Wilkins; 2004.
19. Bynum B, Bynum H. Object lessons: reflex hammer. Lancet. 2017;390:641.
20. Pearce JM. Early days of the tuning fork. J Neurol Neurosurg Psychiatry. 1998;65:728–33.
21. Hill AB. The environment and disease: association or causation? Proc R Soc Med. 1965;58:295–300.
22. Bang OY, Saver JL, Kim SJ, et al. Collateral flow predicts response to endovascular therapy for acute ischemic stroke. Stroke. 2011;42:693–9.
23. Liebeskind DS. Collateral circulation. Stroke. 2003;34:2279–84.
24. Liebeskind DS. Collateral perfusion: time for novel paradigms in cerebral ischemia. Int J Stroke. 2012;7:309–10.
25. Wijdicks EF, Kokmen E, O'Brien PC. Measurement of impaired consciousness in the neurological intensive care unit: a new test. J Neurol Neurosurg Psychiatry. 1998;64:117–9.
26. Geschwind N. Current concepts: aphasia. N Engl J Med. 1971;284:654–6.
27. Damasio AR. Aphasia. N Engl J Med. 1992; 326:531–9.
28. Graff-Radford NR, Damasio H, Yamada T, Eslinger PJ, Damasio AR. Nonhaemorrhagic thalamic infarction. Clinical, neuropsychological and electrophysiological findings in four anatomical groups defined

by computerized tomography. Brain. 1985;108(Pt 2):485–516.
29. Zafar SF, Suarez JI. Automated pupillometer for monitoring the critically ill patient: a critical appraisal. J Crit Care. 2014;29:599–603.
30. Olson DM, Stutzman S, Saju C, Wilson M, Zhao W, Aiyagari V. Interrater reliability of pupillary assessments. Neurocrit Care. 2016;24:251–7.
31. Kramer CL, Rabinstein AA, Wijdicks EF, Hocker SE. Neurologist versus machine: is the pupillometer better than the naked eye in detecting pupillary reactivity. Neurocrit Care. 2014;21:309–11.
32. Maciel CB, Youn TS, Barden MM, et al. Corneal reflex testing in the evaluation of a comatose patient: an ode to precise semiology and examination skills. Neurocrit Care. 2020;33:399–404.
33. Gutierrez S, Warner T, McCormack E, et al. Lower cranial nerve syndromes: a review. Neurosurg Rev. 2020. https://doi.org/10.1007/s10143-020-01344-w.
34. Bhatti MT. Orbital syndromes. Semin Neurol. 2007;27:269–87.
35. Chen PR, Amin-Hanjani S, Albuquerque FC, McDougall C, Zabramski JM, Spetzler RF. Outcome of oculomotor nerve palsy from posterior communicating artery aneurysms: comparison of clipping and coiling. Neurosurgery. 2006;58:1040–6.
36. Yanaka K, Matsumaru Y, Mashiko R, Hyodo A, Sugimoto K, Nose T. Small unruptured cerebral aneurysms presenting with oculomotor nerve palsy. Neurosurgery. 2003;52:553–7. discussion 556–557
37. Scheitler KM, Mustafa R, Wijdicks EFM. Lid and convergence retraction nystagmus in thalamic-midbrain hematoma. Neurocrit Care. 2020. [Online ahead to print].
38. Lee H, Sohn SI, Jung DK, et al. Sudden deafness and anterior inferior cerebellar artery infarction. Stroke. 2002;33:2807–12.
39. Raupp SF, Jellema K, Sluzewski M, de Kort PL, Visser LH. Sudden unilateral deafness due to a right vertebral artery dissection. Neurology. 2004;62:1442.
40. Leigh RJ, Zee DS. The neurology of eye movements. 4th ed. New York: Oxford University Press; 2006.
41. Seemungal BM, Bronstein AM. A practical approach to acute vertigo. Pract Neurol. 2008;8:211–21.
42. Fukuda T. Vertical writing with eyes covered; a new test of vestibulo-spinal reaction. Acta Otolaryngol. 1959;50:2636.
43. Caplan LR. Caplan's stroke. 5th ed. New York: Saunders; 2016.

4 Excess or Paucity: Making Sense of Movements

Injury or abnormal function of the central nervous system may cause involuntary non-paretic slowness or surfeit of movement. When movements are wild and flailing, we notice them more. We see them in the patients in our unit as well as during consults in the emergency department or in other intensive care units. Movement disorders often are presented to us as "tremors, seizures, or cramps"; we as neurologists are asked to better characterize them by name ("can you come and look and figure out what this is?") and certainly make them go away ("we cannot ventilate the patient well due to large tidal volumes").

Examination of abnormal movements is one of the more difficult tasks in neurology, not in the least because of disagreement on diagnosis and classification among colleagues. (see what happens if you show a video in a room full of movement disorder specialists) The requesting team may describe them as "seizures," but often they turn out to be something else. One thing is certain—not all that twitches is a seizure. We may use the term "adventitious" initially, followed by "tremor," "myoclonus," "asterixis," and "chorea." Appearances can be deceiving. On occasion, we diagnose extensor posturing, rigidity, rigors, or simple shivers. Functional (psychogenic) causes are often considered with bizarre, out-of-the ordinary presentations, but these are far less common in intensive care units. Sometimes, it really *is* a seizure, and then, not infrequently, the patient may be in focal status epilepticus with one limb twitching (epilepsia partialis continua) or with repetitive mouth and eyelid movements.

Our descriptions (i.e., semiology) have improved after centuries of confusing terminology. Many movements are reasonably classifiable and can lead to a diagnosis. Some movements remain unclassifiable and just disappear. Some neurologists may just begrudgingly call them "fidgets" or poorly defined pickiness or restlessness. But we observe and settle at the bedside. There are no tests we can order that will help much. Electrophysiology often tells you the movements are out of sync with electroencephalography (EEG) and shows nonspecific slowing as a result of organ dysfunction. We may still have no good explanation, and our phenotypic uncertainty remains.

A movement disorder is not just a matter of discomfort or disability. More worrisome is the reality that movement disorders often lead to a host of secondary complications. Dystonic involvement of the chest wall, neck, and respiratory muscles may cause acute respiratory compromise requiring mechanical ventilation. Excess (hyperkinetic) movements sustained over a period of time increase the risk of muscle breakdown (rhabdomyolysis, potentially resulting in acute renal failure and significant hyperkalemia). Hyperkinetic movement disorders, specifically

E. F. M. Wijdicks, *Examining Neurocritical Patients*, https://doi.org/10.1007/978-3-030-69452-4_4

dystonia, may correlate with significant pain. If patients experience paroxysmal sympathetic hyperactivity, profound dysautonomia (tachycardia and hypertension) often accompanies the uncontrolled movements. It is therefore necessary to recognize these movements and treat some of them urgently. In this chapter, the clinical semiology of movement disorders seen in intensive care units is described to aid recognition. Moreover, we will discuss when they are relevant for day-to-day critical care management.

Movement Disorder Mimickers

First, we should begin by separating abnormal movement from well-known movements such as shivering and chills. Shivering is characterized by involuntary muscle contractions and vasoconstriction in an attempt to increase metabolic activity and heat production. Shivering is very common to regulate temperature and is basically a tremor with a high-oscillating frequency of 200 Hz [1, 2]. Perioperative hypothermia is defined as a core temperature of 33–35 °C, knowing the threshold for this thermoregulatory response in non-anesthetized patients is around 36 °C. Shivering results from hypothermia and, not uncommonly, after the use of anesthetic drugs. General anesthetics inhibit the thermoregulation center of the hypothalamus by lowering the shivering threshold and preventing the compensatory mechanism of shivering. The use of muscle relaxants will also inhibit the body's ability to shiver. There is even a simple, validated shivering score that is helpful to note its severity: 0, no shivering; 1, mild fasciculations of face and neck, electrocardiography (ECG) disturbances in the absence of voluntary activity of the arms; 2, visible tremor in a muscle group; and 3, gross muscular activity involving the entire body. Shivering occurs most often after remifentanil administration [3]. It is an uncomfortable feeling unlike most acute movement disorders. Moreover, untreated shivering increases oxygen consumption sixfold, which, when untreated and prolonged, may cause lactic acidosis, carbon dioxide production, and catecholamine release, resulting in increased cardiac output, heart rate, and arterial pressure. Other adverse effects of shivering include increased intracranial and intraocular pressures. It can be quite a disturbing sight too. A warm blanket may not always do wonders, and radiant heat (e.g., The Bair Hugger ® system) is more effective in reducing oxygen consumption. Moreover, several agents have potent anti-shivering properties; dexmedetomidine is the most effective.

Next to consider is a focal seizure, which is a continuous jerking in face and extremities. It can be seen and felt, and it has the same amplitude in one or two adjacent muscle groups. However, it can be subtle in an eyelid or chin muscle and not easily "picked up." Patients are not alert or fully aware of their environment; they may slip in and out of consciousness if seizures briefly generalize. An EEG is helpful here because the twitch is clearly time locked with a sharp wave or series of sharp waves coming out of a discrete area and with spreading elsewhere on the cortex. The diagnosis is not difficult if there is evidence of cortical irritation, which is very common with acute and subacute subdural hematoma or with a new mass when a glioma is suspected.

Another clinical sign is posturing or intermittent spasmodic stiffness. Convulsion-like, intermittent, spasmodic, clonic contractions [4, 5] may occur in acute brainstem lesions. With good reason, in the distant past, these were called "tetanus-like seizures." There is a wide spectrum of manifestations. They resemble decerebrate posturing but without the full, coordinated, stereotypical movement. These movements may be initially considered "seizures" and treated as such until the neurologist points out there is more to the picture and recognizes clinical signs of a brainstem stroke. Patients exhibiting these movements typically have no significant decline in alertness. The decerebrate rigidity (or better, extensor posturing) seen in comatose patients from a brainstem stroke is different, and examination will be much different with ocular signs (see Chap. 3). With decorticate rigidity, there will be opisthotonus, extension of the lower extremities, and flexion of the upper extremities; forearms are pronated in the wrist and fingers flexed. The legs are extended and abducted with the feet plantar flexed. There may be brief episodes of jaw clenching and extension of all four limbs. When examined, there is sudden yielding with stretching of the muscle ("clasp-knife" phenomenon), and clonus or subclonus tendon reflexes are also present. Conventional wisdom holds that decerebrate rigidity is the clinical expression of a brainstem transection, but this is far from correct. This motor response has been recognized as a strong indicator of a severe structural brain injury. Decerebrate posturing has been observed in brainstem lesions and in lesions involving injury to both hemispheres without evidence of brainstem injury or displacement. More specifically, the site of the responsible lesion can be either corticospinal or striospinal tracts, and the response can be produced by injuries of (1) bilateral forebrain, (2) mesencephalic–pontine injury, and (3) posterior fossa mass lesions compressing the midbrain and rostral pons.

The Major Movement Disorders

Once these common and not so common mimickers have been excluded or are deemed much less likely, we can start categorizing the neurologic movement disorder that eludes some of us. Some simple distinguishing characteristics, based on flow and rhythmicity, are shown in Fig. 4.1. We can get more granular by describing movement initiation, movement velocity, amplitude, and hesitations [6–9]. With structural lesions, movement disorders almost always originate from the basal ganglia or the connected structures, and there are traditional correlations with lesions in the putamen (dystonia), thalamus (unilateral asterixis), subthalamic nucleus (ballism), caudate nucleus (chorea), substantia nigra (hypokinesis), cortex (generalized myoclonus), and spinal cord (segmental myoclonus) [10]. Toxicity-driven movements do not have a well-established topography, and combinations of these major syndromes are common. But generally, the movement falls into

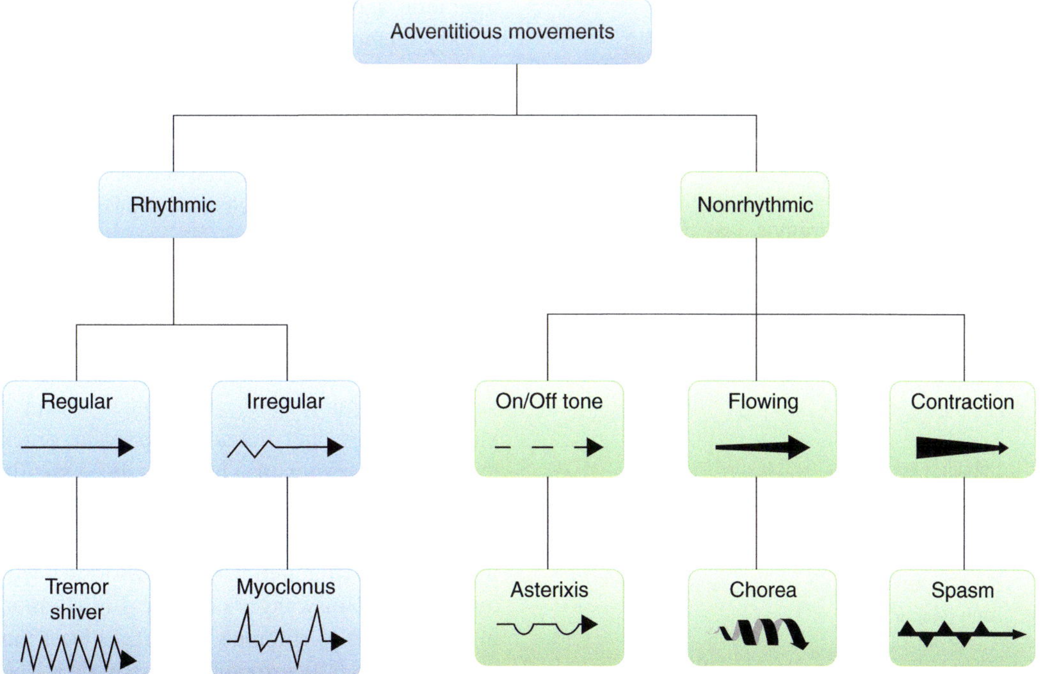

Fig. 4.1 Characterization of acute movement disorder

two main categories—markedly reduced movements (hypokinetic) or markedly increased movements (hyperkinetic) [11–14].

Hypokinetic movements signal abnormally diminished motor activity [14]. Hypokinesia is characterized by a paucity of movement rather than a lack of motor strength. Typical movements, such as pointing to an object, are slow. Hypokinetic disorders present with rigidity and cogwheel phenomenon. These movements are most frequently seen in Parkinson's disease or neurodegenerative disorders [15]. Some causal considerations are shown in Table 4.1. Hypokinetic movements can be expected if the patient has received antipsychotics to control agitation or to establish rapid tranquilization. Psychotropic drugs and dopamine-receptor antagonists can also cause parkinsonism. However, many of these drugs more often cause tremor and not the classic features of Parkinsonism (see Table 4.3). Parkinsonism (as opposed to Parkinson's disease) refers to clinical features associated with other disorders, often very prominently. These include progressive supranuclear palsy, multiple system atrophy, Lewy body dementia, corticobasal degeneration, and the frontotemporal dementias. There is marked loss of facial mimicry and blinking. Cerebellar ataxia and orthostatic hypotension are clinical clues that help rule out typical Parkinson's disease.

Hyperkinetic symptoms are indicative of some very unusual disorders such as chorea, ballism, athetosis, dystonia, tremor, and tics. The causes of dyskinesias seen in the ICU are shown in Table 4.2. Chorea is a fluent, purposeless movement, large in amplitude; when it involves proximal muscles or abdomen or when it becomes more extreme, it is classified as ballism. Patients sometimes describe chorea as a feeling that their limbs are moving uncontrollably, like a marionette's. Dystonia is a sustained, muscle-cramping movement often accompanied by irregular twisting and sustained postures. Tetanus is rare, but the most common presentation is with trismus (lockjaw) resulting from masseter muscle spasms. Profound rigidity and spasm can also affect axial and limb muscles. The spasm results from gamma-aminobutyric acid (GABA) inhibition and glycine release within the spinal cord by tetanospasmin.

Table 4.1 Causes of acute Parkinsonism

Structural
Stroke
Subdural hematoma
Hydrocephalus
Central pontine myelinolysis
Drug-induced
Neuroleptics (including the atypicals)
Antiepileptics
Antidepressants
Chemotherapeutic agents
Amiodarone

Table 4.2 Causes of dyskinesias in ICU

Etiology
Structural
Acute stroke
Autoimmune encephalopathies
Paraneoplastic encephalopathies
Drug-induced/toxic
Neuroleptics
Ondansetron
Metabolic
Extreme electrolyte derangements (rare)

Tremor is defined as a regular, rhythmic, sinusoidal activity of agonist and antagonist muscle groups leading to oscillatory movement around a joint. Tremor can be classically divided into rest tremor (no muscle activation) and action tremor (appearing with muscle contraction). The causes seen in the ICU are shown in Table 4.3. In the ICU, tremor is commonly seen with alcohol or opioid withdrawal and also with recently prescribed and toxic levels of fosphenytoin, valproate, amiodarone, dopamine receptor-blocking drugs, and immunosuppressive drugs; the latter showing a characteristic essential tremor-type picture (although head tremor is rarely apparent). Tremor can be seen in drug-induced or vascular parkinsonism associated with multi-infarct Binswanger's disease. Chin tremors in the chin are so common in critical illness that every consulting neurologist has seen them (but rarely with a reasonable explanation). Patients with prolonged stays in the ICU and multi-organ

Table 4.3 Causes of tremor in ICU

Structural
Cerebellar lesions (intention tremor)
Midbrain lesion (rural tremor)
Drug-induced/toxic
Beta-agonists
Dopaminergic drugs
Valproate
Amiodarone
Cyclosporine
Tacrolimus
Lithium

disease will often have hand tremors that resolve later.

Myoclonus is a repetitive, irregular, quick jerk in several muscle groups often moving the limb. Some have called it "positive" or "negative." When positive, the cause is the contraction of a single agonist and antagonist muscle (or a group of muscles). Negative myoclonus (asterixis) is characterized by transient interruption of tonic muscle tone, with momentary lapses of contraction and, thus, is best seen in active postural muscles (asterixis). The term asterixis, in fact, denotes inability to maintain posture (*a*-privative, *sterixis*-support). Although it strongly resembles flapping, the term "flapping tremor" (used occasionally by internists) is essentially inaccurate—there is neither flapping nor tremor. And it does not describe the movement in feet, face, and tongue.

Myoclonus is a brief, lightning-like, jerky movement, usually less than 100 milliseconds in duration. Myoclonus can be focal, generalized, multifocal, or axial. The causes of myoclonus seen in the ICU are shown in Table 4.4 [16]. Myoclonus is one of the most common movement disorders in intensive care units and is a result of medication or brain or organ injury (Table 4.1). Myoclonus is usually intermittent but can be present continuously, originating from the cerebral cortex (EEG time-locked with EMG) brainstem or spine (EEG not time-locked). Myoclonus may be associated with action tremor (voluntary, with increasing tremor amplitude reaching a target), postural tremor (tremor at beginning and end of movement), or asterixis

Table 4.4 Causes of myoclonus in ICU

Structural
Anoxic–ischemic brain injury
Viral encephalitis (including arthropod-borne virus)
Paraneoplastic or autoimmune encephalitis
Drug-induced
Antibiotics (quinolones, cephalosporins)
Clozapine
Opioids
Toxic
Serotonin syndrome
Lithium
Heavy metal poisoning
Metabolic
Hepatic failure
Renal failure
Hypocalcemia
Hyponatremia
Hypomagnesemia

(bursts of arrhythmic movement brought on by holding hands or limbs in a certain position). It may be seen in the tongue and when puckering the lips. Although both asterixis and myoclonus symptoms are very common, they are infrequently recognized. Patients with long-standing hypercapnia, renal disease, or liver disease may likely have these movements, and they will only disappear with improvement of liver, kidney, or lung function. Spinal segmental myoclonus originates in the spinal cord; jerks affect muscles innervated by one or two neighboring spinal segments (myotomes). Axial myoclonus is myoclonus of the neck and trunk muscles resulting in flexion but also abduction of arms and flexion of hips. It is spinally originated and from the proprioceptive spinal cord fibers. As expected, it is seen mostly with spinal cord injury and, most commonly, after transitioning to sleep. Myoclonus is often stimulus-sensitive and rhythmic (1–3 Hz) and may occur during sleep as well as wakefulness. Myoclonus cannot be controlled by patient effort. When myoclonus occurs with initiation of movement, it is called action myoclonus, a condition most commonly seen in anoxic brain injury after cardiopulmonary resuscitation [17–20].

If they do not resolve quickly, movement disorders lead to significant potential consequences downstream. Prolonged rigidity causes rhabdomyolysis and requires several CPK values. A single CPK may be markedly elevated in the 100s and may seem normal, but follow-up values may be in the 1000s or 10,000s. Extremely high CPK values and dehydration (easily judged by BUN/creatinine ratio) will lead to further worsening of creatinine values and even to acute renal injury and oliguria [21]. Therefore, monitoring urine production (and color) is critical in any patient with hypokinetic emergencies.

Another immediate concern is markedly reduced chest elastance leading to atelectasis and hypoxemia. As a general rule, restrictive pulmonary disease begins at the point of severe rigidity and marked bradykinesia. Pulmonary function studies have convincingly shown improvement, particularly of inspiratory flow, after levodopa administration. Conversely, acute respiratory failure with labored breathing may occur within 24 hours after withdrawal of dopamine agonists. Aspiration in Parkinson's disease is very common and a direct result of delayed swallowing. Marked dysphagia, however, is not a feature of Parkinson's disease and appears more typically in neurodegenerative diseases with parkinsonism. Poor, inadequate swallowing that fails to clear secretions results in acute mucus plugging and frank aspiration, which rapidly lead to pneumonia or ARDS. It is these complications—and not the neurologic manifestations themselves—that make these disorders critical. Prolonged immobilization from protracted recovery can easily lead to deep decubital ulcers, deep venous thrombosis, and pulmonary emboli. Good outcome, therefore, can only be achieved with full support and rapid reversal of the source of complications [22].

Once a movement disorder becomes obvious, a number of considerations must be reviewed. Paroxysmal dyskinesias (e.g., dystonia, chorea, or athetosis) may be due to a secondary cause. Toxicity from overdose is a common occurrence. All drug-induced movement disorders are self-limiting, and there are quite a few. Virtually all of the neuroleptics can cause parkinsonism; clozapine and quetiapine are the only ones without these potential side effects. Parkinsonism has been described with high doses of anti-emetics such as the dopamine-blocking drugs (e.g., metoclopramide, prochlorperazine, and promethazine).

Abnormal Movements of Significant Urgency

Several movement disorders are syndromic and recognizable as a cause of significant new brain injury. Others are clearly drug related: either too much, too little, or a change in a short time period.

Myoclonus Status

Intensivists see severe myoclonus commonly; many patients with severe anoxic ischemic encephalopathy present with severe, sustained myoclonus. It is hard to miss and easy to treat but often misjudged as convulsive status epilepticus. The muscle contractions are brief, of small amplitude, shock-like, and completely different from a tonic–clonic seizure, even when seen in the resolving phase. The movements are usually chaotic and arrhythmic. Myoclonus is due to severe destruction in multiple cortical layers and continuous disinhibition (similar to myoclonus with syncope but continuous). Myoclonus is sensitive to touch and sound and can originate from the basal ganglia, brainstem, or spinal cord; in severe anoxic–ischemic brain injury, all of these locations may be involved [23–25].

Myoclonic status epilepticus in a comatose patient is a strong indicator of poor prognosis because (1) cortical injury is more diffuse and severe and (2) there often is systemic organ injury from prolonged resuscitation to achieve restoration of spontaneous circulation. Myoclonic status should be differentiated from isolated myoclonic jerk, other types of seizures, and an action myoclonus of Lance–Adams. Lance–Adams syndrome frequently follows respiratory arrest, becomes evident after awakening, and has never been associated with coma [26, 27].

Drug-induced myoclonus may involve manifestations at first exposure or appear in toxic doses. The selective serotonin-reuptake inhibitors and lithium can put a patient into myoclonus status. The neurointensivist should inquire specifically about lithium use and perhaps even order a lithium level to exclude this possibility. Patients with lithium become agitated, develop fasciculations and cerebellar dysfunction, and may initially display choreiform movements as part of the clinical manifestation. Seizures and coma rapidly follow when the serum levels reach 3.5 mEq/L, which represents a threefold increase above the normal target range. Lithium intoxication may have ocular myoclonus in all directions and axial myoclonus. Myoclonus is an under-recognized manifestation of lithium intoxication. Toxic exposure to lithium may permanently damage the cortex and basal ganglia. Generalized myoclonus is common in acute metabolic derangements but usually in end-stage organ failure such as hepatic and renal disease. It has also been observed in hypernatremia, hypomagnesemia, and nonketotic hyperglycemia. Less common causes are heat stroke, decompression injury, and pesticide exposure.

Trismus

Trismus is a newly recognized sign (previously associated with tetanus) in autoimmune encephalitis associated with movement disorders [28, 29], but it rarely occurs in isolation and is more commonly seen with dystonic laryngospasm. It is a well-known complication of head and neck cancer [30]. In polytrauma, it can be seen with a fracture or dislocation of the mandible or zygomatic arch. Trismus can be a drug adverse effect (e.g., phenothiazines, metoclopramide, tricyclic antidepressants) or the result of peritonsillar or other pharyngeal abscesses.

Acute Dystonic Reaction

Dystonic movements are characterized by a persistent posture in one extremity. There are sustained, patterned spasms but normal tone in between spasms. The patient may assume bizarre positions in the limbs and trunk. It is useful to distinguish between a generalized or focal dystonia and determine whether dystonia occurs at rest. Ocular deviation or oculogyric crises are a form of dystonia associated with backward or lateral flexion of the neck and, occasionally, protrusion of the tongue. Eyes may deviate upward, sideways, or downward for several minutes at a time and can only be corrected briefly with effort. Drug-induced oculogyric crises can often be successfully resolved by discontinuing the drug. Oculogyric crisis-inducing (and oromandibular dyskinesia-inducing) drugs include phenothiazines and many of the antipsychotic drugs but also carbamazepine, gabapentin, lithium, ondansetron, and, perhaps best known, metoclopramide. However, oculogyric crises also occur in serious neurologic conditions such as bilateral paramedian thalamic infarction, multiple sclerosis, and traumatic brain injury. A physician seeing patients with acute dystonia should consider Wilson's disease, particularly when patients are 20 to 30 years old. Additional findings are artificial grin (retracted lips) and brown iris in previously blue-eyed persons. Diagnostic tests include reduced serum ceruloplasmin level (in 5% of patients, it is normal), Kayser–Fleischer rings under slit lamp, and increased signal in basal ganglia and cortex on magnetic resonance imaging (MRI). Recently, recognized autoimmune encephalitis often displays complex movement disorders such as tremor, myoclonus, chorea, dyskinesias, dystonia, and parkinsonism. Hyperkinetic movement disorders are more common in younger patients, whereas hypokinetic movement disorders, such as parkinsonism, are more common in older patients.

Status Dystonicus

This disorder is recognized clinically as a sustained contraction of multiple muscle groups and may cause repetitive twisting movements or unusual postures [31]. It may lead rapidly to exhaustion and pain from muscle damage or joint overstretching. When it involves shoulder, face, or oropharyngeal muscles, it becomes a

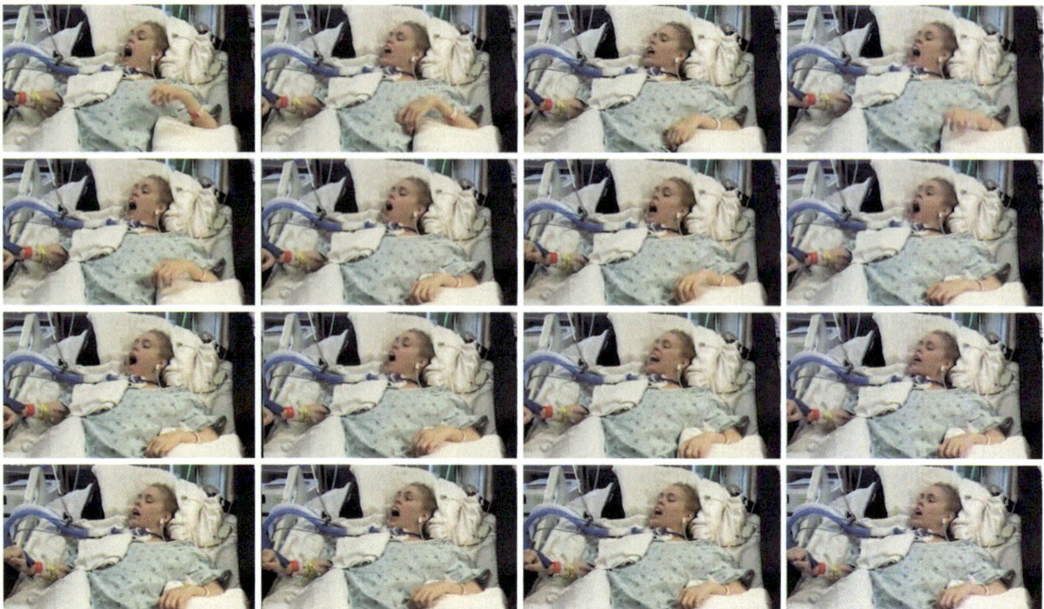

Fig. 4.2 Anti-NMDAR encephalitis Oropharyngeal dystonia and dystonic posturing (still every 20 frames)

neuromedical critical illness [32]. Trismus can severely dislocate the jaw, and often there is lateral flexion or extreme retrocollis. The patient is in distress but not in pain despite frequent extreme grimacing (Fig. 4.2). Jaw dystonia and laryngospasm are common accompaniments of ANNA-2 autoimmunity and are associated with significant morbidity. Tracheostomy may be considered when episodes increase in frequency (several per day), requiring emergency department visits. Following tracheostomy, laryngospasm continues without respiratory compromise [33].

The condition may also occur as the result of a drug reaction or worsening in patients with known focal dystonias (e.g., infection, sudden drug effect withdrawal). Haloperidol, dopamine-receptor blockers, clozapine, and intrathecal baclofen pump failure are all known triggers.

The first line of action is intubation and sedation, particularly if the patient demonstrates pharyngeal, laryngeal, and diaphragmatic dystonia. The next step is fluid resuscitation with multiple boluses of crystalloids and temperature control with cooling blankets or cooling pads. Neuromuscular-blocking agents may be used initially for 24–48 hours. The dystonic contractions are best treated with oral or IV clonidine, 3–5 mg/kg, administered every 3 hours, which may be supplemented with IV midazolam or IV lorazepam. CPK levels are the best monitoring tests, and resolution occurs quickly after treatment. Initial CPK levels may be dramatic and high 10,000s. Gabapentin and baclofen in oral doses have been concomitantly administered [34]. Antiepileptic drugs have not been useful, even in combinations. Acute dystonic reactions are often successfully treated with intravenous or oral administration of anticholinergics (benztropine) or antihistaminic agents (diphenhydramine). Other laboratory tests, including acidosis, hyperkalemia, hypocalcemia, and blood urea nitrogen (BUN), may be abnormal.

Paroxysmal Sympathetic Hyperactivity (Dysautonomic Storming)

The first noticeable sign is extreme new rigidity in paroxysms and overreactivity to stimuli associated with routine nursing care [35, 36]. Tachycardia

(>140/minute) and tachypnea (>30/minute) are well beyond the normal range. Profuse sweating with pearls of sweat may occur. Bed linen becomes soaked. These clinical features of paroxysmal sympathetic hyperactivity or dysautonomic storming occur simultaneously and, if untreated, will occur multiple times per day for several days. Surprisingly, paroxysmal sympathetic hyperactivity (PSH) remains unrecognized and, thus, goes untreated because it is not seen as a separate manifestation of the injury that requires specific management. It looks bad and can be bad for the patients because these episodes can produce marked rises in intracranial pressure. Potentially, when PSH goes untreated, the severity of the dystonia can cause contractures and make later rehabilitation efforts difficult.

These spells, also known as "sympathetic storms," are frequent in patients with severe, acute brain injury [37–41]. They occur most often in young comatose patients with severe diffuse axonal traumatic brain injury, but we have also seen them after severe anoxic–ischemic encephalopathy. They are often seen with severe acute brainstem injury (primary brainstem injury, large ischemic territories with acute basilar artery embolus, or pontine hemorrhage) and ganglionic hemorrhages [42]. Episodes of PSH can begin during the acute phase and in the more severely affected patients. PSH can also become manifest later, after some interval, and thus are more unexpected.

The main clinical features are paroxysms of major autonomic manifestations. Patients become tachycardic, hypertensive (with increased pulse pressure), tachypneic, and diaphoretic, manifesting pupillary dilatation, piloerection, and skin flushing. The rigidity is most evident. The rigidity is smooth (lead pipe) and not jerky (cogwheel), but it may assume a dystonic posture. It is not easy to move the limbs or bend them, and it may also affect body core (axial) muscles.

Neuroleptic Malignant Syndrome

Neuroleptics (e.g., metoclopramide, droperidol, promethazine) cause this syndrome, which closely resembles parkinsonism. Incidence rates for neuroleptic malignant syndrome range from 0.02% to 3% among patients taking these agents, and thus, it is not entirely negligible [43, 44]. NMS is most often seen with high-potency neuroleptic agents (e.g., haloperidol, fluphenazine) but may also occur with the atypical antipsychotic drugs (e.g., clozapine, risperidone, olanzapine) and antiemetic drugs (e.g., metoclopramide, promethazine). Common triggers are an initially high dose, a recently increased major dose, and a reaction to the first dose. Concomitant use of lithium may increase the risk. Fever, rigidity, rhabdomyolysis, and dysautonomic signs, such as tachycardia, tachypnea, and profuse sweating, are key findings. This is nearly identical to parkinsonism–hyperpyrexia syndrome, which is a result of sudden withdrawal of Parkinson's drugs. Patients have been described as warm and stiff. The clinical diagnosis is supported by an expected, marked increase in serum creatine kinase (CK), up to 10,000 IU/L. Fever and dehydration will lead to other laboratory abnormalities including hypocalcemia, hypomagnesemia, and hypernatremia. This disorder has a major overlap with a much less commonly seen entity called malignant hyperthermia, which can often be prevented if a genetic mutation is known [45]. Patients with specific mutations in RYR1 (ryanodine receptor 1) and, less frequently, in CACNA1S (calcium channel, voltage-dependent, L type, alpha 1S subunit) and STAC3 (SH3 and cysteine rich domain 3) are at risk for a life-threatening, malignant hyperthermic reaction. Any patient with a congenital myopathy is at risk [46]. Skeletal muscle cells exposed to triggering anesthetics can cause massive muscle cramping and contracture, massive rhabdomyolysis, hyperkalemia, acute renal failure, and fatal cardiac arrhythmias [47].

Serotonin Syndrome

Serotonin syndrome is a rapidly increasing but still poorly recognized disorder (Table 4.5). The selective serotonin reuptake inhibitors (SSRI) causing this syndrome are currently prescribed

Table 4.5 Drugs known to cause and exacerbate serotonin syndrome in the ICU (combinations are common)

Pain treatment
Opiates (e.g., fentanyl)
Psychiatric treatment
Selective serotonin reuptake inhibitors
Serotonin–norepinephrine reuptake inhibitors
Tricyclic antidepressants
Buspirone hydrochloride
Monoamine oxidase inhibitors
Illicit drugs
3, 4-Methylenedioxymethamphetamine (ecstasy)
Lysergic acid diethylamide
Amphetamines
Cocaine

drugs such as sertraline, fluoxetine, fluvoxamine, paroxetine, and citalopram [48]. Drugs that exacerbate a serotonin syndrome are lithium, valproate, anti-emetics such as ondansetron, metoclopramide, and, most recently recognized, opioids. Other antidepressants such as trazodone or buspirone may worsen serotonin syndrome when symptoms are initially mild. Serotonin syndrome occurs only rarely but is typically attributed to ingestion of bupropion. Bupropion is a commonly used antidepressant and also used for smoking cessation. It selectively inhibits neuronal reuptake of dopamine and norepinephrine and may indirectly affect serotonergic receptors. In high doses, the drug will cause tachycardia, slurred speech, dry skin, ataxia, and seizures but no hyperreflexia or hyperthermia following overdose. The serotonergic effects of bupropion have been debated.

Serotonin syndrome can occur within days of starting a serotonin-reuptake inhibitor as a result of co-ingestion of a drug that reduces its clearance or as a result of a suicide attempt. (Serotonin syndrome is seldom included in the differential diagnosis of an elderly patient because acute agitation is attributed to pre-existing dementia). Mortality can be substantial because the condition can lead to metabolic acidosis, rhabdomyolysis, acute liver failure, renal failure, and, in the most extreme cases, disseminated intravascular coagulation. The patient will not improve unless the drug is discontinued. Myoclonus frequently causes patients to be hyperactive (predominantly in the legs), markedly rigid with hyperreflexia, febrile, and with leukocytosis. Fentanyl is an under-recognized trigger in this syndrome, and some patients even have been treated accidentally with fentanyl, worsening the syndrome. The prognosis of serotonin syndrome is usually very good.

Acute Emergencies in Parkinson's Disease

The causes of acute parkinsonism were shown in Table 4.1, but do not expect it with most acute structural brain lesions. The main acute manifestations are acute akinesia or acute dyskinesia. Acute akinesia is associated with hyperthermia and also known as parkinsonism–hyperpyrexia syndrome or even akinetic crisis [49, 50]. Acute parkinsonism has also been described with acute toxins (organophosphates, sarin, cyanide, and methanol). Common causes are sudden withdrawal of medication or a recent surgery (often in elderly patients undergoing orthopedic surgery). The syndrome also affects patients with olivopontocerebellar atrophy. Clinical features are increased core temperature, which may reach hyperthermic values (40 °C). There is marked rigidity and dysautonomia. The dysautonomia is usually not recognized but includes tachycardia, anhidrosis or sweating, dysperistalsis of the gut, or even adynamic ileus, marked blood pressure changes, all associated with rhabdomyolysis as a result of constantly contracting skeletal muscles [43]. Rhabdomyolysis may lead to diffuse intravascular coagulation. The only way that levodopa withdrawal syndrome, neuroleptic malignant syndrome, and malignant hyperthermia differentiate themselves is by underlying disease (i.e., known Parkinson's disease, known schizophrenia, or a known prior myopathy).

A major controversial issue is parkinsonism-associated hydrocephalus; recorded cases of rapidly worsening parkinsonism and acute hydrocephalus have been published. Akinesia may improve with shunting but not in all patients. Acute gait freezing has also been reported with acute strokes in expected areas such as the substantia nigra, but these are exceptional circumstances.

Acute Torticollis

A nontraumatic subluxation of the atlantoaxial joint due to inflammation from a spreading infection may cause acute torticollis (also known as Grisel's syndrome) [51–56]. It is common in children but may occur in adults; very few neurologists are familiar with this syndrome. On examination, the chin hangs down and to one side. CT will show a significant C1–C2 rotary subluxation. Treatment consists of antibiotics (it often occurs after tonsillectomy for tonsillitis) and muscle relaxants. Cervical spinal tumors and posterior fossa tumors have been associated with torticollis.

Functional Movement Disorders

Surprisingly for some, functional movement disorders are not unknown in the ICU. They invariably occur after elective surgery (sometimes those with dubious indications), after an overdose (as a cry for help), or after a major acting-out episode [57]. Careful examination and consultation with colleagues may be needed to diagnose these abnormalities with certainty. Clues are movement disorders that disappear when patients are not being observed with surveillance video monitoring or with "lying on the hands." Criteria include (1) abrupt onset; (2) peculiar, changing characteristics, and combinations of movements; (3) coexisting fatigue and exhaustion; (4) disappearance with distraction; (5) vivid startle response; (6) response to placebo or suggestion; (7) spontaneous remission; (8) minor trauma to a limb; and (9) prior evaluation for a psychiatric or emotional disorder.

More Reflections

Is this a real thing? A commonly overheard question in any patient with a new (unexpected) movement. Any wildly "flailing and shaking" patient attracts attention, but ironically, acute immobility from hypokinesis is also a movement disorder. Indeed, it is far more dangerous. Movement disorders presenting anew in the emergency department require an immediate neurologic assessment. Many neurologists will first characterize the abnormal movement—hypokinetic or hyperkinetic—and indeed, as noted previously, should ask themselves if it may harm the patient. Several of these abnormal movement disorders can act as a "perfect storm" for the development of a critical (and even life-threatening) condition. Recognition of the urgency is critical, and treatment should be started without hesitation. Treatment is nonspecific but can be diagnostic. Try 50 mg of diphenhydramine IV or benztropine, 4 mg, in severe laryngospasm (after securing airway first); usually, the patient may be extubated within 48 hours. Propofol is most effective in muting myoclonic status; everything else is useless. Escalating anti-epileptics rarely work in this condition and only waste time. In many other conditions, recognizing a drug side effect and discontinuing it can be diagnostic and therapeutic. It will not come as a surprise that sometimes treatment (e.g., escalating benzodiazepines) is worse than the movement disorder.

One final note. We have seen a few cases of neuroexcitation with propofol. The movements associated with these events have strongly suggested convulsive activity but they are not. Usually these are healthy young patients who underwent elective surgery under conscious sedation and emerge from sedation with transient but repetitive violent motor activity and impaired consciousness. These manifestations required considerable mobilization of multiple health care workers to protect the patient from inflicting harm. All patients fully recovered from this event. Brief treatment with dexmedetomidine does wonders [58].

Pointers and Takeaways

- Ask yourself: What twitches? Where does it twitch? Is it rhythmic or non-rhythmic and finally regular? Is it irregular, on/off tone, flowing, or contracting?
- Paucity of movement can be more life-threatening than excess of movement.

- Laryngeal dystonia and trismus may block the airway and, thus, are immediately life-threatening.
- Identify a potential drug adverse effect and discontinue it immediately.
- Recognize that failure to administer antiparkinsonian drugs may lead to a dramatic clinical picture
- Rigidity may cause rhabdomyolyses with CPKs in multitudes of normal.
- Excitation ("propofol frenzy") is occasionally seen in the recovery room and best treated with dexmedetomidine

References

1. De Witte J, Sessler DI. Perioperative shivering: physiology and pharmacology. Anesthesiology. 2002;96:467–84.
2. Eberhart LH, Doderlein F, Eisenhardt G, et al. Independent risk factors for postoperative shivering. Anesth Analg. 2005;101:1849–57.
3. Mathews S, Al Mulla A, Varghese PK, Radim K, Mumtaz S. Postanaesthetic shivering--a new look at tramadol. Anaesthesia. 2002;57:394–8.
4. Rollins N, Pride GL, Plumb PA, Dowling MM. Brainstem strokes in children: an 11-year series from a tertiary pediatric center. Pediatr Neurol. 2013;49:458–64.
5. Saposnik G, Caplan LR. Convulsive-like movements in brainstem stroke. Arch Neurol. 2001;58:654–7.
6. Dewey RB Jr, Jankovic J. Hemiballism-hemichorea. Clinical and pharmacologic findings in 21 patients. Arch Neurol. 1989;46:862–7.
7. Gandhi SE, Newman EJ, Marshall VL. Emergency presentations of movement disorders. Pract Neurol. 2020;20(4):practneurol-2019-002277.
8. Kipps CM, Fung VS, Grattan-Smith P, de Moore GM, Morris JG. Movement disorder emergencies. Mov Disord. 2005;20:322–34.
9. Schaefer SM, Rostami R, Greer DM. Movement disorders in the intensive care unit. Semin Neurol. 2016;36:607–14.
10. Cossu G, Colosimo C. Hyperkinetic movement disorder emergencies. Curr Neurol Neurosci Rep. 2017;17:6.
11. Poston KL, Frucht SJ. Movement disorder emergencies. J Neurol. 2008;255(Suppl 4):2–13.
12. Robottom BJ, Factor SA, Weiner WJ. Movement disorders emergencies. Part 2: hyperkinetic disorders. Arch Neurol. 2011;68:719–24.
13. Robottom BJ, Weiner WJ, Factor SA. Movement disorders emergencies. Part 1: Hypokinetic disorders. Arch Neurol. 2011;68:567–72.
14. Schilder JC, Overmars SS, Marinus J, van Hilten JJ, Koehler PJ. The terminology of akinesia, bradykinesia and hypokinesia: past, present and future. Parkinsonism Relat Disord. 2017;37:27–35.
15. Ahlskog JE. The Parkinson's Disease Treatment Book: Partnering with Your Doctor to Get the Most from Your Medications. 2nd ed. New York: Oxford University Press; 2015.
16. Bhowmick SS, Lang AE. Movement disorders and renal diseases. Mov Disord Clin Pract. 2020;7:763–79.
17. Freund B, Kaplan PW. Differentiating Lance-Adams syndrome from other forms of postanoxic myoclonus. Ann Neurol. 2016;80:956.
18. Gupta HV, Caviness JN. Post-hypoxic myoclonus: current concepts, neurophysiology, and treatment. Tremor Other Hyperkinet Mov (N Y). 2016;6:409.
19. Levy A, Chen R. Myoclonus: pathophysiology and treatment options. Curr Treat Options Neurol. 2016;18:21.
20. Marcellino C, Wijdicks EFM. Posthypoxic action myoclonus (the Lance-Adams syndrome). BMJ Case Rep. 2020;3(4):e234332.
21. Jankovic J, Penn AS. Severe dystonia and myoglobinuria. Neurology. 1982;32:1195–7.
22. Mizuno Y, Takubo H, Mizuta E, Kuno S. Malignant syndrome in Parkinson's disease: concept and review of the literature. Parkinsonism Relat Disord. 2003;9(Suppl 1):S3–9.
23. Callaway CW. Neuroprognostication postcardiac arrest: translating probabilities to individuals. Curr Opin Crit Care. 2018;24:158–64.
24. Freund B, Kaplan PW. Myoclonus after cardiac arrest: where do we go from here? Epilepsy Curr. 2017;17:265–72.
25. Mikhaeil-Demo Y, Gavvala JR, Bellinski II, et al. Clinical classification of post anoxic myoclonic status. Resuscitation. 2017;119:76–80.
26. Aicua Rapun I, Novy J, Solari D, Oddo M, Rossetti AO. Early Lance-Adams syndrome after cardiac arrest: prevalence, time to return to awareness, and outcome in a large cohort. Resuscitation. 2017;115:169–72.
27. Lance JW, Adams RD. The syndrome of intention or action myoclonus as a sequel to hypoxic encephalopathy. Brain. 1963;86:111–36.
28. Blomme L, Van de Velde K. Trismus as a presenting symptom in a case of progressive encephalopathy with rigidity and myoclonus. Case Rep Neurol. 2019;11:132–6.
29. Swayne A, Tjoa L, Broadley S, et al. Antiglycine receptor antibody related disease: a case series and literature review. Eur J Neurol. 2018;25:1290–8.
30. van der Geer SJ, van Rijn PV, Roodenburg JLN, Dijkstra PU. Prognostic factors associated with a restricted mouth opening (trismus) in patients with head and neck cancer: systematic review. Head Neck. 2020;42:2696–721.
31. Allen NM, Lin JP, Lynch T, King MD. Status dystonicus: a practice guide. Dev Med Child Neurol. 2014;56:105–12.

32. Weiner WJ, Goetz CG, Nausieda PA, Klawans HL. Respiratory dyskinesias: extrapyramidal dysfunction and dyspnea. Ann Intern Med. 1978;88:327–31.
33. Pittock SJ, Parisi JE, McKeon A, et al. Paraneoplastic jaw dystonia and laryngospasm with antineuronal nuclear autoantibody type 2 (anti-Ri). Arch Neurol. 2010;67:1109–15.
34. Narayan RK, Loubser PG, Jankovic J, Donovan WH, Bontke CF. Intrathecal baclofen for intractable axial dystonia. Neurology. 1991;41:1141–2.
35. Lemke DM. Riding out the storm: sympathetic storming after traumatic brain injury. J Neurosci Nurs. 2004;36:4–9.
36. Lemke DM. Sympathetic storming after severe traumatic brain injury. Crit Care Nurse. 2007;27:30–7. quiz 38
37. Perkes I, Baguley IJ, Nott MT, Menon DK. A review of paroxysmal sympathetic hyperactivity after acquired brain injury. Ann Neurol. 2010;68:126–35.
38. Perkes IE, Menon DK, Nott MT, Baguley IJ. Paroxysmal sympathetic hyperactivity after acquired brain injury: a review of diagnostic criteria. Brain Inj. 2011;25:925–32.
39. Scott RA, Rabinstein AA. Paroxysmal sympathetic hyperactivity. Semin Neurol. 2020;40:485.
40. Thomas A, Greenwald BD. Paroxysmal sympathetic hyperactivity and clinical considerations for patients with acquired brain injuries: a narrative review. Am J Phys Med Rehabil. 2019;98:65–72.
41. Wijdicks EFM. Brain storming in brain trauma. Neurocrit Care. 2020;32:620–3.
42. Siu G, Marino M, Desai A, Nissley F. Sympathetic storming in a patient with intracranial basal ganglia hemorrhage. Am J Phys Med Rehabil. 2011;90:243–6.
43. Granner MA, Wooten GF. Neuroleptic malignant syndrome or parkinsonism hyperpyrexia syndrome. Semin Neurol. 1991;11:228–35.
44. Guneysel O, Onultan O, Onur O. Parkinson's disease and the frequent reasons for emergency admission. Neuropsychiatr Dis Treat. 2008;4:711–4.
45. Ruffert H, Bastian B, Bendixen D, et al. Consensus guidelines on perioperative management of malignant hyperthermia suspected or susceptible patients from the European Malignant Hyperthermia Group. Br J Anaesth. 2020;125:133.
46. Kynes JM, Blakely M, Furman K, Burnette WB, Modes KB. Multidisciplinary perioperative care for children with neuromuscular disorders. Children (Basel). 2018; Sept 12.
47. van den Bersselaar LR, Snoeck MMJ, Gubbels M, et al. Anaesthesia and neuromuscular disorders: what a neurologist needs to know. Pract Neurol. 2020; Oct 27.
48. Brosen K, Naranjo CA. Review of pharmacokinetic and pharmacodynamic interaction studies with citalopram. Eur Neuropsychopharmacol. 2001;11:275–83.
49. Manji H, Howard RS, Miller DH, et al. Status dystonicus: the syndrome and its management. Brain. 1998;121(Pt 2):243–52.
50. Onofrj M, Thomas A. Acute akinesia in Parkinson disease. Neurology. 2005;64:1162–9.
51. Bhattacharya D, Choudhari KA. 'Idiopathic' torticollis: have you ruled out a spinal tumour? Br J Hosp Med (Lond). 2008;69:412–3.
52. Gourin CG, Kaper B, Abdu WA, Donegan JO. Nontraumatic atlanto-axial subluxation after retropharyngeal cellulitis: Grisel's syndrome. Am J Otolaryngol. 2002;23:60–5.
53. Kumandas S, Per H, Gumus H, et al. Torticollis secondary to posterior fossa and cervical spinal cord tumors: report of five cases and literature review. Neurosurg Rev. 2006;29:333–8. discussion 338
54. Morgante F, Edwards MJ, Espay AJ. Psychogenic movement disorders. Continuum (Minneap Minn). 2013;19:1383–96.
55. Samuel D, Thomas DM, Tierney PA, Patel KS. Atlanto-axial subluxation (Grisel's syndrome) following otolaryngological diseases and procedures. J Laryngol Otol. 1995;109:1005–9.
56. Welinder NR, Hoffmann P, Hakansson S. Pathogenesis of non-traumatic atlanto-axial subluxation (Grisel's syndrome). Eur Arch Otorhinolaryngol. 1997;254:251–4.
57. Chatterjee SS, Das S, Gupta S, Bhattacharya S. "The twisted mind" – psychogenic dystonia in an adolescent, responding to antidepressant therapy. Shanghai Arch Psychiatry. 2018;30:133–4.
58. Carvalho DZ, Townley RA, Burkle CM, Rabinstein AA, Wijdicks EFM. Propofol Frenzy: Clinical Spectrum in 3 Patients. Mayo Clin Proc. 2017;92:1682–7.

5 Mimickers, Misleads, and Confounders

Neurology is perhaps the only specialty where examiners should seriously worry about not getting the clinical examination right because something else is in play. Questions arise when parts of the examination do not fit with what has been traditionally known from the standards of localization. So, which situations throw us for a loop? Some other factor—a drug, a toxin, or a laboratory outlier—might mimic the neurologic findings from a brain lesion, and these mimickers can "muddy the waters." Often, patients are drowsier than expected or not recovering well after some intervention. Drugs administered to the patients may have an obvious sedating effect, but the lingering effect on responsiveness often undermines the neurologist assessment. When it comes to drug-induced neurologic symptoms, the list is endless. Some drugs are closely associated with a movement disorder (tremor, myoclonus, dyskinesia, or less-defined flailing) or with inability to walk, or changes in muscle tone (flaccid or spastic). In other situations, a new metabolic derangement with neurologic manifestations can produce neurologic findings on its own or worsen prior deficits. For example, severe hyperglycemia may worsen a preexisting hemiparesis or aphasia. Cumulative exposures to drugs given by transferring hospitals might cause newly admitted patients to arrive seriously sedated. A commen recent issue is the use of ketamine during helicopter transport, which may only clear after several hours. Recognizing these flukes can lead to correction of the mimicker with resolution often occurring within a number of hours or days (but not always, as we will see).

In essence, we must continually ask ourselves whether our examination accurately reflects what is happening to the patient. Good clinical acumen may improve our intuition that something is not right. And misinterpretations may lead to overly pessimistic predictions that have serious consequences for outcome assessment. Don't let the examination discourage you. When faced with a puzzling finding on examination, we all have to marshal every possible resource to understand it. What I hope to provide in this chapter is the knowledge needed to avoid misdiagnoses. To achieve this, we need to consider pharmacology, neuroanatomy, and how changes in homeostasis affect the brain.

What You Need to Know About ICU Pharmacology

Without explaining the extraordinarily complex field of ICU pharmacology, it is still useful to repeat some principles [1–3] if only to remind ourselves that the three most common reasons for only partial awakening of hospitalized patients are drugs (with an unexpected or unknown interaction), drugs (inadvertently overdosed), and drugs (insufficiently cleared, often due to changes in organ function). Generally,

E. F. M. Wijdicks, *Examining Neurocritical Patients*, https://doi.org/10.1007/978-3-030-69452-4_5

bioavailability of intravenously administered drugs reaches 100%. One can assume that at five times the half-life, the mean plasma concentration of the drug is constant and at a plateau for drugs with zero-order kinetics (i.e., the rate of drug elimination is constant over time). Assuming no organ system dysfunction, the reverse occurs with the plasma concentration decreasing to zero in 5 half-lives. Once administered, drug distribution follows a 2-compartment model. The first compartment is the intravascular volume and rapidly perfused tissues. The second is distribution in the other tissues. Lipophilic drugs (e.g., barbiturates) distribute more readily in fat tissues. Pharmacokinetic changes are common in obese patients, and there are differences in dosing. Total body weight is usually used in most sedatives, anticoagulants, and antimicrobials. Ideal body weight is recommended for other medications (e.g., acyclovir) even when the patient is overweight; taking the true weight could easily lead to substantial overdose. Drug interactions can be caused by pharmacokinetic interactions (drug A affecting the absorption of drug B) or by pharmacodynamics interactions (drug A having an additive effect with drug B).

Additional important principles are 1) additive toxicity (e.g., 2 nephrotoxic agents); additive effect (e.g., similar mode of action); and 3) multiple drugs, unknown patient history, and unstable patients. Drug elimination changes in patients with acute renal failure, and these glomerular filtration changes are mostly seen after sepsis, with or without prolonged hypotension and use of vasopressors.

Ubiquitous use of sedatives and analgesics has been climbing over decades. It is easier to calm the patient than to manage agitation by other means. Admittedly, uncontrolled behavior carries its own risks: (1) self-extubations; (2) increased risk of lung injury, when there is very high minute ventilation from increased drive; and (3) increased myocardial demand, possibly leading to troponin leaks. Fortunately, the intensive-care world has launched a call to action to limit doses to what is minimally needed to achieve effect.

As a necessary additional chore, every neurologist now must carefully scrutinize the medication listing, even going back several days. "The patient is completely off sedation" is not a necessarily satisfactory answer. The statement that "the renal and hepatic functions are normal" may be incorrect, and more specific laboratory tests may be days old or simply have not been examined (such as the seldom-measured serum ammonia). Drugs linger far longer than we appreciate [4]. And then, despite thinking through the absorption, distribution, and the two routes of drug elimination (metabolism and excretion) in the patient we are about to see, we will rarely be able to assess accurately the effect of drug interactions, body physique, or age. Changes in renal function may result in a relative overdose of the drug. An ideal sedative agent—rapid onset, rapid recovery after discontinuation, predictable dose response, lack of drug accumulation, and no toxicity—is hard to find. There are substantial changes in pharmacokinetics in any seriously ill patient (Fig. 5.1).

The Pervasiveness of Analgosedation

The combined use of opioids and benzodiazepines is an ongoing concern. Before we review a few of the usual suspects, it is important to note that the duration of infusion of analogosedative drug combinations affect clearance (Fig. 5.2). How much is not exactly known, and of course, it depends on factors such as severity of critical illness; all factors delaying clearance are simply unknowable.

Physicians routinely switch between the drugs dexmedetomidine and propofol. Dexmedetomidine effects are three-fold: sedation, analgesia and anxiolysis, and peak effect of 15 minutes. Clearance is not altered by organ dysfunction. Dexmedetomidine has linear pharmacokinetics up to doses of 2.5 μg/kg/h but may linger with higher doses to a maximal dose of 5 μg/kg/h [5]. Propofol has a more rapid onset (in the 10- to 60-second range) but does not last more than 20 or 30 minutes (remember that dexmedetomidine lingers on 4 times longer than propofol). Renal or hepatic disease does not significantly alter clearance, but decreased hepatic blood flow can reduce clearance in patients treated for distributive shock. While it is very

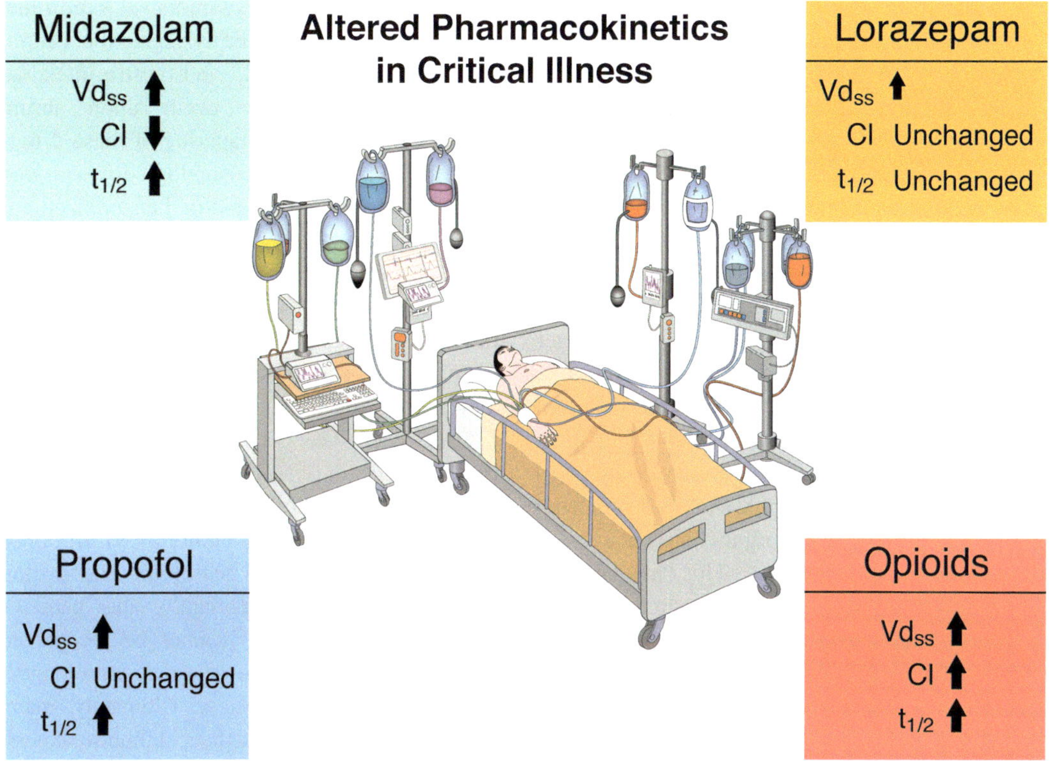

Fig. 5.1 Pharmacologic changes in commonly used analgesics and sedatives in intensive care units (Vd_{ss} volume of distribution steady state, CL clearance, t1/2 half –life) up-arrows means increase, down-arrows means decrease

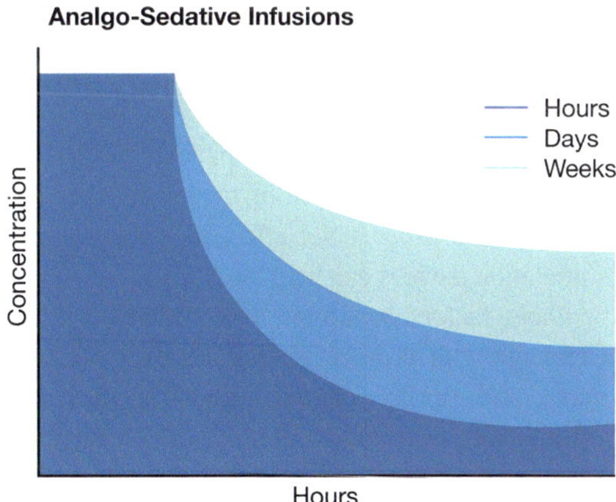

Fig. 5.2 Approximation of clearance rates related to duration of infusion

uncommon for a major drug effect to occur several hours after the infusion has been shut off, we cannot just shut off the infusion, wait for a few minutes, and expect to examine the patient reliably. However, this erroneous practice is not uncommonly seen in many places. Midazolam has an onset in minutes and is short-acting (30–90 sec). This drug becomes significantly more sedative when cirrhosis, obesity, and renal failure prolong the half-life. Active metabolites of midazolam linger much longer in renal and hepatic dysfunction or failure but are also eliminated far more slowly, adding to the sedative mix. Because it is a lipophilic drug, we can expect prolonged effects in obese patients and, certainly, in patients with reduced serum albumin levels.

Lorazepam, which is often used in infusions, is not much better than midazolam; in fact, the half-life is much longer than in midazolam (12 hours) and even doubles in the setting of severe renal failure (old or recent). Lorazepam, however, is metabolized by hepatic conjugation to inactive metabolites. More than a few drugs inhibit the metabolism of benzodiazepines and increase their effects. Calcium channel blockers and cytochrome P450 inhibitors (e.g., erythromycin, fluconazole) all prolong sedation. Drugs prolonging clearance of sedatives and analgesic agents are shown in Table 5.1.

Fentanyl is used very commonly and not always with the right indications. It is almost an anticipatory drug (the "patient must be in pain so let's treat pain" fallacy). As predicted, the effect of opioids (with differences based on their intrinsic properties) is complex, and there is no such a thing as a "tiny dose"—at least not for neurologist who must examine the patient. The bottom line is that the half-life doubles with many analgesics and sedatives after a few hours' infusion, but the effects linger on longer (dexmedetomidine, midazolam, ketamine, propofol). Fentanyl, however, is a different story. Half-lives increase with longer infusions, and clearance in critical illness may be up to 10 times less when compared with patients who have no evidence of organ dysfunction or hemodynamic instability [4]. This "neurologist-unfriendly" (and seriously potent) drug, fentanyl, begets even more problems. Opioids are used in combination with other sedatives for analgesia. Although the onset is immediate and the duration is 1–2 hours, the half-life of fentanyl of (~6 hours) is prolonged with continuous infusions and after repeated doses. Fentanyl accumulates in tissue. Significant accumulation is also expected with renal or hepatic disease. (There is a prolonged sedative effect when used concomitantly with CYP 3A4 inhibitors such as diltiazam.) As noted, the drug is used liberally, and its frequent use over days ("the patient is not waking up after we stopped everything") often comes as a surprise.

Once a combination of sedative and analgesic drugs is started, there is an incentive to continue. Part of the problem with starting analgosedation is the fear of iatrogenic withdrawal syndrome, a combination of autonomic dysregulation, central nervous system arousal, and gastro-intestinal symptoms, which may occur upon abrupt discontinuation or rapid tapering of these drugs. Critically ill patients receiving high doses or who are exposed to opioids and/or benzodiazepines for >72 h are at risk [6]. Self-extubation may increase, and intracranial pressure may abruptly rise in neurocritical patients. Daily interruption trials do not have an effect on ICU stay because many patients stay in some way sedated and do not fully awaken [7].

Table 5.1 Drugs prologing clearance of sedatives and analgesicagents

Drug	Interfering drug
Midazolam	Diltiazem, erythromycin, fluconazole, ranitidine, verapamil
Lorazepam	Clozapine, quetiapine, valproic acid
Propofol	Lidocaine
Morphine	Cimetidine, azithromycin, itraconazole
Fentanyl	CYP 3A4 inhibitors
Diazepam	Cimetidine, erythromycin, fluoxetine

From Wijdicks and Clark [69]

Over recent years, the use of neuromuscular junction blockers (NMB) has decreased—but not completely. They are useful in treating severe forms of acute respiratory failure [8, 9]. But once used, hope of an accurate neurologic examination is lost, and more importantly, the patient's awareness (and seizures) cannot be recognized clinically. Other substantial risks include prolonged, ICU-acquired limb weakness. Ultimately, many patients develop prolonged, flaccid limb weakness, caused by critical illness polyneuropathy or myopathy.

Succinylcholine is the only depolarizing agent. Intravenous pancuronium, vecuronium, atracurium, and cisatracurium all have an ~45- to 60-minute duration, but rocuronium is shorter, at 30 minutes. Rocuronium is less potent than vecuronium, but elimination is prolonged in patients with liver dysfunction. The presence of tendon reflexes is a clue that the drug is not effective. (Train-of-four stimulation devices can help too.)

Other factors contribute to a prolonged effect. Older age, with its decrease in total body water, body mass, and serum albumin, leads to decreased volume of distribution for NMBs leading to increased drug effect and decreased drug elimination. Hypothermia prolongs the duration of action of NMBs due to altered sensitivity of neuromuscular junction and reduced renal or hepatic clearance. Hypokalemia, hypermagnesemia, and prolonged acidosis through a variety of theoretical mechanisms also prolong blockage by nondepolarizing agents and even reduce the ability of neostigmine to reverse blockade. Hypomagnesemia also prolongs the duration of blockade by inhibiting calcium. Acidosis may enhance blockade effects of nondepolarizing agents. Remember that pancuronium half-life is increased in liver failure and that vecuronium and pancuronium prolonged action might be a result of metabolite accumulation in renal disease.

If there are still uncertainties with assessment, it is best to reverse these drugs with several known antidotes (Table 5.2). Both naloxone and flumazenil reverse suspected effects of benzodiazepine and opioid, but there are major caveats. Flumazenil is a nonspecific, competitive antagonist to benzodiazepine receptors. A single dose may not be enough, and repeated doses of 0.2 mg IV over 1-minute intervals up to 1 mg may be necessary to get the optimal effect. In case of re-sedation, the maximal dose is 3 mg in 1 hour. Many physicians do not appreciate how unreliable this drug is; we should not assume that benzodiazepine is reversed and a neurologic examination can proceed without any confounding effect. Furthermore, with a prior history of seizures, flumazenil may precipitate withdrawal seizures. Rarely, benzodiazepine reversal will cause a single seizure, but seizures are more common following an overdose of benzodiazepine, tricyclic antidepressants, and other seizure-inducing agents.

Reversing opioids is more complex; many physicians do not realize that the main goal of naloxone is adequate ventilation, not full return of consciousness. As with flumazenil, dosing is one dose of 0.4 mg naloxone IV after another until effective. If there is no response after 10 mg, something other than opioids is causing the neurologic symptoms.

Table 5.2 Agents for acute reversal of neuromuscular blockage

Reversal with *neostigmine*
Dose: 0.04–0.07 μg/kg
Reversal within 1 min with a peak effect for 9 min
Increases the amount of acetylcholine in the neuromuscular junction
Reversal with *sugammadex* (for urgent rocuronium reversal)
Dose: 1.2 mg/kg for single rocuronium administration
Reversal in about 3 minutes
If no twitch in train-of-four stimulus, then 4 mg/kg bolus (actual body weight)
If spontaneous recovery with second twitch in train of four, then 2 mg/kg bolus
If tendon reflexes are intact, reversal mostly complete

Generally, we must conclude that while reversal agents are somewhat helpful, the effect of benzodiazepines or opioids does not dissipate easily. Patients remaining comatose after suspected benzodiazepine or opioid overdose and use of reversal agents likely have something else going on, which could be another unidentified drug or inhalant.

The Other ICU Drug Effects

Entering the ICU fray is daunting for the consulting neurologist tasked with finding out which drug is potentially responsible for neurologic manifestations. Adverse effects causing neurologic symptoms are most often related to inaccurate dosing or sudden withdrawal. A commonly used drug causing changes in neurologic examination is cefepime. Cefepime crosses the blood brain barrier (BBB), but in patients with sepsis, central nervous system infection, uremia, and previous brain injury, BBB integrity may be further disrupted, causing increased penetration of cefepime (up to 45%), which is much higher than the 10% under normal conditions [10]. Because cefepime is predominantly excreted renally (85% unchanged), reduced renal function proportionally increases the elimination half-life and reduces the rate of total body clearance. On electroencephalography (EEG), periodic sharp waves and triphasic waves are characteristic of cefepime neurotoxicity. In a recent study, periodic epileptiform discharges on EEG were five times more frequent in patients receiving cefepime compared to patients receiving meropenem, but overall the prevalence of this finding was still relatively low (1%) [11].

A recent systemic review found different degrees of reduced consciousness (47%), myoclonus (42%), and confusion (42%) as the most common symptoms [12]. All patients tested with EEG had abnormalities including nonconvulsive status epilepticus (25%), myoclonic status epilepticus (7%), and triphasic and focal sharp waves (40%). Half of the patients were overdosed; however, 26% experienced neurotoxicity despite appropriate dosing. Seizures (and even nonconvulsive status epilepticus presenting as "encephalopathy") are fairly uncommon clinical manifestations of cefepime neurotoxicity. Rather, the more likely clinical scenario is a difficult-to-explain "altered mental status," which may be a depressed level of consciousness, confusion, or disorientation. The key is that myoclonus and asterixis should not necessarily be attributed to renal impairment; they are potentially suggestive of the diagnosis and should prompt the discontinuation of cefepime in favor of an alternative antibiotic [11–14]. We have observed patients with renal failure and severe encephalopathy or coma, who improved dramatically after the cessation of cefepime. During CRRT, limiting a cefepime dose to ≤2 g may decrease the risk of CEF-induced neurotoxicity but may possibly induce treatment failure.

Antibiotics have been associated with a range of movement disorders. Sulfamethoxazole-trimethoprim can cause tremor. Fluoroquinolones, macrolides, and aminoglycosides slow neuromuscular transmission by inhibiting presynaptic acetylcholine release and binding postsynaptic acetylcholine receptors. Given the high likelihood of exacerbated illness, these medications are contraindicated in patients with myasthenic syndromes, orofacial dyskinesia, dystonic reactions, and tic disorders. Chloramphenicol can cause chorea and dystonic reactions. Multiple other antibiotics have been associated with delirium or some undefined encephalopathy. Cephalosporin and penicillin have been linked to seizures and delirium [15]. Sulfonamides, fluoroquinolones, and macrolides have been associated with acute psychosis [16]. Aminoglycosides can also induce cerebellar syndromes. Metronidazole occasionally precipitates an acute cerebellar syndrome. Metronidazole correlates with neurotoxicity and may be coadministered with cefepime to provide additional anaerobic coverage. Metronidazole-associated encephalopathy merits special attention as it can produce an acute cerebellar syndrome and seizures starting several weeks after treatment initiation. Affected patients can have symmetric hyperintense lesions in the dentate nuclei on MR imaging [16].

There are some other oddities. Medication-induced mydriasis has previously been described as a reaction to nebulized treatments with anticholinergic agents such as ipratropium. Nitric oxide is often confused with nitrous oxide. The effect of nitrous oxide gas on pupils is decreased size and altered responsiveness. Nitric-oxide-induced mydriasis has not previously been described. This gaseous medication is used in patients with pulmonary hypertension and right-sided heart failure and acts via relaxation of smooth muscle. Nebulizing through a facemask is a less common administration method; it is typically provided through a microcatheter in conjunction with ventilatory support. During nebulization, condensation moisture on eyelids or eyebrows may lead to direct corneal contact. Nitric oxide generally induces vascular dilatation by acting on smooth muscle cells. This is actually relevant; with this different mechanism of widening of the pupil, the common use of the 1% pilocarpine test (miosis of the normal eye and no response in the pharmacologically dilated eye) will not be helpful despite the usual recommendation for suspected anticholinergic use. The wide pupil resolves with removal of the offending agent, which may take a full day if not longer. If the drug is essential for treatment,

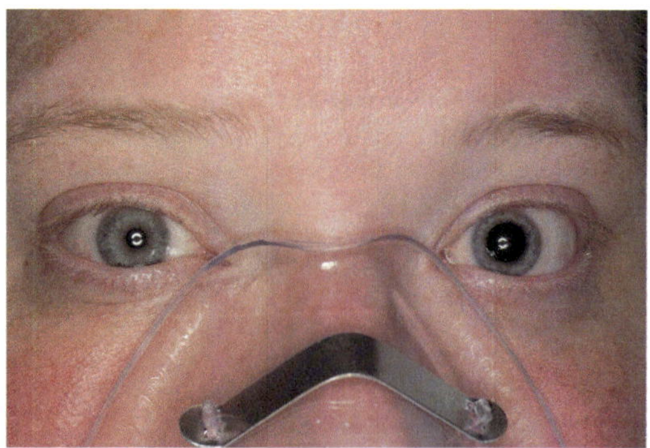

Fig. 5.3 Mydriasis of the left pupil from an inhaled vasodilator. (From *ref. [17] used with permission*)

awaiting full reversal is critical simply to ensure a proper fitting respiratory mask (Fig. 5.3).

Deranged Physiology of Critical Illness and the Brain

Major changes in metabolic function, ventilation, and hemodynamics can suppress brain function. The exact mechanism is unknown, and there is considerable individual variability in how major changes from the norm can affect higher (executive) brain function. Take, for instance, the different presentations of hypoxemia. Some patients will be dyspneic with labored fast breathing and restless but others, calm despite oxygen saturations in the 60% range or below. For many hypoxemic patients, oxygen saturations <60% can be tolerated for some time with only moderate and transient alterations in thinking or other signs and symptoms. Hypoxia stimulates increases in depth and rate of breathing, but the carotid artery baroreceptors are set at quite low levels, and the response is even more blunted by a low PCO_2, This increased breathing is also associated with increases in ventricular output by pumping harder and faster [18, 19]. In patients who can increase breathing and, thus, lower the arterial partial pressure of carbon dioxide, breathlessness is limited. There are limits to cardiovascular compensation to hypoxemia although, again, with wide variations among individuals. Eventually it can lead to pulseless electrical activity type of cardiac arrest. Clinically, a severely hypoxemic patient is tachypneic, restless (not necessarily delirious), and gradually hypotensive, which eventually is a more crucial factor in declining consciousness than hypoxemia alone. Hypercapnia is a different story; climbing numbers of $PaCO_2$ (>60–70 mmHg) lead to stupor rather quickly—assuming the patient is not a "CO_2 retainer" from long-standing COPD.

Hypotension alters consciousness, and the mean arterial pressure when we start becoming drowsy is somewhere at a MAP of 50 mmHg. Cerebral perfusion pressure (CPP) is MAP–ICP or CVP (the highest value counts) and is typically kept at 60 mmHg. CPP can be a moving target with normal autoregulation (constant cerebral blood flow with cerebral perfusion pressures between 50 and 150 mmHg), but in a chronically hypertensive patient, the curve has moved to the right, changing these limits. The upshot is that the higher level provides protection, but the lower level may cause neuronal ischemia more easily. Obviously, this has implications for the association of impaired consciousness and hypotension. Most importantly, a MAP of <50 will cause ischemic zones in watershed areas, and the impaired consciousness may become persistent for days and weeks. (Compare these MAP thresholds with other organ systems—cardiac (coronary) MAP for adequate coronary perfusion pressure of 65 mmHg and MAP of 70–75 mmHg for adequate perfusion of the kidney. Remember that when cardiac and kidney ischemia is present, brain ischemia may follow.)

There are a few other simple observations to keep in mind. First, all acute severe metabolic derangements generally produce reduced alertness, which evolves into drowsiness and, eventually, stupor. Localizing signs, such as aphasia or hemiparesis, should not be expected with diffuse brain dysfunction due to a metabolic derangement [20–22]. A single exception is nonketotic, hyperosmolar hyperglycemia causing transient aphasia or hemiparesis, although the mechanism is poorly understood. Second, correction of the abnormality, particularly when it occurs quickly (albeit a largely undefinable qualification), may cause pontine tract demyelination from rapid osmotic shifts. Long-standing alcoholics are most susceptible to changes resulting from correction of chronic hyponatremia. Some thresholds for neurologic manifestations have been proposed (Table 5.3); let's review the ones that have been best described.

Hyponatremia typically manifests itself with restlessness and confusion, followed by generalized tonic-clonic seizures. Both the rapidity of decline and ultimate level of serum sodium determine symptomatology. Hyponatremia usually is asymptomatic until levels drop below 125 mmol/L. It may cause drowsiness, in some cases interrupted by generalized tonic-clonic seizures. Although the

Table 5.3 Thresholds of laboratory values compatible with altered consciousness

Derangement	Serum
Hyponatremia	≤125 mmol/L
Hypernatremia	≥160 mmol/L
Hypercalcemia	≥3.4 mmol/L
Hypermagnesemia	≥5 mg/L
Hypercapnia	≥70 mm Hg
Hypoglycemia	≤50 mg/dL
Hyperglycemia	≥800 mmol/L

rapidity of sodium change is likely more crucial, the degree of hyponatremia is also an important factor. If there is a precipitous drop to very low levels (<115 mmol/L), cerebral edema may occur and can be rapidly fatal. (Usually it is associated with dramatic increase of free water administration.) Normally, the brain can extrude inorganic solubles within hours of hyponatremia. This is then followed by water loss, reducing the risk of cellular swelling. However, this compensatory system may be overwhelmed, leading to brain edema. This rare syndrome of postoperative hyponatremia in premenopausal women with cerebral edema and respiratory arrest with fatal outcome (many become brain dead from swelling resulting also in diabetes insipidus) has been previously described [23–25]. If a patient presents with a flaccid tetra paresis, bulbar palsy, dysarthria, dysphagia, and ocular motor abnormalities, the central pons is likely involved, and the diagnosis is central pontine myelinolysis. Extrapontine myelinolysis can affect the cerebellum, putamen, caudate nuclei, lateral geniculate bodies, hippocampus, cerebral cortex, and thalamus. The clinical manifestations of CPM can be diverse, but pseudobulbar palsy, facial weakness, inability to swallow or speak, and quadriplegia are prominent. In some patients, presentation is with parkinsonism and other movement-disorder symptoms such as dystonia, chorea, and athetosis. In many (but not all) patients, rapid correction of chronic severe hyponatremia is responsible. In this situation, we define "rapid" as correction of sodium by 0.5 mmol/L per hour. Over-correction to hypernatremia may be an equally important factor [26–28].

It is different from hypernatremia. The level at which hypernatremia confounds the clinical examination is generally defined as 160 mmol/L or more [29]. However, plasma osmolality may correlate better with the level of consciousness than with an absolute value of hypernatremia. Plasma osmolality, >350 mosm/L, results in progressive drowsiness in other situations. As in many other electrolyte disorders, the rapid development of hypernatremia is more important than the absolute value.

It is very rare to see symptoms at potassium levels of <2.0 mmol/L, but when they occur, they include weakness—generalized proximal-greater-than-distal weakness. We may also hear about leg cramps, paresthesia, irritability, and rhabdomyolysis. It is rarely seen, usually in acute admissions to the emergency department and often an unexplained surprise. Our patients, as well as the patients in other intensive care units, have their potassium levels religiously checked and corrected. Hyperkalemia, in itself, does not cause neurologic symptoms, but the disorders causing it (e.g., acute renal disease due to the impaired ability to excrete potassium) can, indeed, induce neurologic symptoms.

Hypocalcemia is far more symptomatic. Neuropsychiatric symptoms seen with severe hypercalcemia include subtle personality changes, difficulty concentrating, confusion, depression, dementia, and anxiety, but proximal muscle weakness is also possible. If levels reach <7.5 mg/dL, a number of neuromuscular manifestations, including oral numbness, tingling of the fingertips, muscle cramps, and very painful muscle spasms or the more dangerous laryngospasm or bronchospasm, are common. Classic examination findings in hypocalcemia are the Chvostek and Trousseau signs. In the Chvostek sign, sharply tapping the skin over the facial nerve just anterior to the ear causes facial muscle contraction. In the Trousseau sign, inflating the blood pressure cuff to 20 mm Hg higher than the patient's systolic blood pressure for 3–5 minutes causes carpal spasm.

Hypermagnesemia reduces neuromuscular excitability due to displacement of calcium by magnesium at the neuromuscular junction. Hyporeflexia and weakness are noted at levels between 7 mmol/L and 9 mmol/L, and areflexia and parasympathetic blockade are seen at levels greater than 9 mmol/L.

Here is the bottom line: check a recent set of electrolytes before starting the evaluation. We have all had patients in whom it was immediately obvious that the electrolyte outlier produced all of the neurologic symptoms. Finally, be sure to order serum ammonia testing for possible new hyperammonemia. Patients with acute hyperammonemic encephalopathy present with progressive drowsiness to coma, seizures, and, often, status epilepticus due to direct toxic effects of ammonia on the cortical structures. (MRIs indeed show associated restricted diffusion abnormalities involving the insular cortex, cingulate gyrus, and cortical mantle [30].) Acute hepatic dysfunction is most frequently implicated, but other causes are portosystemic shunt surgery, drugs (e.g., sodium valproate, asparaginase therapy, or chemotherapy), infections, severe hypothyroidism, multiple myeloma, and post-lung or bone marrow transplantation [31–33]. Serum ammonium levels may reflect the extent of injury. Coma with cerebral edema is seen at high levels of plasma ammonia (at least 4–5 times the normal range), but the first clinical manifestations of hyperammonemia (e.g., irritability, lethargy, vomiting, disorientation) can be seen with serum ammonia levels of 40–60 μmol/L; even at this level, we can find early MR imaging changes [30]. Failure to consider hyperammonemia is quite common.

Illicit Drugs and Neurologic Examination

The topic is too broad for this book. Clinical neurotoxicology books or poison control hotlines are the best sources of information. One problem for neurointensivists is that, despite a large number of confirmed observations, neurologic details are not well delineated, described neurologic findings are limited to pupils, speed of recovery has been poorly studied, and patients often mix up drugs, combining the different toxidromes. Many intoxications lead to neurologic manifestations (Tables 5.4 and 5.5). Ingestion of any toxin in sufficient doses will impair consciousness or even produce coma with total flaccidity. More often, intoxications manifest as jitteriness, agitation, hallucinations, delirium, seizures, muscle rigidity, dystonia, or other movement disorders, but none of these are defining characteristics. The six major toxidromes are cholinergic, anticholinergic, sympathicomimetic, serotonergic, narcotic, and extrapyramidal. Pupil size, blood-pressure value, heart rate, and skin and muscle tone appearance are important determinants for each of these toxidromes, but it is the combination, rather than

Table 5.4 Illicit drugs and neurologic manifestations

Class	Neurologic manifestations	Late concerns
Amphetamines	Mydriasis, paranoia, hallucinations, delirium, focal signs (cerebral hemorrhage)	Cerebral infarcts (vasculopathy)
Cocaine	Seizures (complex partial type) and dystonia, chorea, migraine, coma, subarachnoid hemorrhage, rhabdomyolysis	Anoxic-ischemic brain injury, transient ischemic attack, cerebral infarction, aneurysmal rupture
Barbiturates	Coma, hypoxic-ischemic encephalopathy, shock, respiratory arrest	Cerebral infarct
Hallucinogens	Colored geometric images, catalepsy mydriasis, piloerection, insomnia, hyperthermia, coma	Ischemic stroke, cognitive decline
Opioids	Coma to euphoria, miotic pupils (but normal eye exam does not exclude opioid intoxication), hypothermia, seizures	Anoxic-ischemic encephalopathy

Table 5.5 Common neurologic findings with toxins

Eye findings	
Miosis	Opioids (heroin), organophosphates, clonidine
Mydriasis	Tricyclic antidepressants, MDMA intoxication
Horizontal nystagmus	Ethanol, antiepileptic drugs, ketamine phencyclidine
Vertical nystagmus	Opioids (PCA pump, ketamine, phencyclidine)
Motor findings	
Flaccidity	Benzodiazepines, baclofen
Rigidity	Neuroleptic agents, carbon monoxide, hypoglycemics

one or two findings, that defines the toxidrome. So we can see cholinergic with miosis, hypertension, tachycardia fasciculations, salivation, and sweating; anticholinergic with mydriasis, hypertension, tachycardia, and dry skin; sympaticomiemetic with mydriasis, hypertension, tachycardia, and sweating; serotonergic with mydriasis, labile blood pressures, tachycardia, myoclonus, narcotic with miosis, hypotension, bradycardia, and fasciculations; and extrapyramidal with normal pupils, hypertension, tachycardia, rigidity, and dry skin. The agitated, intoxicated patient often has an anticholinergic or sympathomimetic toxidrome (or some attenuated syndrome). These toxidromes are similar in tachycardia and hyperthermia, but the sympathomimetic patient is wet with very moist axillae, while the anticholinergic toxidrome patient is "dry as a bone." Patients' mouths are dry ("cotton mouth") causing a muffled tone. Moreover, responses of the dilated pupil are often absent (due to inhibited constrictor fibers). Many patients are agitated due to a full bladder.

Truth to tell, we almost always forget ethanol and atypical alcohols, which demonstrate osmolar gaps and ethanol levels. Methanol is found in commercial products such as windshield washer fluids, de-icers, antifreeze, paints, wood stains, glass cleaners, and hand sanitizers. Along with coma, neurologic manifestations of atypical alcohol intoxication also include cranial nerve deficits, nystagmus, ophthalmoplegia, facial palsy, and dysphasia. Tetanic cramps and myoclonus, but also seizures (late in presentation), may occur. Alcohol is typically eliminated at a rate of 20 mg/dL per hour, which depends on coingestion of other drugs. This implies that a patient with acute alcohol intoxication may require a full day to clear alcohol levels. Blood alcohol levels of >400 mg/dL (blood alcohol content, 0.4%) can lead to profound central nervous system depression and respiratory arrest.

Any inner-city-based police officer in the United States knows that pinpoint pupils indicate heroin and fentanyl (or carfentanil); and wide pupil, cocaine [34, 35]. Drugs rarely affect pupils in hospitalized patients; actually, sick patients often have small pupils. Fentanyl analogs have also been identified as adulterants in street heroin or cocaine. Furthermore, synthetic opioids have been sold as other illicit opioids or even as benzodiazepines. Fentanyl analogs pose a high risk of life-threatening poisoning to users, and the high potency of the majority of these compounds may aggravate this risk. Fentanyl is 100 times more potent than morphine [36]. Fentanyl analogs cause profound respiratory depression requiring repeated bolus injection or continuous infusion of naloxone for up to several hours. Fentanyl is known to cross-react with some immunoassays targeting LSD [36].

Then there is a whole world of inhalants. The so-called volatile substance-abuse problem includes products such as aerosols and dry-cleaning fluids, and some contain substances that damage the brain. Injury may also be related to suffocation using plastic bags or as a result of vomit. The acute effects of inhalants are hallucinations, sometimes in the setting of a full psychotic break. In exceptional cases, acute effects involve acute leukoencephalopathy (from sniffing moth balls). The major toxins reported to cause deep coma and loss of all brainstem reflexes in some were ethylene glycol [37, 38], baclofen, bupropion, and several antiepileptics (in most unusual circumstances). Table 5.5 shows the major neurologic signs associated with overdose. Baclofen or bupropion often remain the most common surprise intoxications [39–43]. Of course, routine laboratory toxicological screening tests do not include ethylene glycol, and its indirect detection mainly depends on the calculation of the osmotic gap. The increased osmolar gap also indicates the severity of the atypical alcohol intoxication. Acute renal failure due to dispersing calcium oxalate crystals is a major concern and can be detected in urine. Hospital practice screening is limited to a few substances including ethanol, benzodiazepines, opioids, acetaminophen, salicylates, or a drug if specifically asked.

Misleading (False Localizing) Signs

Localization of the lesion has been a major component of clinical neurologic thinking. Known clinical signs can be traced to a specific location in the brain. It is what we think we know best. We can now, mostly and instantaneously, verify our localization with MRI, and we should honor our colleagues from the past who could not see their triumphs or failures in real time (or had to wait for autopsy). Several were so good at finding new combinations in the brainstem that they earned an eponym (e.g., Weber, Wallenberg, Foville, Millard-Gubler). There are not that many fundamental contradictions. After the explosion of neuroimaging modalities challenged our clinical ability, clinicians have realized that at times MRI scans point toward a different location and even explanation.

We must also tip our hats to the neurologists who reasoned why the signs were misleading and how to easily explain them. The earliest reports were in the early 20th century, when neuroimaging essentially did not exist, and pathologists told the clinician that the recognized signs "falsely" pointed toward another region of the brain than predicted on the basis of anatomical knowledge. Frontocerebellar pathway damage (e.g., as a result of infarction in the territory of the anterior cerebral artery) may result in incoordination of the contralateral limbs, mimicking cerebellar dysfunction. The most common "mistake" occurred when neurosurgeons operated on a presumed infratentorial tumor and were surprised to find none. Or they made a Burr hole and then had to go to the opposite side to find a subdural hematoma. The most commonly reported false localizing signs were lesions of the cranial nerves, particularly palsies of the abducens, trigeminal, oculomotor, and also of the trochlear nerve [44–47].

False localizing signs are relatively common in the acute setting [48, 49]. They may lead to the wrong test being ordered. Neuro-ophthalmologic signs, including a sixth-nerve palsy associated with increased intracranial pressure [50–52] and a fixed pupillary dilatation opposite from the actual pathological site, were frequently encountered. Hemiparesis at the site of a hemispheric mass had also been recognized but more commonly in more slowly expanding tumors. The presence of an abducens paresis in supratentorial tumors fascinated many neurosurgeons. Alternative explanations for stretching of the nerve included its long course, compression at its pontomedullary-junction exit, compression by the anterior-inferior cerebellar artery, and compression at the clivus.

Most puzzling were patients with a fixed pupil opposite to the acute lesion [53–55]. Traditional explanations have included compression of the oculomotor nerve by a herniating uncus of the temporal lobe. (The pupil-constrictor fibers are superficially located and easily injured.) The opposite pupil becomes involved when the brainstem shifts downward and pulls on the opposite oculomotor nerve to cause bilateral fixed and dilated pupils. Dilation of the opposite pupil first is harder to explain. Rotation of the brainstem in the axial plane could slacken the ipsilateral oculomotor and stretch the opposite oculomotor. Ischemia at a nuclear level is another distinct possibility (Fig. 5.4). Both mechanisms may be at play.

False localizing signs have also been described in spinal cord lesions with compressive cervical myelopathy causing false localizing at the thoracic sensory level, largely explained by ischemia of the medial portion of the spinothalamic tracts [56–62]. This partial sensory defect on the contralateral side of the body is due to the somatotopical organization of the spinothalamic tract in the medulla oblongata. A mediolateral lesion involves both the crossed lateral spinothalamic tract and the ventral trigeminothalamic tract, corresponding to sensory loss in the contralateral face, arm, and

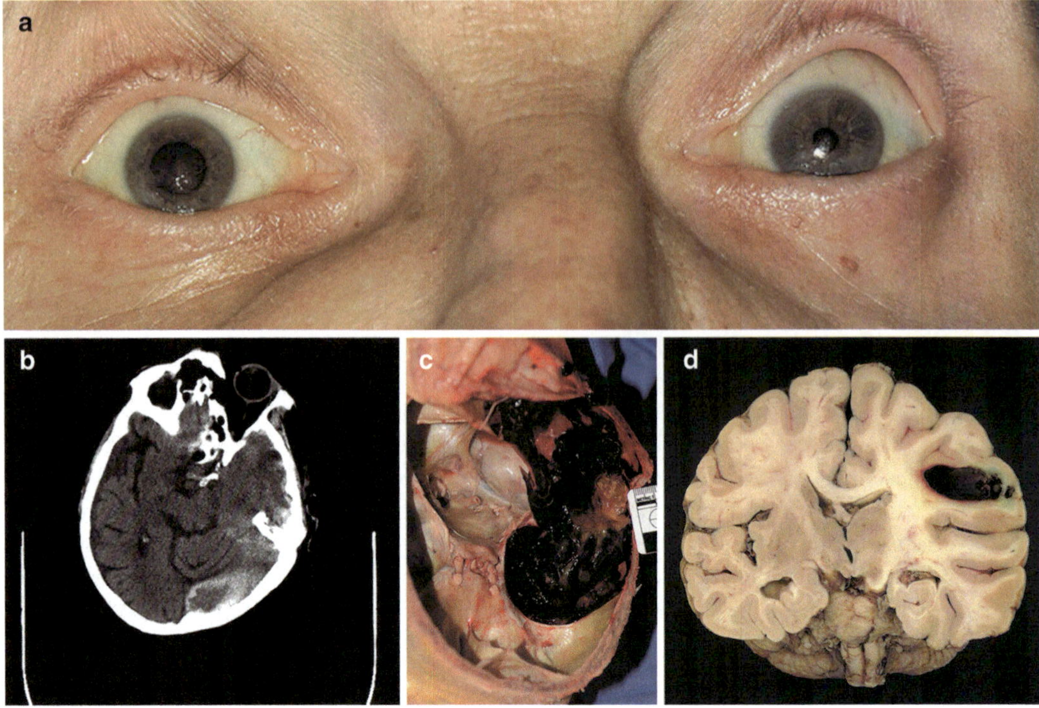

Fig. 5.4 Wrong side pupil. Note the contusional hematoma has rotated and shifted brainstem likely damaging the third nerve. (From *ref. [53] used with permission*)

upper trunk. The ventrolateral lesion of the infarct damaged the far lateral part of the spinothalamic tract, corresponding to sensory loss in the contralateral lower trunk and leg (Fig. 5.5). Precise localization in the spinal cord has always been somewhat unreliable and tricky. Established misjudgments are known with foramen magnum/upper cervical cord mix-ups; tingling in the hands with intrinsic hand muscle wasting and areflexia, with or without long tract signs, suggestive of a lower cervical myelopathy, may occur with lesions at the foramen magnum or higher cervical cord. Moreover, a compressive lower-cervical or upper-thoracic myelopathy may produce spastic paraplegia with a mid-thoracic sensory sensation [60, 63].

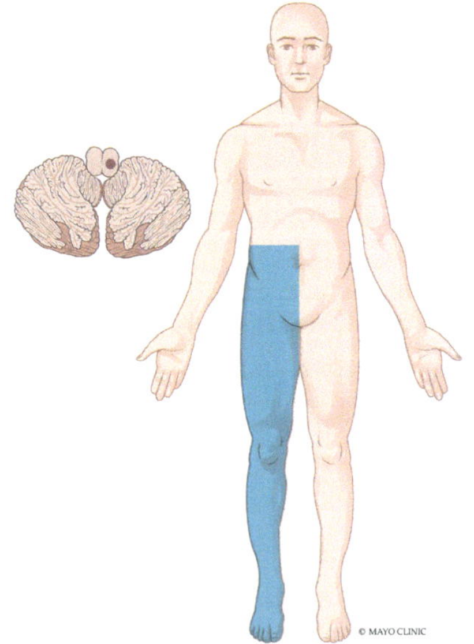

Fig. 5.5 Sensory level from medulla oblongata lesion

Other false localizing signs pertaining to the practice of neurocritical care are pseudo-internuclear ophthalmoplegia—suggesting a brainstem lesion—in myasthenia gravis [64]. (See also Chap. 10 and Fig. 10.1.) It has also been observed in dermatomyositis. Similarly, myasthenic nystagmus may suggest a brainstem lesion. Cerebellar ataxia contralateral to a posterior fossa mass lesion has also been reported [65].

Another question is whether what we see—with neuroimaging or at autopsy—reflects the clinical picture. This is a major issue with the interpretation of "herniation." Traditionally seen at autopsy with displaced tissue, it compresses ("grooves") structures such as the oculomotor nerve (herniation of the uncus of the temporal lobe) or cerebellum (herniation of cerebellar tonsils through the foramen magnum). For many years, it seemed a settled issue until CT and MR scanning found far more shift with "clinical signs of uncal herniation" rather than wedging of tissue under pressure. The earliest anatomical description of uncal herniation appeared in the beginning of the twentieth century with a pathology series. Hemianopia was described as a false localizing sign of uncal herniation, explained by compression of the posterior cerebral artery. Other evidence of displacement was discovered at autopsy. Pathologist Kernohan and neurologist Woltman published a seminal pathology work in 1929 on ipsilateral hemiplegia accompanying cerebral mass lesions. It provided the first comprehensive pathological observation that a groove of the shifted crus cerebri occurred on the side contralateral to a brain tumor. Pathological proof was provided: "...herniation and displacement may be evidenced by a groove sweeping over the uncinate gyrus on the side of the tumor." Since it was now possible to explain homolateral hemiplegia, it could possibly reduce clinical error. Over the years, this observation has become known as the "Kernohan–Woltman syndrome" or "Kernohan's notch" [66, 67].

More deserving of distinction are Groeneveld and Schaltenbrand, who explained the injury as a knife-like stab ("Messerähnlichen Wirkung") from the tentorium and possibly petrosal bone [68]. However, even now, the mechanism by which this V-shaped indentation or groove is produced, whether by displacement of the brainstem at a diencephalic level or pushing by the herniating uncus, remains unclear. Moreover, this "notching" occurs more commonly as a result of gradual tumor growth and compression. We have seen several cases of the Kernohan's phenomenon in acute subdural hematoma (Fig. 5.6). Recovery is expected.

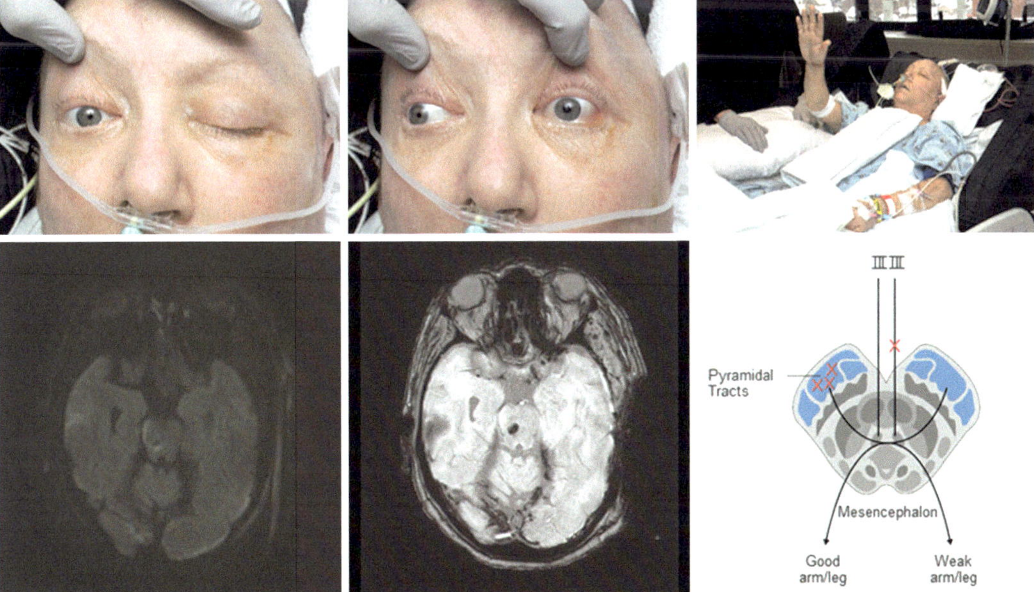

Fig. 5.6 Wrong side hemiparesis (Kernohan's notch injury). Patient with acute subdural hematoma and shifted brainstem. Upper row: Oculomotor paresis (Right arm antigravity and normal as shown, left arm and leg are paralytic). Lower row: MRI shows right peduncular injury and a Duret hemorrhage

More Reflections

We neurointensivists must be vigilant because neurology throws us quite a few "curveballs," and some clinical manifestations require deep thinking. Intuitively, we see that the presentation does not fit traditional teaching. The patient may look different than a CT or MRI indicates. Reconciliation of these discrepancies remains a formidable task, and we may not be able to do it. When drug infusions have been administered, we may be fooled. Patients with initially poor prognoses have recovered when the drugs finally wore off, sometimes significantly so. A prolonged, lingering sedative effect is often not expected, and it can be difficult to quantify, let alone to prove. The consulting neurologist is typically called when the intensivist feels that the toxin is lingering too long, even allowing for decreased elimination or prolonged absorption. We are flummoxed and may fail miserably to find a cause in many cases, but the odds are patients will eventually awaken fully. Unusual explanations for unusual presentations remain a useful exercise (and sometimes can change direction).

Pointers and Takeaways

- Always ask yourself: Does this make sense?
- In this modern era of critical care, we should expect a number of confounders.
- Drugs do not clear well in critical illness, and drug–drug interactions are often underestimated.
- Every obtunded patient needs a new set of electrolytes and a serum ammonia test.
- Only extreme electrolyte derangements affect neurologic examination.
- False-localizing sign is a misnomer; it localizes correctly, but the circumstance is unusual.

References

1. Marengoni A, Pasina L, Concoreggi C, et al. Understanding adverse drug reactions in older adults through drug-drug interactions. Eur J Intern Med. 2014;25:843–6.
2. Roberts DJ, Hall RI. Drug absorption, distribution, metabolism and excretion considerations in critically ill adults. Expert Opin Drug Metab Toxicol. 2013;9:1067–84.
3. Smith BS, Yogaratnam D, Levasseur-Franklin KE, Forni A, Fong J. Introduction to drug pharmacokinetics in the critically ill patient. Chest. 2012;141:1327–36.
4. Tse AHW, Ling L, Lee A, Joynt GM. Altered pharmacokinetics in prolonged infusions of sedatives and analgesics among adult critically ill patients: a systematic review. Clin Ther. 2018;40:1598–615. e1592
5. Iirola T, Aantaa R, Laitio R, et al. Pharmacokinetics of prolonged infusion of high-dose dexmedetomidine in critically ill patients. Crit Care. 2011;15:R257.
6. Cammarano WB, Pittet JF, Weitz S, Schlobohm RM, Marks JD. Acute withdrawal syndrome related to the administration of analgesic and sedative medications in adult intensive care unit patients. Crit Care Med. 1998;26:676–84.
7. Mehta S, Burry L, Cook D, et al. Daily sedation interruption in mechanically ventilated critically ill patients cared for with a sedation protocol: a randomized controlled trial. JAMA. 2012;308:1985–92.
8. deBacker J, Hart N, Fan E. Neuromuscular blockade in the 21st century management of the critically ill patient. Chest. 2017;151:697–706.
9. Murray MJ, Cowen J, DeBlock H, et al. Clinical practice guidelines for sustained neuromuscular blockade in the adult critically ill patient. Crit Care Med. 2002;30:142–56.
10. Durand-Maugard C, Lemaire-Hurtel AS, Gras-Champel V, et al. Blood and CSF monitoring of cefepime-induced neurotoxicity: nine case reports. J Antimicrob Chemother. 2012;67:1297–9.
11. Naeije G, Lorent S, Vincent JL, Legros B. Continuous epileptiform discharges in patients treated with cefepime or meropenem. Arch Neurol. 2011;68:1303–7.
12. Fugate JE, Kalimullah EA, Hocker SE, Clark SL, Wijdicks EF, Rabinstein AA. Cefepime neurotoxicity in the intensive care unit: a cause of severe, underappreciated encephalopathy. Crit Care. 2013;17:R264.
13. Honore PM, Spapen HD. Cefepime-induced neurotoxicity in critically ill patients undergoing continuous renal replacement therapy: beware of dose reduction! Crit Care. 2015;19:455.
14. Payne LE, Gagnon DJ, Riker RR, et al. Cefepime-induced neurotoxicity: a systematic review. Crit Care. 2017;21:276.
15. Grahl JJ, Stollings JL, Rakhit S, et al. Antimicrobial exposure and the risk of delirium in critically ill patients. Crit Care. 2018;22:337.
16. Bhattacharyya S, Darby RR, Raibagkar P, Gonzalez Castro LN, Berkowitz AL. Antibiotic-associated encephalopathy. Neurology. 2016;86: 963–71.
17. Braksick SA, Wijdicks EFM. Moisture and mydriasis Pract Neurol. 2014;14:187–8.

18. Ainslie PN, Poulin MJ. Ventilatory, cerebrovascular, and cardiovascular interactions in acute hypoxia: regulation by carbon dioxide. *J Appl Physiol (1985)*. 2004;97:149–59.
19. Hoffman CE, Clark RT Jr, Brown EB Jr. Blood oxygen saturations and duration of consciousness in anoxia at high altitudes. Am J Phys. 1946;145:685–92.
20. Espay AJ. Neurologic complications of electrolyte disturbances and acid-base balance. Handb Clin Neurol. 2014;119:365–82.
21. Riggs JE. Neurologic manifestations of electrolyte disturbances. Neurol Clin. 2002;20:227–39. vii
22. Diringer M. Neurologic manifestations of major electrolyte abnormalities. Handb Clin Neurol. 2017;141:705–13.
23. Arieff AI, Llach F, Massry SG. Neurological manifestations and morbidity of hyponatremia: correlation with brain water and electrolytes. Medicine (Baltimore). 1976;55:121–9.
24. Fraser CL, Arieff AI. Epidemiology, pathophysiology, and management of hyponatremic encephalopathy. Am J Med. 1997;102:67–77.
25. Gill G, Huda B, Boyd A, et al. Characteristics and mortality of severe hyponatraemia--a hospital-based study. Clin Endocrinol. 2006;65:246–9.
26. Narins RG. Therapy of hyponatremia: does haste make waste? N Engl J Med. 1986;314:1573–5.
27. Sterns RH. The treatment of hyponatremia: first, do no harm. Am J Med. 1990;88:557–60.
28. Sterns RH. Disorders of plasma sodium--causes, consequences, and correction. N Engl J Med. 2015;372:55–65.
29. Adrogue HJ, Madias NE. Hypernatremia. N Engl J Med. 2000;342:1493–9.
30. JM UK-I, Yu E, Bartlett E, Soobrah R, Kucharczyk W. Acute hyperammonemic encephalopathy in adults: imaging findings. AJNR Am J Neuroradiol. 2011;32:413–8.
31. Clay AS, Hainline BE. Hyperammonemia in the ICU. Chest. 2007;132:1368–78.
32. Rimar D, Kruzel-Davila E, Dori G, Baron E, Bitterman H. Hyperammonemic coma--barking up the wrong tree. J Gen Intern Med. 2007;22: 549–52.
33. Takanashi J, Barkovich AJ, Cheng SF, Kostiner D, Baker JC, Packman S. Brain MR imaging in acute hyperammonemic encephalopathy arising from late-onset ornithine transcarbamylase deficiency. AJNR Am J Neuroradiol. 2003;24:390–3.
34. Edlow JA, Rabinstein A, Traub SJ, Wijdicks EF. Diagnosis of reversible causes of coma. Lancet. 2014;384:2064–76.
35. Shannon MW, Borron SW, Burns MJ, editors. Haddad and Winchester's clinical management of poisoning and drug overdose. 4th ed. Philadelphia: Saunders; 2007.
36. Volpe DA, McMahon Tobin GA, Mellon RD, et al. Uniform assessment and ranking of opioid mu receptor binding constants for selected opioid drugs. Regul Toxicol Pharmacol. 2011;59: 385–90.
37. Nahrir S, Sinha S, Siddiqui KA. Brake fluid toxicity feigning brain death. BMJ Case Rep. 2012;2012: bcr0220125926.
38. Tobe TJ, Braam GB, Meulenbelt J, van Dijk GW. Ethylene glycol poisoning mimicking Snow White. Lancet. 2002;359:444–5.
39. Leroy JM, Olives TD, Lange RL, Cole JB. Near death experience with baclofen poisoning and the role of the poison center. Clin Toxicol (Phila). 2015;53:703–4.
40. Ostermann ME, Young B, Sibbald WJ, Nicolle MW. Coma mimicking brain death following baclofen overdose. Intensive Care Med. 2000;26:1144–6.
41. Stranges D, Lucerna A, Espinosa J, et al. A Lazarus effect: a case report of Bupropion overdose mimicking brain death. World J Emerg Med. 2018;9:67–9.
42. Sullivan R, Hodgman MJ, Kao L, Tormoehlen LM. Baclofen overdose mimicking brain death. Clin Toxicol (Phila). 2012;50:141–4.
43. Wang A, Malik N, Shah SO. Bupropion overdose mimicking brain death: a case report. Neurology. 2017;88:P6.058.
44. Halpern JI, Gordon WH Jr. Trochlear nerve palsy as a false localizing sign. Ann Ophthalmol. 1981;13:53–6.
45. Lepore FE. False and non-localizing signs in neuro-ophthalmology. Curr Opin Ophthalmol. 2002;13:371–4.
46. O'Connell JE. Trigeminal false localizing signs and their causation. Brain. 1978;101:119–42.
47. Ro LS, Chen ST, Tang LM, Wei KC. Concurrent trigeminal, abducens, and facial nerve palsies presenting as false localizing signs: case report. Neurosurgery. 1995;37:322–4. discussion 324–325
48. Gassel MM. False localizing signs. A review of the concept and analysis of the occurrence in 250 cases of intracranial meningioma. Arch Neurol. 1961;4:526–54.
49. Larner AJ. False localising signs. J Neurol Neurosurg Psychiatry. 2003;74:415–8.
50. Davis M, Lucatorto M. The false localizing signs of increased intracranial pressure. J Neurosci Nurs. 1992;24:245–50.
51. Dodge HW Jr, Clark EC, Wagener HP, Hustead AP. Certain false localizing signs of intracranial tumor: report of case. Proc Staff Meet Mayo Clin. 1955;30:453–61.
52. Ehni G. "False" localizing signs in intracranial tumor; report of a patient with left trigeminal palsy due to right temporal meningioma. AMA Arch Neurol Psychiatry. 1950;64:692–8.
53. Wijdicks EFM, Giannini C. Wrong sided dilated pupil Neurology. 2014 14;82:187.
54. Chung KH, Chandran KN. Paradoxical fixed dilatation of the contralateral pupil as a false-localizing sign in intraparenchymal frontal hemorrhage. Clin Neurol Neurosurg. 2007;109:455–7.
55. Marshman LA, Polkey CE, Penney CC. Unilateral fixed dilation of the pupil as a false-localizing sign with intracranial hemorrhage: case report and literature review. Neurosurgery. 2001;49:1251–5. discussion 1255–6.

56. Adams KK, Jackson CE, Rauch RA, Hart SF, Kleinguenther RS, Barohn RJ. Cervical myelopathy with false localizing sensory levels. Arch Neurol. 1996;53:1155–8.
57. Matsumoto S, Okuda B, Imai T, Kameyama M. A sensory level on the trunk in lower lateral brainstem lesions. Neurology. 1988;38:1515–9.
58. Phan TG, Wijdicks EF. A sensory level on the trunk and sparing the face from vertebral artery dissection: how much more subtle can we get? J Neurol Neurosurg Psychiatry. 1999;66:691–2.
59. Soffin G, Feldman M, Bender MB. Alterations of sensory levels in vascular lesions of lateral medulla. Arch Neurol. 1968;18:178–90.
60. Sonstein WJ, LaSala PA, Michelsen WJ, Onesti ST. False localizing signs in upper cervical spinal cord compression. Neurosurgery. 1996;38:445–8. discussion 448–449
61. Vaudens P, Bogousslavsky J. Face-arm-trunk-leg sensory loss limited to the contralateral side in lateral medullary infarction: a new variant. J Neurol Neurosurg Psychiatry. 1998;65:255–7.
62. Vuilleumier P, Bogousslavsky J, Regli F. Infarction of the lower brainstem. Clinical, aetiological and MRI-topographical correlations. Brain. 1995;118(Pt 4): 1013–25.
63. Ochiai H, Yamakawa Y, Minato S, Nakahara K, Nakano S, Wakisaka S. Clinical features of the localized girdle sensation of mid-trunk (false localizing sign) appeared in cervical compressive myelopathy patients. J Neurol. 2002;249:549–53.
64. Bakheit AM, Behan PO, Melville ID. Bilateral internuclear ophthalmoplegia as a false localizing sign. J R Soc Med. 1991;84:627.
65. Maurice-Williams RS. Multiple crossed false localizing signs in a posterior fossa tumour. J Neurol Neurosurg Psychiatry. 1975;38:1232–4.
66. Jones KM, Seeger JF, Yoshino MT. Ipsilateral motor deficit resulting from a subdural hematoma and a Kernohan notch. AJNR Am J Neuroradiol. 1991;12:1238–9.
67. Carlstrom LP, Perry A, Puffer RC, et al. A puzzling exam: Kernohan's notch reimaged. J Clin Neurosci. 2020;80:290–1.
68. Dammers R, Volovici V, Kompanje EJ. The history of the Kernohan notch revisited. Neurosurgery. 2016;78:581–4.
69. Wijdicks EFM, Clark S. Neurocritical care pharmacotherapy: Oxford University Press; 2018.

6 Detecting Worsening

Detecting clinical worsening is taxing for a neurointensivist and for any other involved healthcare worker at the bedside. By all means, neurocritically ill patients must be monitored clinically and closely, but what exactly are we looking for? Do we know what to expect in each patient with a specific acute problem? Do we need to recognize different types of presentations or is there some fixed pattern? Have these clinical indicators of potential deterioration been appropriately communicated (Chap. 12)? Detecting deterioration basically requires assigning an experienced observer to a patient to recognize the major manifestations of illness and to call upon physicians not only when changes occur but specifically when the changes are enough to draw notice. Thus, this undeniably difficult burden falls on the neurosciences nursing staff. Their assessment will lead physicians to rush to the bedside, rapidly confirm the change, and carry out new tests and clinical interventions to prevent further worsening. Clinical worsening does not occur in a vacuum; there is always a reason, and we should always try to explain it.

We only have a clinical examination to guide us; all other means are surrogates for (or approximations of) what we see at the bedside. Brain function can currently be monitored with intracranial pressure (ICP) monitoring, brain tissue oxygenation, and microdialysis; each of these devices may signal neuronal distress. Electroencephalogram (EEG) video monitoring may be used in patients with changing degrees of consciousness to diagnose non-convulsive status epilepticus and to titrate antiepileptic drugs. Some neurointensive-care teams put patients "under" sedation (to reduce brain metabolism, reduce pain, and minimize ventilator struggles), apply multimodal monitoring (a blanket term meaning different things and different combinations of devices), and then hope these "numbers" tell a story. But critical care teams and NICU personnel must definitively rely on repeated expert neurologic examination and interpretation of critical, changing neuroimaging features. We should not function otherwise [1–3].

Strangely enough, clinical monitoring is not standardized and based on generalities. We have introduced parlance such as "stuttering stroke," "stroke in progress," "clinical signs of herniation," "coning," and "seizing" without much further clarification [4, 5]. A worsening motor response, new or worsening weakness, new confusion or agitation, or a change in pupil size often does not provide enough information because deterioration in a patient with an acute brain injury is disease-specific. Our clinical tools may be insufficient and even useless if we do not convey the key findings to the person who monitors or is asked to interpret a change. One fact is clear. There are no "stable" acutely ill neurologic patients, and something will change ineluctably. Furthermore, clinical worsening may have reversible causes.

E. F. M. Wijdicks, *Examining Neurocritical Patients*, https://doi.org/10.1007/978-3-030-69452-4_6

Will we detect them? It all depends, as with so many other things, on how hard we look or if we take sufficient time to think it through. I have seen unexpected hypercarbia (due to failure to check an arterial blood gas), sudden severe hyponatremia (due to failure to calculate significant free water intake), and nosocomial meningitis (due to missing nursing records of a postoperative cerebrospinal fluid [CSF] leak). Nevertheless, wholly unexplained periods of deterioration exist, but most are brief (and easily forgotten). We also need to ask ourselves how we define clinical deterioration and what we might consider normal fluctuations. We do not want to send patients for a CT scan every time there is a change, only to find out that CT is unchanged and the patient is back to baseline (the so-called "therapeutic CT scan").

A stable patient is stable in the present, and we know how quickly that can change. (A teacher of mine said, with some pride, that he loved neurology because neurology fluctuated so much and nobody could claim to have done a better exam.) Do we have the dexterity and panache to judge deterioration? This chapter is an attempt to tackle head on the definition of deterioration, the art of recognizing deterioration, and what causes to consider in a number of key diseases.

Recognizing Clinical Worsening

To better understand the clinical diagnosis of deterioration requires asking five basic questions. First, what do we mean by the word fluctuation, and what does it mean for the patient? Second, how bad should a fluctuation be to convince us that something new has happened? Third, how consequential is it? In other words, should we do anything different? Fourth, what are the neurologic causes of deterioration beyond the obvious, and fifth, are there non-neurologic causes that might play a role? In our day-to-day practice, we may not have satisfactory answers to all these questions, but they should be asked with each event. There are a few more salient points to make. There are two major components to consider with a clinically worsening patient: (1) what is part of the anticipated trajectory of evolving signs of the disorder and (2) what is not and unexpected. Moreover, there are the age-old differences between a good and not-so-good observation and between looking and seeing. Inevitably, there are questions about whether the deterioration represents new findings or pre-existing clinical signs that were not truly examined earlier. It is easy to list the ones we tend to forget or sometimes do not do. Neurologic examination in neurocritical illness can be very focused and quick, but we should not forget what is important in certain clinical settings. Not examining for nystagmus and stance in a patient on a gurney with sudden and progressively worse vertigo is unacceptable; not testing for loss of sensation in buttocks in a lying patient with extreme back pain may delay diagnosis. A few things to watch for are shown in Table 6.1. However, subtleties exist: a patient with paralysis may develop worsening neglect; new gaze preference or gaze limitation may occur, and pupils may change their symmetry. On the other hand, patients may have lost so much hemispheric and brainstem function that further deterioration is not possible. Clinical examination remains critical; not all worsening CT scans of the brain indicate a recognizable deterioration (Table 6.2). Best known are patients with enlarging hematomas or contusions not reflecting in a notably worsening weakness.

Table 6.1 Deterioration or detection

Ataxia (never walked)
Neglect (not tested)
Aphasia (reading and naming not tested)
Visual field cut (missed or not tested)
Abdominal reflexes (not tested)
Sensation in the back and buttocks (not tested)
Two-point sensory discrimination (not tested)

Table 6.2 New changes on CT scan, often with no observable deterioration

Hemorrhage in tumor
Hemorrhage in cerebral infarct
Cortical subarachnoid hemorrhage
Frontal lobe hemorrhage
Subdural hematoma (less than 0.5 cm)
Enlarging frontal contusion

So we have to distinguish between expected, completed injury (with or without progression), and worsening. In acute neurologic disorders, can we reliably say that they can be maximal at onset, are stepwise (stutter), gradually declining, or fluctuating? The clinical trajectories mentioned in Chap. 1 exemplify how certain disorders present themselves and how trajectories can differ in the same disease. It may depend on when a patient is seen during the disease course. Seen early, many worsen; seen late, many are stable only to deteriorate later.

Many of these designations apply to acute stroke, which provides a key illustration of why the timing of interventions remains difficult. A "stuttering" onset (weak, then not weak, and then weak again) is very common in a thrombotic process and often in thrombotic occlusion in penetrating-branch disease (lacunar stroke). Acute major deficits are seen with an embolus to a large cerebral artery, but then improvement and worsening may be a result of a challenged collateral circulation trying to take over the acute perfusion challenge. In the extreme, we have seen the word "malignant" creep into the medical lexicon such as in malignant cerebral edema; fulminant may be a more appropriate term. Caplan has categorized the early course of stroke (Table 6.3) but acknowledges that biologic processes are not always crystal clear [6].

When considering causes of deterioration, a number of common clinical conditions should be considered, and they are summarized in Fig. 6.1.

Table 6.3 Deterioration in stroke

Acute stroke type	*Most common course*
Intracerebral hemorrhage	Gradual smooth
Ischemic stroke	Maximal at onset with some stepwise
Lacune	Stepwise, fluctuating, gradual
Aneurysmal SAH	From good to bad and vice versa

Adapted from Caplan's Stroke [6]

The Neurology of Clinical Worsening

What are unambiguous clinical clues for deterioration? There must be a persistent decline in consciousness, a change in pupil size and light response, newly absent brainstem reflex(es), changes in eye movements or new spontaneous eye movements, new profound limb weakness where there previously was none, a new acute speech impediment or acute muteness, and a number of associated system signs signaling brain tissue shift such as acute bradycardia, a hypertensive surge, or sudden loss of vascular tone if someone becomes brain dead after a catastrophic injury. Changed motor or verbal response has always been seen as a good marker but with some caveats. Changed speech (sudden muteness or aphasia) is extraordinarily important and always signals a new problem. The exception is dysarthria, which can easily occur as a result of administered medication. It requires some skill to differentiate "confusion" from acute global aphasia; usually, asking the patient to repeat a sentence or read aloud will bring out major deficiencies. Word-finding difficulties may not always be due to aphasia. These difficulties could be easily an effect of ICU-related sleep exhaustion, certainly possible after many days. Aphasia may occur with or without preservation of motor function. It is far more common than appreciated after traumatic brain injury with acute subdural hematoma or frontal/temporal contusions. It may appear after evacuation. In the case of subdural hematoma evacuation, aphasia may appear and disappear multiple times a day or several days in a week. The clinical substrate remains to be determined and currently is interpreted as spreading cortical depression.

Generally, we should note that many causes are potentially at play when a patient stops responding to voice or does not seem to understand simple tasks such as pointing at the ceiling. Fluctuating consciousness may present

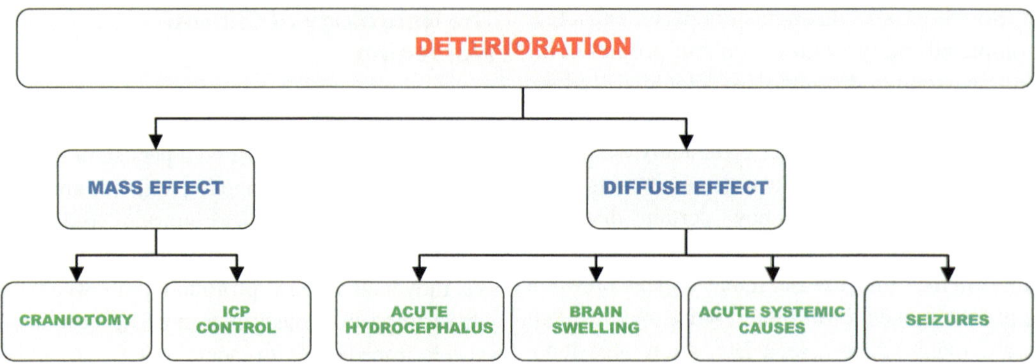

Fig. 6.1 The main trajectories with deterioration

itself as daytime sleepiness. I am still surprised by how often sleepiness is attributed to "delirium" or poor rest due to frequent "neurochecks." New sleepiness in patients who were not previously sleepy may be symptomatic of something ominous that we need to find. Fluctuating consciousness with transient eye deviation and, often, eye fluttering indicates seizures. Seizures may be more common than appreciated, and in any such patient, continuous EEG monitoring is warranted [7]. Patients with cortical ischemic and hemorrhagic lesions, encephalitis, and major tumor surgery are at particularly high risk for seizures. Nonconvulsive seizures or seizures with minimal clinical presentations are seldom causes for deterioration outside these circumstances. A spot EEG (20- to 30-minute recording) may also miss brief seizures; unsurprisingly, prolonged monitoring increases detection. We can categorically state that patients with more severe acute brain injury are more likely to have transient multifocal epileptiform discharges and be more prone to seizures. Clinicians are particularly bothered by "electrographic seizures" with no overt ictal clinical signs. Unfortunately, many EEG patterns share common features with classic electrographic seizures but are not uniformly considered to be "ictal." These patterns include lateralized periodic discharges (LPDs) and generalized periodic discharges (GPDs), patterns that may produce ictal-like behavior such as fluctuation alertness and, at other times, no recognizable clinical signs even with prolonged close personal or video observation. The conceptual framework called the *ictal-interictal continuum* [7] is not particularly useful. Even with discharges confined to a single hemisphere (periodic lateralized epileptiform discharges or LPDs), it may not be meaningful despite that hemisphere having the highest association with clinical and electrographic seizures.

As has been pointed out many times before, a changed motor response does not always coincide with a changed level of consciousness, although it often does. A changed motor response may also be due to changed pain stimulus. Some are much stronger (temporomandibular joint) than others (sternal rub, nail bed, or shoulder squeeze); I have seen major variability in the reaction of patients with different physicians and nursing staff. However, not all changes from one reflexive motor response to another (decorticate vs. decerebrate) necessarily represent deterioration because both can be present simultaneously. Clinically, decerebrate responses (hyperpronated, extended arms often with tight closed fists or a thumb in fist) are far more obvious than decorticate responses (flexion in arms, adduction of the shoulders, rotation inwards). The difference between decorticate responses and withdrawal flexion is often not as clear as everyone wants to believe. We also do not know the origin of these responses. They are indicative of a major bilateral corticospinal-tract involvement, but we cannot simply assign them to some level in the brainstem. (In the original well-known Sherrington's Law, experimental cuts [between the superior and inferior colliculi in the midbrain] induced spasmodic reflexes.)

Clinical Patterns of Tissue Shift and Brainstem Injury

Brain swelling can cause secondary injury. Brain swelling displaces brain tissue, compresses the diencephalic structures, and injures the brainstem in a secondary manner. Clinically, patients become comatose, require ventilator assistance, and show a new syndromic pattern.

How do we understand volume and pressure within the skull and craniospinal axis? Four volume compartments determine the total volume, which is encompassed by a rigid, non-expandable (at least in adults) skull, and volume is constant. These volume compartments include venous and arterial volume (both 5%), CSF (10%), and brain parenchyma (80%). Both CSF and intracranial blood volume contain 75 mL. Cerebrospinal fluid (CSF) pressure is equal throughout the CSF pathways but only if CSF is not obstructed somewhere. The Monro–Kellie doctrine states that change (increase or decrease) in one compartment should be followed by a change (increase or decrease) in one or two other compartments. Up to 150 mL of new volume can be "tolerated," but this depends on the volume of the brain parenchyma. The brain parenchyma cannot be compressed; it behaves as water and, as such, is essentially incompressible. Once a maximum pressure is reached (which depends on the degree of brain atrophy, prior encephalomalacia, or congenital defects), part of the brain dislocates out of its compartment (as if squeezing toothpaste out of a tube).

The clinical effects of increased intracranial pressure are largely unpredictable, but a major tenet is that brain tissue shifts with increased intracranial pressure, which leads to compression of eloquent structures that shows as new neurologic findings. Increased intracranial pressure from acute hydrocephalus compresses the thalami (third ventricle), subcortical tracts connecting to the cortex (lateral ventricle), and brainstem tegmentum (fourth ventricle). Cranial nerve VI is particularly at risk of injury. The clinical signs of supratentorial herniation are less immediately consequential than herniation of the cerebellar tonsils, which compresses the medulla oblongata and attached vascular structures. The brainstem perfusion, therefore, is quickly in jeopardy, leading to difficulty in the respiratory drive and apneic periods.

Clinical findings in a deteriorating patients lapsing into coma can be localized to the bi-hemispheric or intrinsic brainstem regions (Table 6.4). Localization is important to establish a baseline. If the mass is in one hemisphere, we can monitor it for lateral movement [8]. Lateral brain tissue shift compresses the thalamus and its connected mesencephalon, and this is reflected in clinical presentation. These patients present with declining consciousness (from thalamus-mesencephalon involvement). The pupils change due to pulling of thc oculomotor nerve or change

Table 6.4 Syndromic patterns in deteriorating patients

Location	Clinical pointers
Bihemispheric	Spontaneous eye movements (roving, dipping, ping-pong)
	Upward or downward eye deviation
	Intact oculovestibular responses
	Intact pupil corneal reflexes
	Variable motor responses
	Myoclonus status epilepticus
Intrinsic brainstem	Skew deviation
	Internuclear ophthalmoplegia
	Vertical nystagmus or bobbing
	Miosis
	Variable pupil or corneal reflexes (or anisocoria)
	Absent oculocephalic or oculovestibular responses
	Extensor flexion motor responses
Brainstem displacement (from hemispheric or cerebellar mass)	Anisocoria or unilateral fixed wide pupil (lateral displacement from hemispheric mass)
	Midposition fixed pupils (downward central displacement from hemispheric mass)
	Absent corneal reflexes and intact pupil reflexes (displacement from cerebellar mass)
	Extensor or withdrawal motor responses

in perfusion of its nucleus in the brainstem due to traction. These early manifestations typically originate from the mesencephalon, which is the first structure with tracts that interrupt motor function. This course, however, can be mimicked by acute lesions in the thalamus that suddenly extend asymmetrically to the mesencephalon (e.g., a thalamic hemorrhage).

I should point out that mass lesions in the cerebellum can also compress the brainstem but more typically at the pontine level due to proximity. Most notable is predominance of pontine signs with possible bilateral miosis, loss of both corneal reflexes, and oculocephalic reflexes. Frequently, episodes of bradycardia occur with or without hypertension (Cushing signs). Therefore, lateral brainstem-displacement syndromes (due to a mass moving the brain laterally) can occur above or below the tentorium. Above the tentorium, hemispheric lesions compress the thalamus and mesencephalon. In these patients, a unilateral fixed and dilated (varying from 5 to 7 mm) pupil is seen and, if untreated, is mostly followed by bilateral, fixed, mid-position pupils. As noted, the fixed, dilated pupil occurs as a result of a stretch injury to the oculomotor nerve or ischemic damage to its nucleus due to displacement of the brainstem. However, its mid-position narrowing is due to sympathetic tract involvement, a sign of worsening brainstem injury. (Its true mechanism remains unknown; experiments have raised doubts about oculomotor nerve compression as the main and overriding mechanism [9].) Alternatively, pontine reflexes initially remain intact but may gradually disappear if the syndrome progresses.

Brainstem displacement from a lesion below the tentorium is usually in the cerebellum. In these patients, bilateral miosis and loss of both corneal reflexes and oculocephalic reflexes from pontine compression are more predominant and found early in the clinical course.

A more centrally located mass will compress and distort the thalamus and mesencephalon in a vertical plane causing fixed mid-position (4–6 mm) pupils initially. Asymmetric compression of the mesencephalon with anisocoria and a larger, oval-shaped pupil may be seen at the site of the lesion. Motor responses vary from decorticate to extensor, sometimes even with variation throughout the day and no other evidence of deterioration. In patients with a gaze preference toward the expanding mass, thalamic compression may cause the gaze to reverse. Brief periods of periodic lateral gaze may occur. Further vertical displacement of the entire thalamic/mesencephalic pontine structure may occur but only after direct compressive destruction of the upper brainstem. It may occur with bilateral thalamic compression from diffuse brain edema. Patients who lose all brainstem reflexes usually lose ponto-mesencephalic reflexes at onset and medulla function later. A common progression is flaccidity, no motor response, loss of ponto-mesencephalic reflexes, and, finally, failure to trigger the ventilator (i.e., brain death).

Bihemispheric syndrome is characterized by the absence of any specific localizing finding. Patients may have a gaze preference. Brainstem reflexes are intact. Pupils may be small or normal; motor responses are usually withdrawal to pain, extensor responses, or none. Localization to a noxious stimulus may occur if patients are more "at the surface" or recovering from coma. In some patients, myoclonus or focal twitches can be seen. The breathing patterns can be periodic but are mostly regular. Nothing points to a specific area in the brain hemispheres, and lesions may be in the cortex, white matter, or thalamus.

As previously noted, brainstem syndromes fall into intrinsic brainstem syndromes and brainstem-displacement syndromes, both of which may overlap in a clinical sense. An intrinsic brainstem syndrome (due to a primary lesion inside the brainstem) is characterized by extensor posturing or flexion of the motor response, skew deviation, miosis or anisocoria, and, often, abnormal oculovestibular responses. A 50-cc ice-water injection into the ear can detect or rule out internuclear ophthalmoplegia. The eye on the same side as the injected ear moves toward the ear while the opposite eye fails to move to the injected ear. The finding of internuclear ophthalmoplegia is specific for a lesion of the medial longitudinal fiber that courses through the brainstem connecting the III and VI cranial-nerve nucleus. It is a useful clue for an acute brainstem lesion.

Central brainstem displacement is due to a midline-located lesion or lesion that exerts its force more centrally rather than peripherally downward compressing the thalamus and causing a torque that buckles the mesencephalon. Early fixed, mid-position pupils from mesencephalic interruption are seen with variable motor responses. In general, pontine reflexes remain intact. Figure 6.2 and Table 6.4 summarize the main findings of these syndromes.

Neuroimaging (in most cases by comparing CT scans done within a short time period) can confirm these clinical syndromes. Coronal images may add to the three-dimensional image of brain-tissue shift, but MR imaging will be

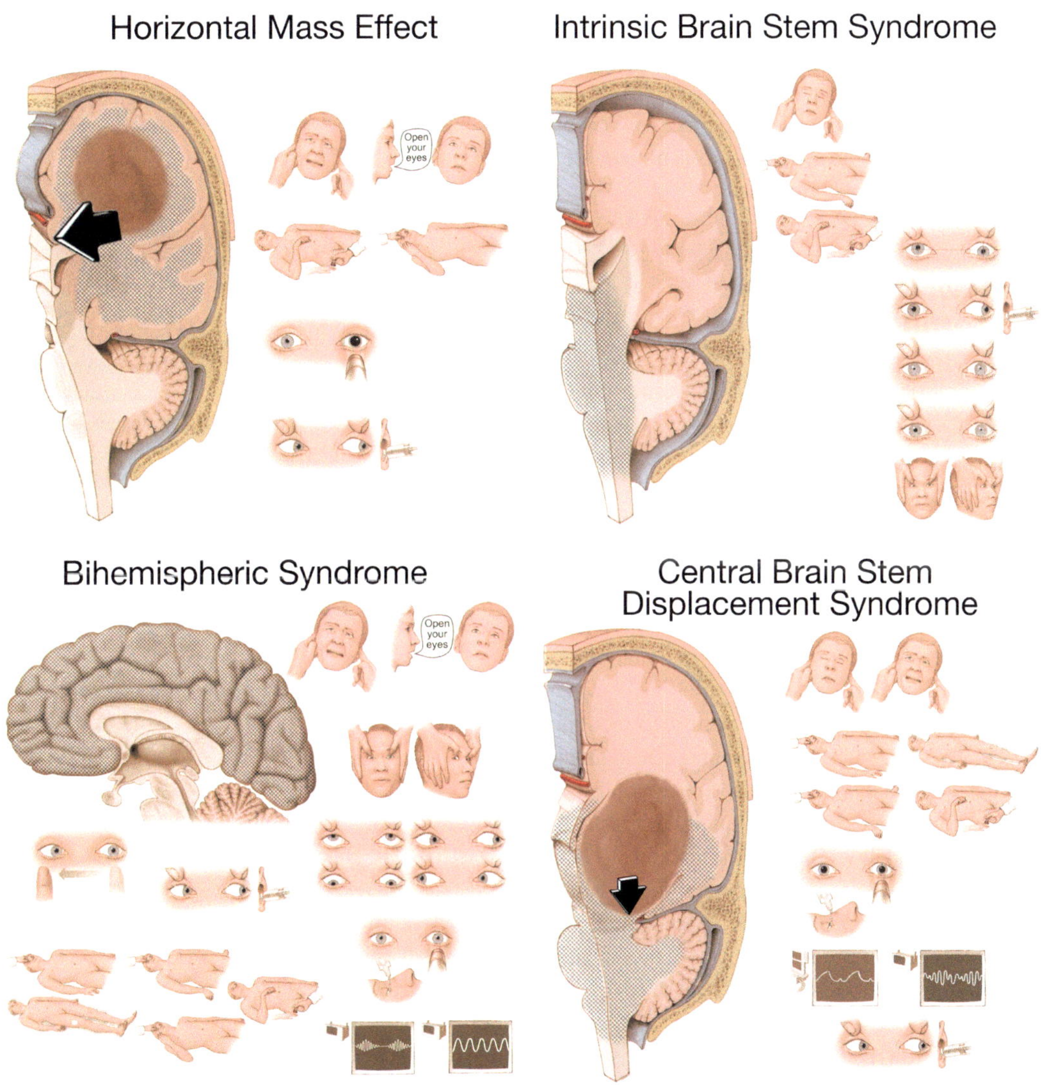

Fig. 6.2 Patterns of brain tissue shift and brainstem injury with associated clinical findings

much more informative. MR imaging requires a considerable time commitment and currently cannot be used in a rapidly deteriorating patient in need of neurosurgical intervention. Clinically, mass effect may develop too fast, and MR imaging at the very moment of emerging clinical signs may thus be impossible. Neuroimaging can show new enlargement of ventricles from obstruction and signs of mass effect with further displacement of the pineal gland and septum pellucidum, obliteration of basal cisterns, sulcal effacement, compression of the brainstem, and intraventricular extension of hemorrhage among other signs of a changing lesion. The red nuclei are recognizable structures in the midbrain (best visualized by spin-echo sequences on coronal MR images), and downward shift is noticeable by tilting of their paired presence.

Deterioration in Specific Neurocritical Conditions

Deterioration can be disease-specific but not necessarily. The principles of brain-tissue shift remain the same whatever the injury, but there are different probabilities of presentation, different time courses, and perhaps different consequences. In any patient with a central nervous system (CNS) hemorrhage (traumatic, spontaneous, or from a vascular anomaly), deterioration depends on clotting status (prior anticoagulation, thrombocytopenia, and genetic clotting disorder). Control of clotting is urgent. Failure to recognize this major variable in outcome (and, consequently, failure to treat it promptly and adequately) has major consequences. It is equally important to correct abnormal vital signs immediately. (Nearly half of the patients with severe traumatic brain injury have episodes of hypoxemia and hypotension.) The main categories of neurocritical illness are discussed here. Causes of clinical worsening are summarized in Table 6.5. (Deterioration in acute neuromuscular diseases is discussed in Chap. 10).

Table 6.5 Causes of clinical deterioration in selected disorders

Disorder	Causes of clinical deterioration	Clinical signs of deterioration
Aneurysmal subarachnoid hemorrhage	Acute hydrocephalus Delayed cerebral ischemia Rebleeding Expanding lobar hematoma Seizures	Decline in consciousness Upward gaze palsy Pinpoint pupils Sudden loss of upper brainstem reflexes and transient apnea
Ganglionic or lobar hemorrhage	Expanding volume Rebleeding	New aphasia or hemiparesis Eye deviation and eyelid twitching
Cerebellar hematoma	Compression of fourth ventricle and acute hydrocephalus Displacement of pons	Decreased consciousness Worsening hemiparesis Acute anisocoria or wide fixed pupil and extensor posturing
Hemispheric infarct	Hemorrhagic conversion Brain swelling	New Cushing signs Pinpoint pupils and downward gaze Comatose and need for intubation
Traumatic brain injury	New contusional lesions Uncontrollable cerebral edema Extension of subdural or epidural hematoma	Sudden coma with extensor posturing and mid position pupils Gradual decline in consciousness New onset cerebral ptosis New fixed dilated pupil New decerebrate or decorticate responses

Traumatic Brain Injury

Minor traumatic brain injury should be set apart from severe traumatic brain injury. Severe traumatic brain injury is best defined by (1) persistent, impaired consciousness to coma; (2) any CT-scan abnormality; and later (3) no recollection of the impact and post-traumatic amnesia of at least a week. The most severe injury leaves the patient in immediate coma from catastrophic white-matter impact that disconnects white matter from the cortical mantle. Further swelling may accelerate deterioration such that patients quickly lose all brainstem function. These patients are often poly-trauma cases with many additional contributing injurious factors such as poor oxygenation, massive transfusion, and unstable blood pressures from other bleeding sites that must be identified and stopped. The initial CT scan in patients susceptible to diffuse brain swelling is more often abnormal, revealing diffuse shear injury and subarachnoid hemorrhage (SAH) [10]. Uncontrolled brain swelling often leads to brain death, and virtually no therapeutic option is available other than aggressive ICP management.

Gradual deterioration may result from enlargement of hemorrhagic contusions, new emerging contusions with mass effect, diffuse cerebral edema, or further expansion of an extra-axial hematoma. Patients with traumatic (or even seemingly trivial) subarachnoid hemorrhage are at risk of deterioration due to severe impact to the skull.

Often, deterioration occurs from enlargement of a hemorrhagic contusion. In contusions initially larger than 20 ml, the risk of progression is increased fivefold [11]. Allison et al. have proposed a four-point CT predictive score with three radiologic features: presence of SAH (2 points), presence of subdural hematoma (SDH, 1 point), and presence of skull fracture (1 point) [12]. Moreover, CT growth >5 cm^3 is seven times more likely to require surgery. A bifrontal or temporal-lobe hematoma increases the chance of marked deterioration. These patients, known by the moniker "talk and deteriorate" [13–16], may look pretty good initially with physicians questioning their admission to the intensive care unit, but then they change dramatically in front of our eyes. Often, they are whisked into the operating room at the first signs of deterioration. If the temporal lobe is damaged, they undergo lobectomy in the inferior segment of the damaged temporal lobe, while other severely damaged brain tissue is resected.

Treatment for subacute subdural hematoma may remain conservative if the thickness of the hematoma is similar to the thickness of the skull bone and no midline shift is noted. However, if enlargement is accompanied by clinical deterioration, a timely craniotomy is necessary for a potentially successful outcome. Delayed epidural hematoma has been noted in hypotensive patients whose condition deteriorated after correction of hypotension, which most likely increased cerebral perfusion pressure and thus caused recurrent bleeding.

Excluding those with a rapid tissue shift, acute wide pupils, and deteriorating motor responses, deterioration in traumatic brain injury is always a combination of clinical signs including agitation (perhaps related to toxins), a focal seizure (often becoming focal status), and decline in responsiveness (from following commands with a motor response to barely localizing a stimulus). Unintubated patients may stop speaking; intubated patients may stop responding to simple directions.

Emerging paroxysmal sympathetic hyperactivity is a common cause of deterioration and a sign of major trouble to come [17, 18]. These spells are now more commonly termed "paroxysmal sympathetic hyperactivity syndrome"to highlight three characteristics. The spells are paroxysmal manifestations of excessive sympathetic activity. Patients become tachycardic, hypertensive (with increased pulse pressure), tachypneic, febrile, or diaphoretic and often develop markedly increased muscle tone, which may result in "dystonic" postures. Most vital systems become rapidly overwhelmed by the excessive sympathetic activity, which increases the metabolic demand and, potentially, intracranial pressure.

Aneurysmal Subarachnoid Hemorrhage

Patients with aneurysmal subarachnoid hemorrhage are prone to deterioration from cerebral ischemia due to delayed cerebral vasospasm, rebleeding, acute hydrocephalus, and enlargement of a temporal lobe hematoma [19,

20]. Despite many years of caring for these patients, we still cannot assuredly recognize cerebral vasospasm in a timely fashion. We must constantly ask ourselves if this sleepy patient is sleepy from diffuse distal vasospasm, if this quiet patient is abulic from anterior cerebral narrowing vasospasm, and if this disinhibited patient is agitated from vertebrobasilar artery narrowing. The problem is that the clinical presentation of delayed cerebral ischemia (delayed because it rarely happens before 4 days after aneurysmal rupture) is a continuum, meaning that there is considerable variation in how patients present, and this is often a combination of disordered brain functions. It is not just a hemiparesis. It is much more. Delayed cerebral ischemia or symptomatic vasospasm manifests by a gradually decreased level of consciousness in most patients and, in some, is associated with hemiparesis, mutism, and, less frequently, apraxia. Unusual presentations, such as paraparesis, have been described. Patients with delayed cerebral ischemia may become apathetic, cut short answers to questions, and have initial weakness of one leg or both legs, indicating infarction in both territories of the anterior cerebral arteries. Patients have fluctuating levels of consciousness: days with daytime sleep and minimal wakefulness intermingled with days of appropriate behavior and better responsiveness.

The typical clinical features of rebleeding are loss of consciousness associated with loss of several brainstem reflexes including pupillary light response and oculocephalic responses. In most patients, respiratory arrest or gasping occurs, necessitating immediate endotracheal intubation and mechanical ventilation. Recovery from rebleeding is difficult to predict, but many patients begin to trigger the ventilator within hours; a return of brainstem reflexes also signals recovery. Patients may improve rapidly up to the point of self-extubation. Rebleeding can be much less dramatic in patients presenting with acute headache alone. In some fortunate patients, rebleeding begins with sudden emergence of fresh blood in the collection bag of the ventricular drain, and rapid evacuation of intraventricular blood is often life-saving.

The clinical presentation of acute hydrocephalus is characterized by progressively impaired consciousness. Patients become more drowsy, tachypneic, and unable to protect the airway or cough up secretions. Most cannot follow complex commands; only vigorous pain stimuli will open the eyes and cause localization of a pain stimulus. Pinpoint pupils and downward deviation of the eyes seldom occur but are more common in patients with a severely enlarged ventricular system. The diagnosis of acute hydrocephalus becomes clear when the CT scan shows further enlargement of the ventricular system. Seizures may cause sudden deterioration, but most are observed at the initial rupture or during rebleeding.

Failure to awaken fully after a generalized tonic-clonic seizure may point to non-convulsive status epilepticus, but this cause of deterioration is very unusual. Acute deterioration in SAH remains unexplained in 20 to 30% of patients. Unwitnessed seizures, drug effects (e.g., from large doses of opioids for pain management), new hyponatremia, or swelling surrounding a parenchymal hematoma may be implicated in some instances, but the causes often remain elusive.

Cerebral Hematoma

Approximately 30% of patients with lobar and ganglionic hemorrhages deteriorate to a more significant neurologic deficit [21]. In ganglionic hemorrhage, the cause is expansion of the hemorrhage; in lobar hematoma, expansion is followed by worsening peri-hematoma edema [22–29]. We can expect one in four patients with a lobar hemorrhage to deteriorate; this is more common with already large clots, early shift, and compression. Worsening hemiparesis and decreased levels of alertness are most common; in others, a new neurologic deficit, such as speech and language difficulties, may become apparent. Pupil dilatation on the side of the hematoma indicates lateral brainstem displacement with third-nerve damage in a temporal-lobe hematoma but may also indicate extension of the thalamic hemorrhage to the midbrain.

In the first 12 to 24 hours, enlargement of the hematoma causes deterioration. Later, acute hydrocephalus is a possible cause when CSF circulation is interrupted. We cannot differentiate expanded hematoma from a ballooning ventricle, and we see worsening responsiveness with patients lapsing into a permanent stupor. Patients who present with an intracranial hemorrhage-related coma with ventricular hemorrhage and dilatation seldom benefit from ventriculostomy, although it often seems the only rational option. A particularly difficult clinical situation is fluctuating consciousness in a patient with a moderate-sized hematoma and some shift but superficial localization in the frontal or temporal lobe. The threshold for evacuation should probably be low, but this is a neurosurgical judgment call and depends on your neurosurgeon's approach (tentative, aggressive, conservative or whatever is logistically feasible).

Further deterioration in patients with cerebellar hematoma can be predicted with certainty. The odds are higher when uncontrolled systolic blood pressure at admission exceeds 200 mm Hg, corneal reflexes are abnormal, and oculocephalic reflexes are impaired. These signs reflect brainstem displacement. Neurologic deterioration is also likely in patients with vermian hemorrhage or hemispheric hemorrhage tracking into the vermis and in patients with early hydrocephalus on initial CT scan. Risk of further clinical deterioration is low in patients without brainstem displacement, upward herniation, or, particularly, compression of the fourth ventricle. Most patients with a cerebellar hematoma deteriorate from direct brainstem compression rather than hydrocephalus. The clinical features are progressive limitation, loss of upward gaze, and deepening coma. Pupils become asymmetrical in size and, soon after, pinpoint. Spontaneous hyperventilation with considerable respiratory alkalosis often occurs at the same time as marked deterioration in consciousness. Often, blood pressure increases with widening of the pulse pressure, and we have found this a helpful clinical sign of early deterioration. Deterioration from hydrocephalus alone may occur in patients with cerebellar hematoma, usually in those with only marginal compression of the fourth ventricle, and has a more gradual clinical course. CT scanning confirms further enlargement of the lateral horns, bulging of the third ventricle, and, on lower CT scan slices, clearly visualizes the temporal horns. These patients gradually become drowsier and are unable to protect the airway. New neurologic signs or symptoms seldom develop.

Ischemic Stroke

Large vessel occlusion in usually symptomatic and with considerable deficits. Patients with minimal deficits (for example from M1 or M2 MCA occlusions) may become far more symptomatic quickly and roughly in a third of the patients (within 24 hour time span and despite intravenous thrombolysis.) Clot retrieval is usually only performed after deterioration and not preemptively. Deterioration in patients with a large hemispheric stroke most often is caused by brain swelling [30]. Brain swelling invariably occurs to some degree in patients with complete middle cerebral artery (MCA) territory occlusion, usually after a 2- to 7-day interval (median, 4 days). The development of anisocoria ($\geq$2 mm) and bilateral ptosis are important initial clinical signs of brain swelling [31–33]. Very often, a Cheyne-Stokes breathing pattern becomes sustained hyperventilation, but periodic breathing may follow when coma deepens. This must prompt endotracheal intubation and mechanical ventilation. A unilateral fixed, dilated pupil may be observed early, but small, constricted-but-reactive pupils from downward diencephalic displacement are more common. In most patients, the clinical course usually worsens gradually over days [34–38]. However, drowsiness from brain swelling can be transient, and the level of consciousness can improve.

Another neurologic cause for deterioration in patients with hemispheric stroke is hemorrhagic conversion of the infarct. This can be explained by either petechial hemorrhage or small hematomas without significant mass effect. Most patients do not have clinical deterioration. Similar to hemispheric infarction, the most reliable clinical symptom of tissue swelling in cerebellar stroke is a decreased level of consciousness and arousal. In addition, pontine compression may lead to ophthalmoparesis, breathing irregularities, and cardiac dysrhythmias. In most patients, neurological deterioration usually occurs within 72 to 96 hours. Some patients may experience deterioration within 4 to 10 days when previously at-risk penumbral tissue progresses to infarction, followed by delayed swelling, and in some cases, hemorrhagic transformation. However, the exact mechanism of this clinical course is still unclear.

Fluctuating deficits, some of which are profound, are typical in basilar artery occlusion but not fully explained pathophysiologically. Failing collaterals with hypotension and recovery with blood pressure augmentation may be considered, but in others, a propagating clot knocks off a series of penetrations leading to clinical progression with fits and starts.

A maximal deficit is present on the first day of admission in approximately one-third of the patients. In another third, deterioration from further clot outgrowth results in successive occlusion of perforating arteries originating from the basilar artery. Many patients, devastated by occlusion of the basilar artery, remain in coma with retention of some brainstem reflexes.

Acute basilar artery occlusion rarely results in complete loss of brainstem reflexes, and some function of the medulla oblongata often remains. Patients with vertebral artery dissection rarely have deterioration, and in most, the event is monophasic. Patients with dissection and recurrent, transient ischemic attacks in the posterior circulation can be successfully treated with balloon occlusion of the vertebral artery if sufficient collateral circulation is present. This usually implies retrograde flow from the contralateral vertebral artery to the ipsilateral posterior. Patients should remain relatively stable despite development of cerebral infarction and impressive occlusion. In addition, hemorrhagic cerebral infarcts located in the occipital lobes may not produce a significant mass effect. Clinical neurologic deterioration is probably a consequence of further thrombosis in the venous system. Impaired consciousness may indicate that the thrombosis has extended into the tributary cortical veins. Increased sagittal sinus pressure may result in hemorrhagic infarction, often bilaterally in the parietal lobes, and increased capillary filtration, causing cerebral edema.

Deterioration from an enlarging hematoma with progressive brainstem compression may require evacuation of the hematoma by craniotomy or decompressive craniotomy. Progression is rapid, usually beginning soon after admission. In advanced cerebral edema, pupils become sluggish and dilated, and papilledema appears. CT scanning may show signs of cortical effacement, but this may be very difficult to appreciate in young patients with less prominent sulci. Progressive effacement of the sylvian fissure and compression of the ventricles and basal cisterns are more diagnostic for cerebral edema on CT. Patients with acute brain edema require intubation and mechanical ventilation to a respiratory alkalosis with PaCO2 in the low 30s.

Acute CNS Infections

Encephalitis is suspected in a patient who presents with headache, fever, confusion, and, in more advanced cases, changes in the level of consciousness. Seizures (focal or generalized) are a common complication. Examination may show neck stiffness or focal deficits, but their absence is not infrequent. In fact, the diagnosis may not even be considered when the patient is seen early in the course of the disease and is just "confused" [39]. Permanent brain injury can occur rapidly after an overwhelming infection. Cortical infarctions are common and widespread and rarely lead to swelling or mass effect. Vasculitis (or thrombotic vasculopathy) may lead to ischemia. In a rapidly worsening patient, subdural empyema should be considered for what initially seems to be acute bacterial meningitis. The clinical presentation of headache and worsening coma can be analogous to acute bacterial meningitis. Important clues are prior paranasal sinusitis and recent sinus surgery, both of which correlate with subdural empyema in a considerable number of patients. To complicate matters further, CT may not have been performed initially or, if done without contrast, can yield normal results. Magnetic resonance imaging may demonstrate multilocular collections not observable on CT, particularly those localized at the convexity. In most patients, however, CT scanning with contrast demonstrates the subdural pus pocket. A large craniotomy is needed to salvage the patient. Outcome is poor if patients are comatose.

Another important cause of clinical deterioration is therapeutic failure. In recent years, penicillin-resistant *Staphylococcus pneumoniae* strains have increased in frequency. Any patient with bacterial meningitis, rapid clinical deterioration, persistent high fever despite antibiotic treatment, and diplococci in a repeated CSF culture after several days of treatment may have a penicillin-resistant strain of *S. pneumoniae*. As mentioned earlier, adding vancomycin to the initial empirical antibiotic treatment should reduce therapeutic failures. Rarely, in addition to systemic therapy, intrathecal vancomycin with monitoring of CSF levels may be needed.

Seizures may cause deterioration in a patient's condition, but the low incidence (10%) does not justify prophylactic treatment. Non convulsive status epilepticus is a rare cause of deterioration in patients with meningitis.

For neurosurgeons (and consulted neurointensivists), meningitis from a recent craniotomy is a major complication and invariably associated with dural CSF leaks [40, 41]. Some noted an increased incidence of meningitis in patients who presented with CSF leak after transsphenoidal surgery and open craniotomy, Factors contributing to postoperative CSF leaks in endoscopic skull base surgery included the disease pathologies, surgical approaches, flap issues, reconstruction techniques, and obesity [42]. Patients with delayed onset of CSF leak are more likely to present with a history of radiation treatment. Surgical closure of skull-base defects could prevent the development of serious neurologic sequelae and ascending infections. Another serious complication is ventriculitis-associated meningitis (~1–3%), for which risk factors have been identified [43, 44]. These are not surprising and include duration of catheter placement, catheter replacement, multiple catheters, increased CSF output, and CSF leaks. In one study, routine CSF collection (every other day or every 3 days) or CSF collection when needed did not increase the risk of infection [44], but the main disadvantages of routine CSF sampling are system breach. Practice adjustments, including increasing length of the tunneling, using antibiotic-coated catheters, maintaining superb hygienic care, and only keeping ventriculostomy when clinically relevant, may all help to reduce infection. A patient with a ventriculostomy in place for more than 15 days has an increased risk of developing infection seven fold [44].

Deterioration from Non-Neurologic Causes

Most patients with an acute neurologic injury have other medical issues, which often flare up in the days after admission. A number of causes require immediate assessment and are best done with a systematic review (Fig. 6.3). Neurologists assigned to a Neuro ICU have all seen worsening congestive heart failure, new cardiac arrhythmias, poorly controlled diabetes mellitus, and unstable blood pressures. Fever may have many infectious and noninfectious causes, and diagnostic tests should be cost effective. Launching a broad

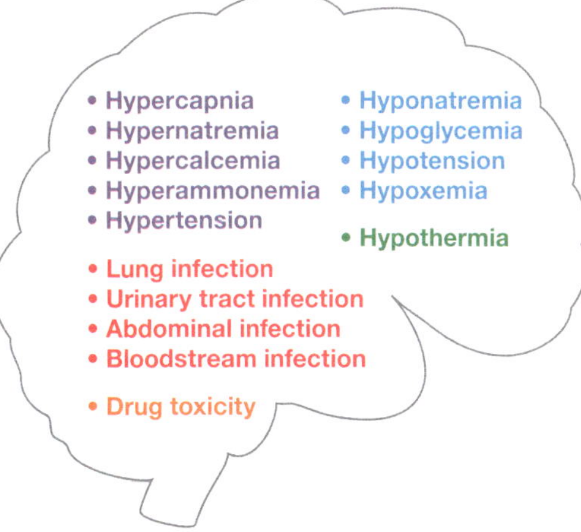

Fig. 6.3 Non-neurologic causes of deterioration

evaluation with every fever spike is nonproductive. Yet, when hypotension accompanies the fever, patients should be rapidly treated for early sepsis following a comprehensive protocol. Any delay in reversing the situation may cause irreparable secondary brain damage. It is a well-established clinical observation that untreated high fever with shock may suddenly (and permanently) change the patient's neurologic outlook. Increased brain temperature can worsen the prior brain injury. Fever not only worsens ischemic injury but exacerbates cerebral edema, increases intracranial pressure, and in general, confounds the general examination of the patient. Markedly febrile patients will have a decreased level of consciousness that can rapidly improve with control of fever.

The medical care of neurologic patients is closely linked to evaluation and treatment of pulmonary problems [45]. Respiratory failure leads to hypoxemia and/or hypercapnia; this abnormality can be further classified as hypercapnic ($PaCO_2$ > 45 mm Hg) or hypoxemic ($PaCO_2$ < 55 mm Hg) respiratory failure. Respiratory failure is defined by hypoxemia (PaO_2 < 50 mm Hg) or hypercapnia ($PaCO_2$ > 50 mmHg) but only if the patient is on room air and had a prior normal blood gas. Hypoxemic respiratory failure is mostly due to alveolar hypoventilation or ventilation-perfusion mismatch (or right-to-left shunt). Acute neurologic disorders are frequently associated with acute hypoxemic respiratory failure and ventilation–perfusion mismatch such as atelectasis or large lung field collapse, aspiration pneumonitis, pulmonary embolism, and pulmonary edema (cardiac or neurogenic). The clinical features of acute hypoxemia are fairly consistent and include impaired arousal, restlessness, tachypnea, tachycardia, and, sometimes, hypertension and peripheral vasoconstriction.

Acute injury to the brain or brainstem may cause massive ("flash") pulmonary edema. Conditions associated with neurogenic pulmonary edema include ganglionic hemorrhage, aneurysmal subarachnoid hemorrhage (SAH), primary brainstem-lesion hemorrhage, status epilepticus, and penetrating traumatic brain injury. The clinical picture is very specific, almost always appearing soon after the initial brain injury. The clinical entity may be mistaken for other pulmonary conditions, such as massive aspiration pneumonia or pulmonary contusion.

Stress cardiomyopathy frequently follows a brain injury but is often overlooked. Stress cardiomyopathy may present with "flash" pulmonary edema and minimal hypotension progressing to severe hypotension. In these instances, neurogenic pulmonary edema is incorrectly diagnosed and cardiac failure not considered. New-onset atrial fibrillation with rapid ventricular response is a common cardiac arrhythmia after acute brain injury and very responsive to IV diltiazem. Discontinuation of beta blockade for "permissive hypertension" in acute ischemic stroke may lead to tachycardia and demand ischemia. Acute bradycardias occur with acute mass effect, any large hemispheric stroke, or acute dysautonomia and seldom need pacing.

Acute bowel ischemia may occur from severe vasoconstriction of splanchnic vessels from a catecholamine surge, which is also responsible for tako-tsubo cardiomyopathy. Another potential pathogenic mechanism is thrombosis-induced ischemia during the pro-thrombotic state following aneurysmal rupture [46].

More Reflections

We often must assess whether deterioration is taking place in our patient. Sometimes worsening is not apparent in the patient's appearance or behavior; for example, the CT scan shows worsening, but the patient seems stable. That occurs commonly in patients with expanding cerebral hematoma – there are increases in volume and mass effect but without any notable change. Conversely, sometimes CT scan or other tests remain unchanged, but the patient seems much worse, pointing to other factors. Detecting deterioration is about making choices and how they matter. There is no question that technology could greatly help the field of neurocritical care if studied rigorously but with a healthy dose of skepticism. It is not difficult to imagine a patient-and-computer interface that provides detailed, sophisticated online information on secondary neuronal stress resulting from a major injury. Automatic adjustment of parameters may follow with digital alerts to smartphones or tablets to inform the nursing staff and neurointensivist. Functional imaging would be commonplace. We will know in the next decades if this will remain out of reach or if our means and ways will be completely different. Still, nothing will (or should) entirely replace clinical observation and detailed examination of the neurologic patient.

Pointers and Takeaways

- Fluctuating and stable are the most commonly used words in the NeuroICU (but not always understood what they mean)
- Clinical worsening may cause secondary brain injury. It remains important, therefore, to monitor changes and to reverse them quickly
- Clinical determination of brain-tissue shift takes priority because it leads to neurosurgical intervention
- Neurocritical disorders have overlapping causes of not only clinical worsening but also disease-specific ones that providers must recognize and treat appropriately
- Neurocritical disorders may lead to secondary organ involvement, which can compromise clinical examination. Tests should be repeated in any worsening patient.

References

1. Wijdicks EFM. Neurocritical care: It's what we do and what we do best. Neurocrit Care. 2006;5:81.
2. Wijdicks EFM. Neurology of critical care. Semin Neurol. 2016;36:483–91.
3. Cheng Y, Wu S, Wang Y, et al. External validation and modification of the EDEMA score for predicting malignant brain edema after acute ischemic stroke. Neurocrit Care. 2020;32:104–12.
4. Broocks G, Elsayed S, Kniep H, et al. Early prediction of malignant cerebellar edema in posterior circulation stroke using quantitative lesion water uptake. Neurosurgery. 2021;88(3):531–7.
5. Fabritius MP, Thierfelder KM, Meinel FG, et al. Early imaging prediction of malignant cerebellar edema development in acute ischemic stroke. *Stroke*. 2017;48:2597–600.
6. Caplan LR. Caplan's Stroke: a clinical approach. 5th ed. Cambridge: Cambridge University Press; 2016.
7. Hirsch LJ, LaRoche SM, Gaspard N, et al. American Clinical Neurophysiology Society's standardized critical care EEG terminology: 2012 version. J Clin Neurophysiol. 2013;30:1–27.
8. Ropper AH. Lateral displacement of the brain and level of consciousness in patients with an acute hemispheral mass. N Engl J Med. 1986;314:953–8.
9. Wijdicks EFM. Through the eyes of monkeys: questions about uncal herniation. Neurocrit Care. 2020;
10. Kinoshita K. Traumatic brain injury: pathophysiology for neurocritical care. J Intensive Care. 2016;4:29.
11. Adatia K, Newcombe VFJ, Menon DK. Contusion progression following traumatic brain injury: a review of clinical and radiological predictors, and influence on outcome. *Neurocrit Care*. 2021;34(1):312–24.
12. Allison RZ, Nakagawa K, Hayashi M, Donovan DJ, Koenig MA. Derivation of a predictive score for hemorrhagic progression of cerebral contusions in moderate and severe traumatic brain injury. Neurocrit Care. 2017;26:80–6.
13. Tan JE, Ng I, Lim J, Wong HB, Yeo TT. Patients who talk and deteriorate: a new look at an old problem. Ann Acad Med Singap. 2004;33:489–93.
14. Rockswold GL, Leonard PR, Nagib MG. Analysis of management in thirty-three closed head injury patients who "talked and deteriorated". Neurosurgery. 1987;21:51–5.
15. Reilly PL, Graham DI, Adams JH, Jennett B. Patients with head injury who talk and die. Lancet. 1975;306:375–7.
16. Lobato RD, Rivas JJ, Gomez PA, et al. Head-injured patients who talk and deteriorate into coma. Analysis of 211 cases studied with computerized tomography. J Neurosurg. 1991;75:256–61.
17. Perkes I, Baguley IJ, Nott MT, Menon DK. A review of paroxysmal sympathetic hyperactivity after acquired brain injury. Ann Neurol. 2010;68:126–35.
18. Baguley IJ, Perkes IE, Fernandez-Ortega JF, et al. Paroxysmal sympathetic hyperactivity after acquired brain injury: consensus on conceptual definition, nomenclature, and diagnostic criteria. J Neurotrauma. 2014;31:1515–20.
19. Wijdicks EFM, Kallmes DF, Manno EM, Fulgham JR, Piepgras DG. Subarachnoid hemorrhage: neurointensive care and aneurysm repair. Mayo Clin Proc. 2005;80:550–9.
20. Rabinstein AA, Lanzino G, Wijdicks EFM. Multidisciplinary management and emerging therapeutic strategies in aneurysmal subarachnoid haemorrhage. Lancet Neurol. 2010;9:504–19.

21. Xi G, Keep RF, Hoff JT. Mechanisms of brain injury after intracerebral haemorrhage. Lancet Neurol. 2006;5:53–63.
22. Arima H, Wang JG, Huang Y, et al. Significance of perihematomal edema in acute intracerebral hemorrhage: the INTERACT trial. Neurology. 2009;73:1963–8.
23. Chen L, Xu M, Yan S, Luo Z, Tong L, Lou M. Insufficient cerebral venous drainage predicts early edema in acute intracerebral hemorrhage. Neurology. 2019;93:e1463–73.
24. Gusdon AM, Nyquist PA, Torres-Lopez VM, et al. Perihematomal edema after intracerebral hemorrhage in patients with active malignancy. Stroke. 2020;51:129–36.
25. Levine JM, Snider R, Finkelstein D, et al. Early edema in warfarin-related intracerebral hemorrhage. Neurocrit Care. 2007;7:58–63.
26. Murthy SB, Moradiya Y, Dawson J, et al. Perihematomal edema and functional outcomes in intracerebral hemorrhage: influence of hematoma volume and location. Stroke. 2015;46:3088–92.
27. Sun W, Pan W, Kranz PG, et al. Predictors of late neurological deterioration after spontaneous intracerebral hemorrhage. Neurocrit Care. 2013;19:299–305.
28. Volbers B, Giede-Jeppe A, Gerner ST, et al. Peak perihemorrhagic edema correlates with functional outcome in intracerebral hemorrhage. Neurology. 2018;90:e1005–12.
29. Wu TY, Sharma G, Strbian D, et al. Natural history of perihematomal edema and impact on outcome after intracerebral hemorrhage. Stroke. 2017;48:873–9.
30. Campbell BCV, Khatri P. Stroke. Lancet. 2020;396:129–42.
31. Wijdicks EFM, Sheth KN, Carter BS, et al. Recommendations for the management of cerebral and cerebellar infarction with swelling: a statement for healthcare professionals from the American Heart Association/American Stroke Association. Stroke. 2014;45:1222–38.
32. Ropper AH, Shafran B. Brain edema after stroke. Clinical syndrome and intracranial pressure. Arch Neurol. 1984;41:26–9.
33. Hacke W, Schwab S, Horn M, Spranger M, De Georgia M, von Kummer R. 'Malignant' middle cerebral artery territory infarction: clinical course and prognostic signs. Arch Neurol. 1996;53:309–15.
34. Kimura K, Iguchi Y, Shibazaki K, Aoki J, Terasawa Y. Hemorrhagic transformation of ischemic brain tissue after t-PA thrombolysis as detected by MRI may be asymptomatic, but impair neurological recovery. J Neurol Sci. 2008;272:136–42.
35. Krieger DW, Demchuk AM, Kasner SE, Jauss M, Hantson L. Early clinical and radiological predictors of fatal brain swelling in ischemic stroke. Stroke. 1999;30:287–92.
36. Pallesen LP, Barlinn K, Puetz V. Role of decompressive craniectomy in ischemic stroke. Front Neurol. 2018;9:1119.
37. Zha AM, Sari M, Torbey MT. Recommendations for management of large hemispheric infarction. Curr Opin Crit Care. 2015;21:91–8.
38. Hwang DY, Matouk CC, Sheth KN. Management of the malignant middle cerebral artery syndrome. Semin Neurol. 2013;33:448–55.
39. Singh TD, Fugate JE, Rabinstein AA. The spectrum of acute encephalitis: causes, management, and predictors of outcome. Neurology. 2015;84:359–66.
40. Ter Horst L, Brouwer MC, van der Ende A, van de Beek D. Community-acquired bacterial meningitis in adults with cerebrospinal fluid leakage. Clin Infect Dis. 2020;70:2256–61.
41. van de Beek D, Drake JM, Tunkel AR. Nosocomial bacterial meningitis. N Engl J Med. 2010;362:146–54.
42. Tang R, Mao S, Li D, Ye H, Zhang W. Treatment and outcomes of iatrogenic cerebrospinal fluid leak caused by different surgical procedures. World Neurosurg. 2020;143:e667–75.
43. Bari ME, Haider G, Malik K, Waqas M, Mahmood SF, Siddiqui M. Outcomes of post-neurosurgical ventriculostomy-associated infections. Surg Neurol Int. 2017;8:124.
44. Sweid A, Weinberg JH, Abbas R, et al. Predictors of ventriculostomy infection in a large single-center cohort. J Neurosurg. 2020:1–8.
45. Della Torre V, Badenes R, Corradi F, et al. Acute respiratory distress syndrome in traumatic brain injury: how do we manage it? J Thorac Dis. 2017;9:5368–81.
46. Chakraborty T, Wijdicks EFM. A punch to the gut from a ruptured cerebral aneurysm. Neurocrit Care. 2021;34:343–4.

7 Unraveling Unconsciousness

We have reached the chapter in which we discuss one of the most important clinical problems in the acute clinical neurosciences – the examination of the unresponsive patient. Causes can be elusive, certainly never easy, and we may be unable to give a satisfactory answer. Patients are often first described as "altered," "altered mental status," "unresponsive," or "encephalopathic," which do nothing other than satisfy our innate desire for labels (Table 7.1). Many have uncritically repeated these sobriquets, but none of the descriptors fully satisfy. Neurologists understand the difficulty of ascertaining what is going on in the brain and with the patient but, for many years, also had the tendency to use an umbrella term "multifactorial toxic-metabolic encephalopathy"— a label that is nothing more than a bunch of words strung together [1]. Some would then list the abnormalities that comprised the patient's critical illness but not always with a careful reasoning as to why these abnormalities lead to brain dysfunction.

We now know that "metabolic encephalopathy" may mimic structural lesions, and organ dysfunction can be associated with poor clearance of medication resulting in toxicity and resolution after stopping the drug. Neurology is full of fanciful terms, and terms to delineate a confused patient abound. "Clouding of consciousness" was used in the first version of DSM-III and then disappeared. None of these terms have brought us closer to the problem of understanding. Generally, how major illness affects neuronal circuitry is not understood. It is the biggest question of all. We should stop looking for structural injury in delirium – it is not there, not even in the worst manifestations of alcohol-withdrawal delirium [2].

Table 7.1 Trying to describe disturbances of consciousness

Altered mental status
Encephalopathy
Global confusional state
Hypokinetic (quiet) delirium
Hyperkinetic (agitated) delirium
Acute brain failure
Toxic delirious reaction
Toxic psychosis
ICU psychosis
Somnolent
In a coma

As we will note, there has been a serious attempt to reduce the number of designations of types of delirium; some have reduced them to three: hyperactive, hypoactive, and mixed. Others, dissatisfied with this terminology, have added subsyndromal delirium, a state in between no delirium and clinical delirium. Simplified terminology may have its benefits, but it reduces the recognition of clear neurologic symptomatology. The proposed definition of "hypoactive delirium" is a patient status characterized by less attention and paucity of movement as opposed to "hyperactive delirium," which is characterized by increased attention, agility, and exaggerated

E. F. M. Wijdicks, *Examining Neurocritical Patients*, https://doi.org/10.1007/978-3-030-69452-4_7

response to a simple stimulus. Mixed forms are defined as a combination of hypoactive and hyperactive delirium. Studies have found hyperactive delirium to be far less common than hypoactive delirium or a mixed form. So-called hypoactive delirium has also been called "quiet delirium," and this remains one of the most problematic designations for any neurologist. It is difficult to determine how much drowsiness or less responsiveness is categorized as hypoactive delirium; many neurologists will ask who in the intensive care would not fulfill these simple criteria? Any drowsy patient lacks concentration and attention and is unable to think clearly. The term encephalopathy may not be ideal but is much better understood than "hypoactive delirium." Introducing the latter, finessed term will dramatically increase the prevalence and, perhaps, unnecessary treatment. This mangling of terms, while minimizing neurologic symptomatology, should certainly raise an eyebrow among seasoned neurologists. With different forms of acute confusion, the main challenge is to recognize one of the more defined agnosias presenting as "disorientation." There may be impairment of visuospatial tasking (right posterior parietal cortex), anosognosia (unawareness for hemiparesis), or severe amnesia (parahippocampus).

The two primary components of abnormal consciousness – reduced vigilance and altered conscious thought – are what we are seeking. The word "engaged" might sum up these two components. Being conscious (in a neurologic but not philosophical sense) is determined by the level of vigilance (the highest level, dozing off; the lowest level, coma). Somewhere in there is delirium. These consciousness (or rather unconsciousness) conundrums are owned by neurologists, and not a day goes by without seeing patients in different states of attentiveness.

How do we unravel altered consciousness and unconsciousness? I will explain the clinical findings at the bedside to bring some clarity with suggestions on how to interpret certain signs – with the caveat that we may never get it right all the time. Our clinical observations may have taken us as far as they can, and our binary classifications of neurologic states – you are either delirious or not; you are either in coma or you are not – may be over-imaginative.

The Spectrum of Altered Consciousness

Neurocritically ill patients – by the nature of their condition – lose their perceptual abilities leading to disorientation in time. Worsening changes of consciousness in a patient with acute brain injury should be viewed within a wide spectrum. Within this spectrum fall sleepiness and agitated confusion. At the end of the clinical spectrum is persistent unconsciousness (Fig. 7.1) [3]. Coma is rarely persistent and most (over 95%) patients awaken. Those who do not will be entering a permanent comatose state or vegetative state, and in a handful of cases, decades may pass and some improvement in awareness (and even speech after being silent) is seen. These cases are exceedingly hard to interpret in retrospect and never followed by the same physician that long, and we can never know if there is true improvement

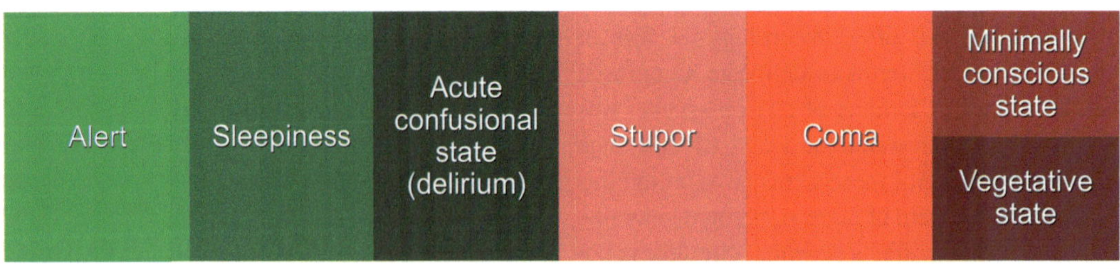

Fig. 7.1 Spectrum of alteration of consciousness

(*recovery*) or the diagnosis was incorrect and was never revisited (*discovery*). A vegetative state is a destroyed brain with vital functions intact (the brainstem provides adequate respiratory drive and vascular tone). In comparison, brain death is the only irrevocably permanent comatose state with loss of vital functions (apnea and no vascular tone) due to absent brainstem function.

In contemplating this spectrum of abnormal consciousness, two questions immediately arise: why and how? Let us start with sleepiness. We do not know why sleepiness occurs, but often it is a need to rest. We see increased sleepiness in patients with acute brain injury – a neurocritical care truism. Patients who are awake and talking throughout the day may easily fall into a deep sleep after receiving nursing care, undergoing tiresome transfers out of bed but also physical therapy, which may be strenuous in any illness, and we can best interpret this as a sign of early exhaustion [4]. Patients who are difficult to awaken with voice or gently sternal rub may awaken with a brief noxious stimulus and stay awake for some time but then drift off to sleep. This excessive sleepiness is not associated with changes in content. Some fluctuation is expected, and at some point, physicians introduced a new vapid (astronomy-derived) term: "waxing and waning." As described in Chap. 6 in more detail, each of these manifestations may indicate a new deterioration (e.g., new cerebral vasospasm, new hydrocephalus, more shift of a new mass, new sodium derangement, or creeping up hypercarbia), or they may remain unexplained.

The next level in the downward spiral is an "acute confusional state," a term used interchangeably with "encephalopathy" or "quiet delirium." None of these terms is satisfactory, and they have been thrown around for years without serious thought given to how they fit into this spectrum. Acute confusional state is both a disorder of content and wakefulness or arousal. In other words, patients are unclear what is happening, slow in thinking, and distracted. To stay in metaphoric descriptions, the patient appears foggy and clouded. We have traditionally associated this condition with the elderly admitted to hospitals; that is, taken out of their familiar environment [5–7]. Within the definition of acute confusional state is a more defined clinical state called abulia. It signifies a lack of initiative, spontaneity, and drive and brief, monosyllabic but often appropriate responses to questions. The patient is completely detached from everybody and everything. All action, ideas, and emotions are blunted. Within a timespan of several minutes (if the time has been taken to observe the patient closely), the patient may correctly answer a question or not at all. The clinical picture is best summarized as demotivated state and a result of frontal lobe executive dysfunction and disruption of the subcortical tracts. Neurointensivists see it most often in aneurysmal subarachnoid hemorrhage with cerebral vasospasm of the anterior cerebral artery, which leads to ischemia or infarction. (It may also occur with thalamic and midbrain lesions.) It is unclear how to differentiate abulia from quiet delirium, if we want to continue to use the latter term.

Delirium traditionally, for many years, has been equated with a hyperactive sympathetic (fight-like) state. Patients are wild and thrash around; they move their heads side to side, pick, buck, bite, and yell at imaginary people. They generally appear very uncomfortable or even terrified [8]. We now see these extreme presentations mostly with autoimmune encephalitis with few options other than to intubate and sedate patients heavily. Intensivists around the world know that alcohol-withdrawal delirium remains common and may include auditory (voices) and visual (mice mostly) hallucinations. For many years, the consensus was that delirium was a state of confusion accompanied by autonomic symptoms such as profuse sweating, muscle twitching, fidgety movements, and even more focused movements such as removing lines and catheters.

As alluded to earlier, some have suggested discarding encephalopathy altogether and substituting the term "hypoactive delirium" for less attentive patients with paucity of movement as opposed to "hyperactive delirium," a condition characterized by increased attention, agility, and exaggerated response to a simple stimulus. Still, one can also easily imagine hypoactive delirium harboring a new frontal lesion, CNS infection, nonconvulsive status epilepticus, a new metabolic derangement such as hyperammonemia, or a major side effect of an administered drug. Some would not accept the more recent term "quiet delirium," having seen delirium only in states of hyperactivity [9]. Others have suggested the term *delirium disorder* ("a plurality of physiologically distinct delirium disorders with a convergent phenotype defined by essential features"). This new term only suggests the need to look for precipitating factors and a causative pathophysiology, which we already do. It seldom resolves the elusive cause of delirium unless it is due to alcohol withdrawal, baseline dementia, or a newly introduced drug [10].

How does a psychiatrist or neurologist define delirium, and how does their definition differ from an intensivist's definition? [11–13] Delirium is very common in medical and surgical intensive care units, and prolonged stay has everything to do with it. [14–16] Many patients with renal or hepatic encephalopathy show daytime sleepiness and nighttime agitation, also known as sundowning. Typically, delirium has several disturbed components (arousal, language, perception, orientation, mood, and sleep). Restlessness is associated with pallor, sweating, and tachycardia but also wide pupils. This occurs most often after withdrawal of stimulants or CNS depressants such as alcohol. Most patients are hyperaroused, with incoherent, rambling speech, and little comprehensible verbal output. Orientation to time is first impaired, followed by place. Patients do not know the building they are in or where they live. Disorientation of place has always been nonlocalizing, but some place it in the nondominant parietal occipital cortex [17]. Eventually, mood changes may include anger and aggression (Fig. 7.2).

Fig. 7.2 Delirium at the bedside

Some patients are delirious as a transitional clinical stage before progressing to stupor or coma. It is underrecognized that patients can progress this way. I have seen it in acute worsening hydrocephalus and, of course, with seizures – agitation followed by deep stupor. A well-known example is rapidly worsening hepatic encephalopathy. The appearance of jaundice marks the development of delirium, convulsions, and coma, but clinicians have noted early phases of irritable temper and restlessness, paroxysms of manic behavior to reduced awareness of surroundings and their stimuli, alternating with yawning, and a tendency to doze off. Patients with fulminant hepatic failure experience a very condensed process of clinical development of hepatic encephalopathy. Fluctuations in attention and slow responses to requests are typically present. Patients are incapable of registration, retention, and recall. The immediate memory span for digits is severely reduced. Overactivity and unrest, disorientation for place, delusions, and repetitive picking movements become evident in later stages.

Clinicians must focus on identification of precipitating causes. Most delirious patients rarely have an acute structural lesion. Acute strokes may cause severe agitation as predominant signs and then usually acute pontine or nondominant hemispheric strokes. Anticholinergic drugs do increase delirium, and it is probable that acetylcholine pathways are involved. Similar observations are found with the use of gamma-aminobutyric acid (GABAergic) or dopaminergic drugs that enhance alertness and, in high doses, cause delirious states. Another way to "prove" the involvement of neurotransmitters in delirium is through the use of certain drugs that improve agitation either through stimulation or toning down of the neurotransmitters. Examples are quetiapine or olanzapine. Haloperidol is an effective antidopaminergic drug.

The Examination of a Confused Patient

How can we best summarize findings on examination? Several nursing scales have been developed, and the most useful is the CAM-ICU scale or the ICDSS (Table 7.2) [18, 19]. The CAM ICU scale has been validated, but many of the tests focus on language comprehension to the

Table 7.2 Monitoring of agitation in ICU using nursing scales

CAM-ICU			ICDSC			
	Criteria	Positive score if		Criteria	Yes	No
1	Acute onset <u>or</u> fluctuating course	One or both features present	1	Altered level of consciousness *YES: RASS score other than zero* *NO: RASS = 0 or recent sedative use*	1	0
2	Inattention *Read the following series of 10 letters and let patient squeeze on the letter "A"* *S A V E A H A A R T*	More than two errors	2	Inattention *Read the following series of 10 letters and let patient squeeze on the letter "A"* *S A V E A H A A R T*	1	0
3	Altered level of consciousness *Based on the RASS score*	RASS score other than zero	3	Disorientation *Disoriented in name, place, and/or date*	1	0
4	Disorganized thinking *<u>Ask the following questions:*</u>* *Will a stone float on water?* *Are there fish in the sea?* *Does one pound weigh more than two pounds?* *Can you use a hammer to pound a nail?* *<u>Command</u>* *Say to patient: "Hold up this many fingers" (hold two fingers in front of the patient), "Now do the same with the other hand" (without repeating the number of fingers)*	More than 1 mistake on the questions or commands	4	Hallucination, delusion, or psychosis *Ask if present*	1	0
			5	Psychomotor agitation or retardation *Either hyperactive, requiring sedatives or restraints, or hypoactive*	1	0
			6	Inappropriate speech or mood *If present*	1	0
			7	Sleep–wake cycle disturbance *Either frequent awakening/ <4 hours of sleep or sleeping during most of the day*	1	0
			8	Symptom fluctuation *Fluctuation of any of the above symptoms over a 24-hour period*	1	0
Positive CAM-ICU: Criteria 1 <u>plus</u> 2 <u>and</u> either criteria 3 <u>or</u> 4			Positive ICDSC score: Subsyndromal delirium: 1–3 points Delirium: 4–8 points			

CAM-ICU: Confusion Assessment method for ICU
ICDSC: Intensive Care Delirium Screening Checklist

exclusion of a detailed neurologic examination. Neurologic examination in a confused patient should, at the very least, include testing recall, naming of three unrelated objects (e.g., car, Mr. Johnson, and tunnel), attention, repeating a series of digits or telephone number, calculation (e.g., counting down from 100 by subtracting 7), but also writing and reading a complete sentence. Ideally, it would also include copying a cube and following a more complex command (e.g., "before you point to the ceiling, point toward your nose"). Neurologic examination is definitely more difficult in agitated patients who may thrash, moan, speak incoherently, or swear. This presentation often occurs in the setting of alcohol or drug withdrawal and examination will be truncated if it can be done at all. The information that a full mental status examination provides is very useful, but a concern remains whether a full mental status examination can even be used to examine the patient in a confusional state. Still, most bedside screening tests nothing more than attention and distractibility and may vary from counting backwards from 20, having the patients name months forwards and backwards, or indicating the letter A in a random list of letters or a specific sentence. Obviously, impaired attention markedly influences testing of other cognitive domains. Nonetheless, there is an obligation to test for clear neurologic abnormalities that may not have been examined with sufficient detail. When aphasia is present, the physician should test fluency, naming, repetition, comprehension, and praxis. This may include recognition of paraphasic errors (it can be phonemic or semantic), naming of an object (e.g., a wedding ring), and repetition of a phrase (e.g., *It is a nice day today*). Notice comprehension by interspersing yes-and-no questions with commands (e.g., *Am I wearing glasses? Point to the clock in the room.*), reading and writing, and praxis (e.g., *Show me your teeth. Puff your cheeks. Wave goodbye.*).

Patients with an apraxia are unable to copy or pantomime. Abulia can be tested using grasp, snout reflex, or motor persistence of the paratonias found. Whether the patient is impaired in a visuospatial or a perceptual region is important and can sometimes only be tested by more complex tests such as copying geometric designs or drawing a clock. A simple test is the executive function test or oral trail-making test, which asks the patient to name the letter of the alphabet and add a number (e.g., A1, B2, C3, D4, and E5). Neglect in a patient who is agnostic is important and must be identified as left-sided neglect, cortical blindness, or prosopagnosia. Each of these abnormalities localizes to specific regions of the brain and may indicate an infarction, hemorrhage, or less clearly identifiable lesions on the CT scan (e.g., PRES).

The Examination of a Comatose Patient

Once there is a major decline in responsiveness, a neurologic examination of the "unresponsive" or comatose patient is needed; carrying it out remains a major obstacle for many physicians. As mentioned in Chap. 2, the initial examination is often relegated to the Glasgow Coma Scale, and overall, this does not provide the necessary details to diagnose the patient. Introduction of the FOUR Score, which adds assessment of eye movements, brainstem reflexes, and breathing patterns, improves the clinical assessment greatly, but there is much more to note.

In essence, the neurologic examination of a patient in coma requires three key components. These are observation and inspection; assessment of responsiveness to stimuli, and, most importantly, localizing to either a presumed lesion in the hemispheres or brainstem. Further granularity in localization (e.g., thalamus, pons) may follow but is rarely important in acute situations. A general physical examination is also helpful in the evaluation of comatose patients. Some combination of changes in vital signs may point to what is commonly called a toxidrome (a combination of toxic and syndrome), which may indicate an illicit or prescribed drug. Extremes in temperature and blood pressure, sweating or dryness, cardiac arrhythmias, or EKG abnormalities may be diagnostic or at least important pointers (Table 7.3). Still, it is not often possible to differentiate clinically between a diffuse structural brain lesion and acute physiologic brain dysfunction. CT or MRI may provide answers.

Table 7.3 Physical signs and considerations

Sign	*Consideration*
Hyperthermia	Endocarditis, sepsis, drug ingestion (cocaine, amphetamines, cyclic antidepressants, salicylates), malignant catatonia, neuroleptic malignant syndrome
Hypothermia	Hypothyroidism, drug ingestion (barbiturates, opioids)
Hypertension	Pheochromocytoma, eclampsia, drug overdose (phencyclidine, cocaine, amphetamine)
Hypotension	Antihypertensives, sepsis
Tachycardia	Alcohol, amphetamine, ethylene glycol
Bradycardia	Uremic coma, myxedema coma
Hypoventilation	Heroin, fentanyl, alcohol
Hyperventilation	Ethylene glycol, salicylate, diabetic ketoacidosis
Sweating	Thyroid storm, hypoglycemia, organophosphate exposure

Readers may first find it helpful to see a summary of the functional anatomy. Several different locations can cause coma. The key structures that underpin awareness and being fully awake are (1) the cortical mantle, (2) the thalamus, and (3) the reticular formation (ascending reticular activating system) in the posterior part of the upper brainstem. With this simple knowledge of key structures, we can deduce structural lesions that cause unconsciousness. Bihemispheric cortical injury should be diffuse in order to cause coma. Destructive damage involving the entire cortical mantle can occur after anoxic–ischemic injury, although the parieto-occipital regions are most severely affected. The white-matter core can be damaged mostly from acute demyelinating disorders or from acute hydrocephalus, and this will interrupt the thalamocortical circuits. The thalamus can be preferentially affected, at times due to ischemic injury from both arterial and venous occlusions, but in most instances, function is impaired due to compression from a new mass. Thalamic injury leads to disconnection from the cortex and the reticular formation. Involvement of the dorsal parts of the mesencephalon, pons, and pontomedullary border interrupts arousal by disconnecting with the thalamus, hypothalamus, and cortex [3, 20]. The hemispheres in comatose patients can only be examined by gross assessment of responses to sound, touch, and noxious stimuli. The thalamus and upper brainstem may produce more localizable signs.

CT scan of the brain often indicates a cause in comatose patients, but a normal scan creates concern that something else is ongoing but not recognized or not yet apparent on CT. In the emergency department, it is either an unrecognized sedative drug, alcohol (including the highly toxic methanol and ethylene glycol), or drug toxicity. In some patients, an embolus to the basilar artery should be considered, requiring an emergent vascular study with a CT angiogram. In the medical and surgical ICU, it could be anoxic–ischemic encephalopathy not yet showing its later cerebral edema.

The neurologic examination of the comatose patients – once their vital signs are secured – is different from a "traditional" (that is, clinical outpatient) neurologic examination. It involves inspection for abnormal muscle movements and twitches, eye position and movement, brainstem reflexes, and tone and limb reaction to a painful stimulus. Tendon, exteroceptive (corneomandibular), or corticospinal (Babinski signs) reflexes follow. The brainstem reflexes are tested (Chap. 3) and are rarely abnormal, but abnormal reflexes have major significance for diagnosis and prognosis. Suggestions that could be helpful and diagnostic when approaching a comatose patient are found in Table 7.4.

When patients are approached first they have there eyes closed and the lids need to be lifted. It is not uncommon for patients to open their eyes briefly and partly, but most of the day they are closed. (the consistent "sleep- wake" cycles, as seen in a persistent vegetative state may come as early as one month after onset. Eyes may open (or more likely twitch) from a loud hand clap but not after asking or yelling. A patient with eyes open but unresponsive could have a locked in syndrome. So, we begin with a careful assessment of pupils, eye position, and eye movements (if any). Pupil size and light reflex should be examined. Pupils can become small or pinhole (pons), midsized (midbrain tectum), asymmetric, or maximally dilated (the oculomotor nuclear complex in mesencephalon or peripheral fibers).

Most pupils in comatose patients are relatively small (2–4 mm). The most commonly known cause of very narrow pupils (1–2 mm) is prior use of opioids. On the street, this is typically heroin; in the hospital, it is patient-controlled analgesia pumps or opioid patches. Midsize, light-fixed pupils are seen in severe midbrain lesions that are often secondary to vertical shift of the brainstem and frequently the first sign of loss of all brainstem reflexes. Dilated (>8 mm) pupils are due to disruption of cranial nerve III secondary to either a mesencephalic injury or lesion of the peripheral portion. Drugs and other toxins, such as amphetamines or lidocaine, can also cause dilated pupils and must be excluded through the patient's history and laboratory testing.

In addition to pupil size and reactivity, any deviation of one or both eyes should also be noted. Spontaneous eye movements, including "ping-pong" eyes (Fig. 7.3) or ocular dipping, generally indicate hemispheric dysfunction. Ocular bobbing, described as a rapid downward deviation of the eyes followed by a slow upward movement, typically implies a pontine lesion. Bihemispheric dysfunction can also lead to "roving eye movements," though it can also be expected with acute metabolic derangements and various neurotoxicities. Rather than supporting particular localization or disease processes, roving eye movements indicate that the brainstem is relatively intact.

Table 7.4 Approaching a comatose patient

• Try to see a patient before the airway is secured (drugs used with intubation will mask degree of consciousness).
• Always consider the possibility that the patient is in a locked-in syndrome; ask them to blink and to look up and down.
• Check the main brainstem reflexes first. An acute brainstem lesion is less common but requires acting urgently.
• Eye movements or forced position is diagnostic albeit nonlocalizing. Vertical nystagmus is brainstem. Roving eye movements or ocular dipping indicate hemispheres. Horizontal gaze is toward a large hemispheric lesion.
• Look for twitches. Eyelids and mouth twitching indicate possible nonconvulsive status epilepticus. Myoclonus is often post-CPR anoxic–ischemic injury.
• Look for muscle tone. Marked rigidity may be drug related (SSRIs). Flaccidity has much less value in determining etiology.
• Tendon reflexes are not clearly discriminatory signs.

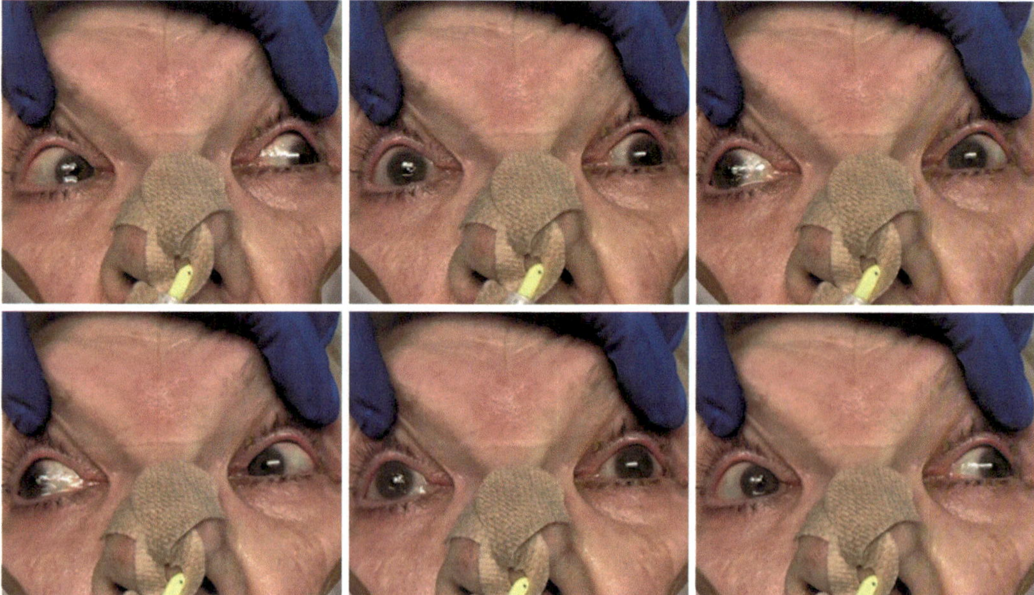

Fig. 7.3 Ping-Pong eye movements (read from upper row left to right to lower row left to right; about 1 second between images)

The oculocephalic reflex deserves special mention as several brainstem structures are assessed simultaneously. Dysfunction of any of these reflexes implies a lesion involving the cranial nerves or the pathways connecting them. In a patient with a normally functioning brainstem, eyes move in a direction opposite the head movement and appear to remain fixated on a point in space (much like a doll's eyes). If a lesion disrupts any of the structures or pathways involved in the reflex, including cranial nerves III, IV, and VI and the medial longitudinal fasciculus, the eyes will move along with the head, remaining in mid-position with respect to the orbits.

Tonic deviation of the eyes, typically in the horizontal plane, may indicate an ipsilateral hemispheric lesion affecting the frontal eye fields or a contralateral pontine lesion (Chap. 3). Horizontal deviation can also be seen in nonconvulsive status epilepticus and may be one of the few signs indicating the need for an emergent electroencephalogram (EEG). However, the oculocephalic maneuver will be able to overcome a gaze preference from a cortical lesion as the pontine and midbrain structures responsible for the reflex remain intact.

An eye position that shows skew (vertical misalignment) implies a brainstem injury. Skew deviation is thought to result from acute damage to areas in the brainstem responsible for vertical control of the oculomotor system. Mostly a structural defect, it can be seen in hepatic coma (and possibly other comas due to severe metabolic derangements).

The assessment of horizontal responsiveness to oculovestibular testing is unfortunately a test rarely done (except in brain death determination Chap. 8), but it is highly informative. Positioning the head at 30° will put the horizontal canals in a vertical position. Cold water drags down the endolymph stimulating the hair cells moving eyes toward the stimulus. Absence of caloric stimulation (and oculocephalic reflexes) often correlates with no recovery. Another advantage of oculovestibular testing is that it can bring out an internuclear ophthalmoplegia (Chap. 3).

A funduscopic examination should also be performed once the external examination of the eyes has been completed. (There have been instances where a funduscopy immediately clinched the diagnosis in ambiguous presentations and before a CT scan was done, often due to active resuscitative efforts for instability.) Retinal and vitreous hemorrhages are indicative of subarachnoid hemorrhage. Funduscopy can reveal subhyaloid hemorrhages (in subarachnoid hemorrhage or asphyxia) or papilledema (increased intracranial pressure or severe hypertensive crisis causing posterior reversible encephalopathy syndrome). The presence of venous pulsations suggests normal intracranial pressure, but their absence is less informative. Ocular ultrasonography of the optic-nerve sheath with a high-frequency probe reportedly estimates intracranial pressure reasonably well. Validation is needed to determine whether this tool is a noninvasive way to detect raised intracranial pressure. High sensitivity rules out increased intracranial pressure in low-risk patients, and high specificity indicates elevated intracranial pressure in those at high risk [21].

The rest of the upper brainstem reflexes can be tested and include assessment of patient mimicry to pain and blinking to threat. However, their presence or absence is not helpful in determining the cause of coma. Inspection of the mouth and oropharyngeal function follows. The mouth can be discolored by corrosives due to a suicide attempt, and a tongue bite may have occurred as a result of a seizure. Other brainstem reflexes involve the lower brainstem and medulla oblongata. Cough reflexes and breathing drive are the key functions to test. The oropharyngeal function can be tested through the gag reflex. However, the pharyngeal reflexes are better tested by using suctioning with catheter in the endotracheal tube. Finally, the breathing drive is noted in intubated patients and may require adjusting of the ventilator or brief disconnection at normal arterial PCO_2.

Motor system testing in an unresponsive patient typically involves applying a noxious stimulus to the supraorbital nerve, the nail bed, or the temporomandibular joint and assessing the patient's reaction. Possible responses include a localizing response, in which the patient reaches toward the stimulus; reflexive responses, such as decorticate or decerebrate posturing; or no response at all. Decorticate posturing involves a slow flexion of the elbow, wrists, and fingers, whereas decerebrate posturing is defined as adduction and internal rotation of the shoulder with arm extension and wrist pronation. Although posturing reflexes have purportedly been useful in lesion localization, these responses can be seen with either focal lesions or global conditions affecting the nervous system. Both responses can be present in the same patient at the same time, and there is no guiding pattern to help with the interpretation of deterioration (Chap. 3).

In addition to stimulation-induced movements, unresponsive patients also make spontaneous movements, the classic example being generalized myoclonus. Although a predictor of an extremely poor outcome following cardiac arrest and due to anoxic ischemic injury, generalized myoclonus status can also be seen with various intoxications, including lithium, cephalosporins, and pesticides.

Failure to Emerge from Coma

Most patients in coma awaken, and that is often greeted as a major improvement. However, increased wakefulness brings its own set of problems as patients demonstrate lingering significant effects of their brain injury. Deeply comatose during the acute phase, some patients remain comatose but transition to a different clinical state in which they regain awake-and-sleep cycles. This clinical syndrome – named persistent vegetative state (PVS) in the early 1970s – described patients with no evidence of a functioning mind [22, 23]. More recently, this state has also been referred to as "unresponsiveness wakefulness syndrome" because of the allegedly negative connotation of the word "vegetative" in PVS, which according to some authors, invites the use of the word "vegetable." [24]. The term is problematic because patients do respond (albeit reflexively), and it does not adress awareness. The term is also close to the older French term *coma vigil* (from vigilance), which was equally nondescriptive. There is no need to discredit Jennett and Plum's original, carefully chosen, and reasoned term "persistent vegetative state." Plum noted that he could have called it "persistent autonomic state" but felt that it was less accommodating. It is a term our hospital colleagues and other professionals understand. The word vegetable is simply an erroneous conjecture brought up by families and should be addressed in a family conference [25]. Their understanding of "being a vegetable" includes any patient requiring full, extensive, and prolonged nursing home care. After prolonged coma, patients have periods of spontaneous eye-opening but do not visually fixate or track objects with their eyes. The key feature is that patients show "no evidence of sustained, reproducible, purposeful, or voluntary behavioral responses" to external stimuli [26, 27]. A patient's eyes may open wide, but consistently demonstrated visual pursuit and fixation are absent. A large mirror held in front of the patient – to track his or her face – is a useful test – and probably the best stimulus – to assess whether visual fixation and pursuit occur. A startle response is often present and may manifest as myoclonus, head flexion, or a decorticate response [28]. Primitive reflexes, including snout, glabella, and palmomental reflexes, may be easily elicited. Random movements of the limbs and trunk, occasional grunts, and even occasional tears or smiles are all consistent with PVS but may provoke uncertainty in family members or inexperienced clinicians. Preserved autonomic and brainstem functions allow patients to maintain adequate circulation and breathe spontaneously without difficulty. The clinical picture fits with what is seen pathologically (i.e., the majority of brains at autopsy showing extensive damage to the subcortical white matter or thalamus, with sparing of the brainstem [29]).

At what point can a vegetative state be considered permanent? With hindsight, we can view autopsy material or MRI findings and find an immense degree of injury [29, 30]. When is there a high degree of clinical certainty that the clinical state is irreversible and the chance of regaining consciousness is exceedingly unlikely? The clinical course of a vegetative state depends in large part on the underlying etiology and the duration of unconsciousness. The most common causes are traumatic brain injury (TBI) and hypoxic–ischemic brain injury. The traditional approach to prognostication has been that patients in posttraumatic PVS are unlikely to regain consciousness after 12 months, while those with anoxic brain injury have even less potential for improvement and very rarely recover consciousness after 3 months. [27] While this is true for the majority of patients, a minority of patients may recover from PVS beyond these cutoffs (from anoxic - schemic brain injury). A small prospective study of patients with PVS found that 7/43 patients (16%) recovered responsiveness and were living at 2 years, 12 (28%) remained vegetative, and 24 (56%) died [31]. All responsive survivors had preserved pupillary light reflexes and present cortical responses with somatosensory-evoked potentials during the acute phase of injury. Notably, those who regain awareness often find themselves severely disabled.

Age also plays a key role, particularly in TBI, with younger patients showing better recovery rates. The American Academy of Neurology, in collaboration with the American Congress of Rehabilitation Medicine and the National Institute on Disability, Independent Living, and Rehabilitation Research, formed a committee to develop a guideline on prolonged disorders of consciousness [32]. The committee advocated more caution in prognostication and warned against clear time-cutoff points beyond which no recovery occurs. The guideline suggests late (after 1 year) improvements primarily occur in younger patients and in approximately 20% of patients initially meeting PVS criteria.

The clinical assessment of a patient already out of the initial critical period and who remains unconscious can be challenging [3]. The examination may need to be repeated at different times of the day because of fluctuations in awareness and circadian oscillations affecting arousal. Some studies suggest a misdiagnosis in a substantial minority of patients in PVS and a reclassification of 10–25% of supposedly vegetative patients using formal scales such as the Full Outline of Unresponsiveness (FOUR) Score or the Coma Recovery Scale-Revised (CRS-R) [33–35]. Diagnostic error rates of 40% have been cited by others, but this frequency may reflect poor reporting standards and insufficient follow-up [36].

The diagnosis of PVS, which requires a prolonged focus of clinical attention, is rarely cemented in acute intensive care settings. It is difficult to be certain. Moreover, whether future brain function studies may find clinical shortcomings remains to be established. What to do with patients who have brain activation on EEG? – about 10% improve, and we cannot tell which patient is the fortunate one [37]. What should we do with patients who show responsiveness on functional MRI? [38–41] Recent MRI findings have suggested different types of responsiveness in what we consider unresponsiveness or coma. The subtypes are more often identified in patients in a minimally conscious state.

Diagnostic neurologic examinations for the purpose of establishing a definitive diagnosis of PVS should perhaps be postponed for at least a month. However, a presumptive diagnosis can be made earlier. Neurologic examination cannot be reliably performed if the patient had recent evidence of bacteremia or early sepsis, situations that significantly confound assessment of consciousness. It also requires exclusion of sedating drugs – a staple of modern intensive care units – the effect of which is commonly underappreciated.

The examination of PVS must be detailed before a presumptive diagnosis is allowed. There are unmistakably characteristic findings in PVS, and they need to be specifically looked for. Most important is the absence of visual pursuit of objects. Patients may open their eyes wide when touched, but visual pursuit – smoothly following an object – is absent, momentary, or irreproducible. Visual fixation is absent, although it can appear later, and mostly at random, without other signs of improvement. A visual orienting reflex may occur with head turning when family members or nursing staff move in the room. Large objects or persons suddenly approaching may cause patients to turn their eyes briefly, suggesting target focusing, but the response extinguishes quickly. However, actions, such as placing the front page of a newspaper or an optokinetic tape before the patient and moving it sideways or tilting a large mirror held in front of the patient, do not consistently elicit visual scanning, optokinetic nystagmus, or tracking. Eye movements are often disconjugate; they typically rove back and forth, interrupted by nystagmoid jerks. Response to sound is complex. It may be present in some rudimentary form, and many patients may show a startle myoclonus. A sudden, loud hand clap may briefly, but only partly, open eyes and move the patient's head toward the stimulus. Characteristically, it occurs only the very first time, and the response is not found with a salvo of hand claps. Consistently looking toward the origin of sound is incompatible with the diagnosis of PVS. Blinking to threat does not occur, because these responses require cortical feedback loops. However, spontaneous blinking or blinking induced by tapping the area between the eyebrows (glabella) does occur because that sign only requires brainstem circuits.

Noxious stimuli can produce some "grimacing," albeit inconsistently, but usually it is entirely absent. Spontaneous grimace may be observed with chewing, yawning, and bruxism and may remain as primitive brainstem reflexes. Grimacing may persist after the stimulus has stopped, but it has an uncertain significance. (Earlier studies have hinted that this sign indicates a better prognosis.) The snout, glabella, palmomental, and corneomandibular reflexes are all easily elicited. The jaw reflex is brisk. In other patients, a tongue depressor may cause forceful biting with the ability to raise the head. Nontracheostomized patients do not talk but may make sounds with different vowels. They may moan, groan, or squeal. Spontaneous, episodic screaming has been reported but is highly unusual. More often, patients may express a muffled cry, often painful to hear for the family. An involuntary swallowing gag reflex may remain. Swallowing movements of saliva do occur, but the coordinated stages of oropharyngeal passage are impaired. An ice chip placed inside the mouth can elicit a few primitive chewing movements. Any food placed in the mouth will likely be inhaled.

Motor response to nail bed or temporomandibular compression is absent or limited to pathologic flexion or extensor responses. Motor responses are muted due to the overriding spasticity and contractures. Thumbs may be buried in balled fists or wedged between the index and middle finger. Nailbed compression induces an increasing pulse rate and tachypnea, with some limb flexion or extension. Reflexes are difficult to elicit because of these contractures, including an equinovarus, but clonus is often present in all extremities. Spontaneous, undirected, choreiform fidgeting of the limbs or extreme opisthotonus (arc-de-cercle) may occur in patients in a persistent vegetative state. These movements are unresponsive to neuroleptic agents, benzodiazepines, or dopaminergic drugs and may persist for weeks. Manifestation of dysautonomia includes increased bronchial secretions, hypertensive surges, tachycardia, and tachypnea.

The "vegetative" manifestations may be pronounced within the first week but then settle down to stable blood pressure, regular breathing, regular pulse, normal sweating, and skin temperature. Sleep–wake cycles are preserved in PVS, possibly due to retained, tonically active mesencephalon synapsing through sympathetic tracts to the pineal gland. Circadian sleep–wake cycles may be absent; the eyes of the patient remain mostly closed with brief episodes of opening, which are more often observed when brainstem lesions are found.

In the 1990s, clinicians involved in the care of brain-injured patients began to recognize that some patients previously diagnosed as PVS showed subtle and partial awareness of their environment. This emerging clinical state was characterized and defined as the minimally conscious state (MCS) by expert consensus of members of a multidisciplinary work group [32, 33, 42]. The presence, albeit partial, of conscious awareness distinguishes MCS from PVS. Reproductible (albeit inconsistent) responses represent another characteristic of MCS. A patient must demonstrate one or more specific reproducible behaviors to meet diagnostic criteria for MCS. These include following simple commands, responding with verbal or gestural yes/no answers, understandable verbalization, and purposeful behavior. In this state, the patient has minimal but clearly noticeable behavioral evidence of awareness of self and environment. However, there is significant inconsistency in responses, which are mostly prolonged and delayed. There may be vocalization or gestures that occur in response to questions. Some patients may reach for an object. Some may touch or hold an object as a purposeful behavior. In general, the signs present in minimally conscious state but absent in persistent vegetative state include eyes holding attention momentarily, looking at a person briefly, turning head in the direction of or establishing eye contact with the person speaking, and mouthing words in response to pain stimuli. The eyes may follow a person's movements and localize to pain. There may be some intelligent verbalization. A recent study [43] found that continuation (habituation) and not extinction of the blink reflex (a series of 10 hand claps every 2 seconds) could be a differentiating feature (albeit with a far from ideal 75% sensitivity and 65% specificity). Correlation of habituation of the blink reflex with MCS was strengthened by PET brain metabolism. The boundaries of chronic disorders of consciousness (DOC) are arbitrarily defined on a broad continuum. When patients previously considered in MCS reliably demonstrate functional interactive communication or can functionally use two different objects, they are generally recategorized as severely disabled. Coma is typically a transient state that evolves to one of the several distinct clinical states within days to weeks. This may be in response to rehabilitation physicians, who often reject the PVS diagnoses of patients referred to them and change them to a minimally conscious state (MCS) or better.

MCS has some awareness and interaction and is, of course, more common than PVS. Some rehabilitation physicians have subclassified MCS into MCS *with* (MCS+) or *without* language (MCS-). Additionally, functional MRI scan-based categorization, including cognitive motor dissociation (CMD), has revealed a patient subset fulfilling all PVS criteria but showing command-following responses on functional MRI [39]. Another patient subset, higher-order cortex-motor dissociation (HMD), exhibits cortical response to auditory stimuli, again without evident awareness (Fig. 7.4).

The clinical course of patients with MCS is still being refined. It is important to differentiate PVS from MCS because the prognosis for recovery appears to be different, with MCS patients recovering at much better rates than those in PVS [33, 44]. A small study showed that the majority of patients in MCS eventually "emerge" and that duration of MCS and age are the best predictors of outcome as measured by the disability rating scale (DRS) [44]. Patients after TBI are more likely to progress to better awareness than patients with nontraumatic disorders. For those with traumatic injuries, recovery can continue for 2 to 5 years [45]. Emergence from MCS is likely first the demonstration of consistency of responses. Only then can we anticipate possible recovery that may include interactions. Next comes understanding these interactions and to remembering them. These not-so-simple functions are needed to recover a state of functionality.

More Reflections

At first glance, consultants may consider the evaluation of the delirious or comatose patient a nearly impossible task, but there can be some "method to the madness." Finding the best additional label of an acute confusional state is not as critical as finding a potentially new structural

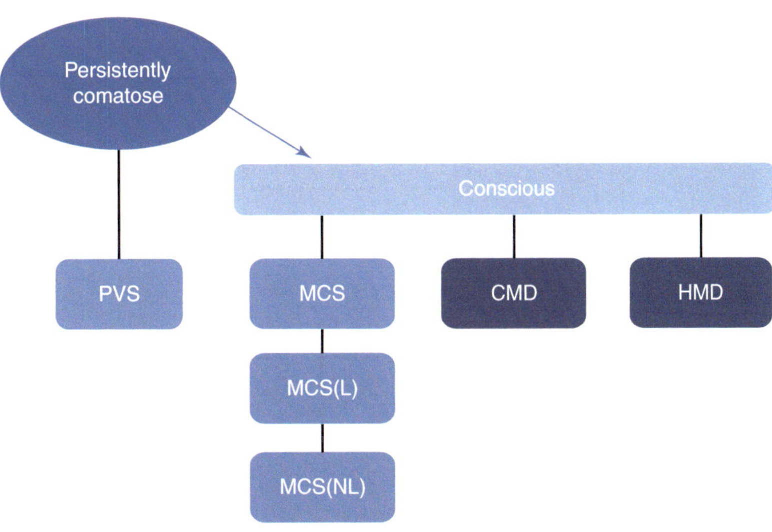

Fig. 7.4 Subsets of prolonged comatose states based on functional MR imaging. These conceptional frameworks cannot be recognized at the bedside and require research settings

lesion, drug effect, or ongoing seizure underlying the agitation. Localizing the lesion responsible for the coma is important; however, localization is the first step in management. The FOUR score assessment and a neurologic examination yield a set of clinical findings, all of which can be helpful, although we need only a few observations to get us to a rapid diagnosis. Let us assume a few things first. We must be absolutely certain that the patient is not in a locked-in state, which will be picked up by the FOUR score. Patients may be "pseudo locked-in" (i.e., no residual vertical eye movements) as a result of neuromuscular blockers and insufficient sedation. I cannot emphasize enough that an unconfounded examination is essential. Once all of this is decided, we should anticipate that most comatose patients have either nonexistent or some responsiveness to a stimulus, so looking for a motor response is the way to start. Next, look for abnormal brainstem findings. As Fig. 7.5 shows, this combination of motor responses with a fixed and dilated pupil or abnormal brainstem reflexes determines subsequent directions. A new anisocoria and a "normal" CT in coma point to an embolus to the basilar artery, and thus a CTA is needed to find it. (In retrospect, we often can see a great white dot in front of the brainstem, also known as an hyperdense basilar artery sign.) The recognition of coma from an acutely occluded basilar artery remains an enormous challenge for many physicians, but following this sequence of thinking, you will be more likely to find it. We recognize an increased probability that the problem is an acute brainstem lesion, a hemispheric lesion, or physiologic global brain dysfunction. Once that is known, a simple differential diagnosis list can be drawn up. All other clinical findings will become supportive or invalidating. This simple method will assist in the initial clinical assessment of a patient presenting in coma. We can then predict what we might see on CT scan. If we already have a CT scan, we will need to correlate the two and resolve discrepancies, which may require repeating the CT scan, performing urgent lumbar puncture, or MRI.

Finally, it should be noted, we are struggling with terminology, and renaming does not solve the fundamental problem. We can only describe what we see, but we only see some of it. Still, what we say matters, and we must be very careful not to exaggerate, diminish, or become unnecessarily vague [46]. The terminology has always been controversial, and physicians can use terms very differently. A recent consensus [47] stated "Acute encephalopathy can lead to a clinical presentation of subsyndromal delirium, delirium, or in case of a severely decreased level of consciousness," a misleading explanation that creates even more ambiguity. In comatose patients, there must be great emphasis on eye findings – static or dynamic,

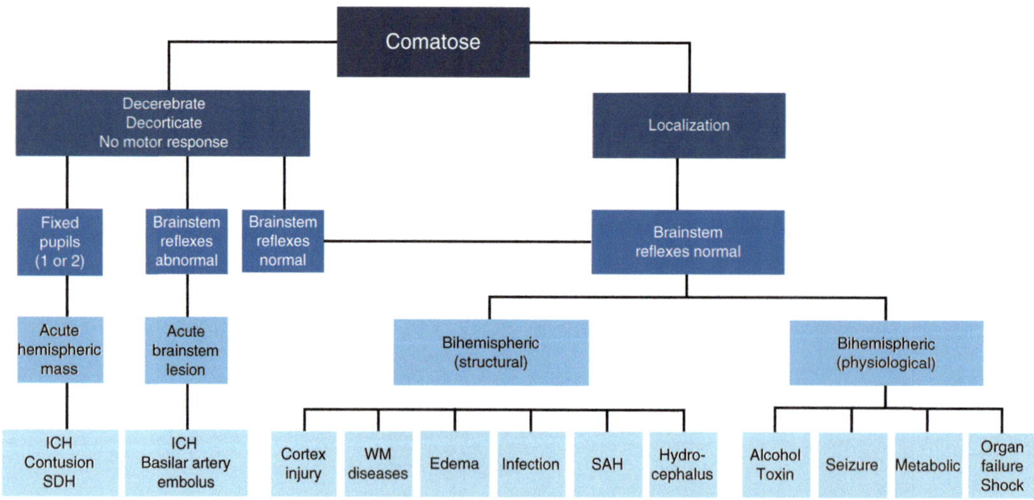

Fig. 7.5 Narrowing down clinical causes of coma. Examination assumes no confounders and exclusion of a locked-in syndrome. ICH intracranial hematoma, SDH subdural hematoma, WM white matter, SAH subarachnoid hemorrhage

how brain shift affects the examination, and the patients' actual state when coma is prolonged.

It should be again emphasized that all of the mentioned neurologic states of abnormal consciousness are determined clinically, a realization that opens up an easy debate on reliability and limitations. Controversies with prolonged states of coma have a long history and have reached religious and philosophic realms. Some categorically state that "the status quo is under [grossly] insufficient treatment, which is equal to therapeutic nihilism" and "substandard treatment can mean that patients are deprived of interventions that might restore their ability to functionally communicate" and basically may violate the Americans with Disabilities Act [48]. Nothing could be further from the truth. Saying treatment should be offered does not make it so. Anything other than meticulous medical care of complications of bed rest and serious intercurrent infections (with often-repeated ICU admissions) will keep patients in this state. Neurologists are very aware of the difficulties these disorders pose and will always try to find reasons to give patients the best care and the best fighting chance. But CT scans often tell another story, showing the ravages over time with emerging liquefactive necrosis [30].

Coming to terms with the uncomfortable reality of their loved one's progressively worsening neurologic state takes a toll on the patient's family as well as the treating neurologist. Families, however, often call a halt to aggressive care (and rightly so) when they anticipate their loved one would be very different from the one they knew. Moreover, family members often ask themselves, "Would you want to live like this?" or declare "He would never ever want this." There are usually no other motivating forces at play, and it is a genuine call to look at a person's quality of life.

Pointers and Takeaways

- The terminology of confused, restless, disoriented patients with autonomic hyperactivity has not helped to categorize these patients; for now, it seems that anything goes.
- Categories of unconsciousness can be recognized clinically, but the definitional borders of each of these designations are less well defined.
- Coma from lesional causes can be traced down to a few locations in the brain.
- Coma from nonlesional causes suppresses brain function more diffusely and thus lacks any localizing findings on examination.
- Patients in a comatose state often improve, and only a very small proportion becomes persistently vegetative. Its rarity is not commonly appreciated, which leads families to envision worst-case scenarios more readily than clinicians.
- Neurologists in their heart of hearts look for signs in the patient to recover to an independent functional state with return to life's simple pleasures.

References

1. Wijdicks EFM. Metabolic encephalopathy: behind the name. Neurocrit Care. 2018;29:385–7.
2. Janz DR, Abel TW, Jackson JC, Gunther ML, Heckers S, Ely EW. Brain autopsy findings in intensive care unit patients previously suffering from delirium: a pilot study. J Crit Care. 2010;25:538.e7–12.
3. Wijdicks EFM. The comatose patient. New York: Oxford University Press; 2011.
4. Kamdar BB, Niessen T, Colantuoni E, et al. Delirium transitions in the medical ICU: exploring the role of sleep quality and other factors. Crit Care Med. 2015;43:135–41.
5. Postoperative delirium in older adults: best practice statement from the American Geriatrics Society. J Am Coll Surg 2015;220:136–148 e131.
6. Liston EH. Delirium in the aged. Psychiatr Clin North Am. 1982;5:49–66.
7. McNicoll L, Pisani MA, Zhang Y, Ely EW, Siegel MD, Inouye SK. Delirium in the intensive care unit: occurrence and clinical course in older patients. J Am Geriatr Soc. 2003;51:591–8.
8. Lipowski ZJ. Delirium: Acute confusional states. New York: Oxford University Press; 1991.
9. Laureno R. Foundations for clinical neurology. New York: Oxford University Press; 2017.
10. Oldham MA, Holloway RG. Delirium disorder: integrating delirium and acute encephalopathy. Neurology. 2020;95:173–8.
11. Dubois MJ, Bergeron N, Dumont M, Dial S, Skrobik Y. Delirium in an intensive care unit: a study of risk factors. Intensive Care Med. 2001;27:1297–304.
12. Ely EW, Margolin R, Francis J, et al. Evaluation of delirium in critically ill patients: validation of the con-

fusion assessment method for the intensive care unit (CAM-ICU). Crit Care Med. 2001;29:1370–9.
13. Pun BT, Ely EW. The importance of diagnosing and managing ICU delirium. Chest. 2007;132:624–36.
14. Pandharipande P, Cotton BA, Shintani A, et al. Motoric subtypes of delirium in mechanically ventilated surgical and trauma intensive care unit patients. Intensive Care Med. 2007;33:1726–31.
15. Pandharipande P, Cotton BA, Shintani A, et al. Prevalence and risk factors for development of delirium in surgical and trauma intensive care unit patients. J Trauma. 2008;65:34–41.
16. Pandharipande PP, Girard TD, Jackson JC, et al. Long-term cognitive impairment after critical illness. N Engl J Med. 2013;369:1306–16.
17. Fisher CM. Disorientation for place. Arch Neurol. 1982;39:33–6.
18. Bergeron N, Dubois MJ, Dumont M, Dial S, Skrobik Y. Intensive care delirium screening checklist: evaluation of a new screening tool. Intensive Care Med. 2001;27:859–64.
19. Neto AS, Nassar AP Jr, Cardoso SO, et al. Delirium screening in critically ill patients: a systematic review and meta-analysis. Crit Care Med. 2012;40:1946–51.
20. McClenathan BM, Thakor NV, Hoesch RE. Pathophysiology of acute coma and disorders of consciousness: considerations for diagnosis and management. Semin Neurol. 2013;33:91–109.
21. Ohle R, McIsaac SM, Woo MY, Perry JJ. Sonography of the optic nerve sheath diameter for detection of raised intracranial pressure compared to computed tomography: a systematic review and meta-analysis. J Ultrasound Med. 2015;34:1285–94.
22. Jennett B. The vegetative state. Medical facts, ethical and legal dilemmas. New York: Cambridge University Press; 2002.
23. Jennett B, Plum F. Persistent vegetative state after brain damage. A syndrome in search of a name. Lancet. 1972;1:734–7.
24. Laureys S, Celesia GG, Cohadon F, et al. Unresponsive wakefulness syndrome: a new name for the vegetative state or apallic syndrome. BMC Med. 2010;8:68.
25. Wijdicks EFM. We believe your father is in a vegetative state. You mean he is a vegetable? He never wanted that. Intensive Care Med. 2021;47(3):363–4.
26. Multi-Society Task Force on PVS. Medical aspects of the persistent vegetative state (1). N Engl J Med. 1994;330:1499–508.
27. Multi-Society Task Force on PVS. Medical aspects of the persistent vegetative state (2). N Engl J Med. 1994;330:1572–9.
28. Wijdicks EF, Cranford RE. Clinical diagnosis of prolonged states of impaired consciousness in adults. Mayo Clin Proc. 2005;80:1037–46.
29. Adams JH, Graham DI, Jennett B. The neuropathology of the vegetative state after an acute brain insult. Brain. 2000;123(Pt 7):1327–38.
30. Wijdicks EFM. Liquefactive necrosis of the brain. Intensive Care Med. 2020;46:1466–7.
31. Estraneo A, Moretta P, Loreto V, et al. Predictors of recovery of responsiveness in prolonged anoxic vegetative state. Neurology. 2013;80:464–70.
32. Giacino JT, Katz DI, Schiff ND, et al. Practice guideline update recommendations summary: disorders of consciousness: report of the Guideline Development, Dissemination, and Implementation Subcommittee of the American Academy of Neurology; the American Congress of Rehabilitation Medicine; and the National Institute on Disability, Independent Living, and Rehabilitation Research. Neurology. 2018;91:450–60.
33. Giacino JT, Kalmar K. Diagnostic and prognostic guidelines for the vegetative and minimally conscious states. Neuropsychol Rehabil. 2005;15:166–74.
34. Schnakers C, Vanhaudenhuyse A, Giacino J, et al. Diagnostic accuracy of the vegetative and minimally conscious state: clinical consensus versus standardized neurobehavioral assessment. BMC Neurol. 2009;9:35.
35. Wijdicks EFM, Bamlet WR, Maramattom BV, Manno EM, McClelland RL. Validation of a new coma scale: The FOUR score. Ann Neurol. 2005;58:585–93.
36. Wade DT. How often is the diagnosis of the permanent vegetative state incorrect? A review of the evidence. Eur J Neurol. 2018;25:619–25.
37. Claassen J, Doyle K, Matory A, et al. Detection of brain activation in unresponsive patients with acute brain injury. N Engl J Med. 2019;380:2497–505.
38. Monti MM, Vanhaudenhuyse A, Coleman MR, et al. Willful modulation of brain activity in disorders of consciousness. N Engl J Med. 2010;362:579–89.
39. Owen AM, Coleman MR, Boly M, Davis MH, Laureys S, Pickard JD. Detecting awareness in the vegetative state. Science. 2006;313:1402.
40. Schiff ND. Uncovering hidden integrative cerebral function in the intensive care unit. Brain. 2017;140:2259–62.
41. Schiff ND, Giacino JT, Kalmar K, et al. Behavioural improvements with thalamic stimulation after severe traumatic brain injury. Nature. 2007;448:600–3.
42. Giacino JT, Ashwal S, Childs N, et al. The minimally conscious state: definition and diagnostic criteria. Neurology. 2002;58:349–53.
43. Hermann B, Salah AB, Perlbarg V, et al. Habituation of auditory startle reflex is a new sign of minimally conscious state. Brain. 2020;143:2154–72.
44. Katz DI, Polyak M, Coughlan D, Nichols M, Roche A. Natural history of recovery from brain injury after prolonged disorders of consciousness: outcome of patients admitted to inpatient rehabilitation with 1–4 year follow-up. Prog Brain Res. 2009;177:73–88.
45. Nakase-Richardson R, Whyte J, Giacino JT, et al. Longitudinal outcome of patients with disordered consciousness in the NIDRR TBI Model Systems Programs. J Neurotrauma. 2012;29:59–65.
46. Wijdicks EF. Being comatose: why definition matters. Lancet Neurol. 2012;11:657–8.
47. Slooter AJC, Otte WM, Devlin JW, et al. Updated nomenclature of delirium and acute encephalopathy: statement of ten societies. Intensive Care Med. 2020;46:1020–2.
48. Fins JJ, Wright MS, Bagenstos SR. Disorders of consciousness and disability law. Mayo Clin Proc. 2020;95:1732–9.

Declaring Brain Death

8

Given the stakes, assessment of brain death must be conclusive; anything less is just inconceivable. We should proceed with the following principles in mind: exclude major confounders, establish the cause of coma, ascertain the futility of any medical or neurosurgical intervention, test the absence of any motor response, accurately test reflexes at all levels of the brainstem, determine the lack of a respiratory drive, and pharmacologically manage a falling blood pressure and increased urinary production. Brain-dead patients need a warming blanket, are on vasopressors because vascular tone has been compromised, and most likely, are on antidiuretic hormone because the pituitary stalk has been damaged by brain tissue shift with most of its function lost.

The examination of a comatose patient who has likely progressed to brain death requires practice. There are only a few neurologists or neurosurgeons who can honestly claim that they do more than 5 or 10 brain death examinations annually (and pediatricians may do even less). Training is obviously warranted, and simulation centers could be potentially useful in recognition of major confounders and how to do an apnea test. However, every intensive care physician should be able to perform a full clinical brain death examination. For now, we must trust that the examiner is fully knowledgeable and competent in all aspects.

This chapter addresses the full examination of a suspected patient suspected of being brain dead and provides suggestions from the American Academy of Neurology guideline for adjusting to known difficult situations [1]. Ultimately, the physician determining brain death must use his own best judgment. Sequential systematic steps in examination are useful to achieve comprehensiveness and to avoid having an important test slip one's mind. But first, we have to confront some important matters head-on.

Maintaining Professional Norms

It is my medical conviction that the clinical determination of brain death is the terrain of neurointensivists. This is hardly a revelation. In fact, in our medical center, we have been doing them nearly exclusively for the last decades [2]. The other specialties (who, of course, *could* do them) have acquiesced without much debate. There are good reasons for the neurointensivist's involvement. First, our principal contribution rests in our ability to distinguish between different comatose states and to recognize the consequences of catastrophic brain injury. Not surprisingly, brain death is often misinterpreted (and misdiagnosed) as a vegetative state or deep coma. It is neither. Brain death is the permanent loss of life because the brainstem – by all accounts, the most vital center of the brain [3] – has irrevocably stopped functioning. Complete loss of brainstem function occurs mostly after

E. F. M. Wijdicks, *Examining Neurocritical Patients*, https://doi.org/10.1007/978-3-030-69452-4_8

hemispheric destruction. It is due to a new mass (hematoma) or diffuse brain swelling above the tentorium that transfers injury to the brainstem situated below the tentorium. The opposite – brainstem injury leading to bi-hemispheric necrosis – occurs too [4]. Major injury throughout the brainstem leads to a very recognizable clinical picture with an areflexive, hypothermic, polyuric, apneic, and comatose patient in distributive shock. Second, and not irrelevant, we have a very good working relationship with organ transplant coordinators, and we understand their needs. We work separately and have separate goals, but we contact each other when needed. This is not universally the case, and attempts to overstep boundaries may occur, potentially triggering oversensitivity, mistrust, or conflicts. It came to my attention that some physicians have, surprisingly, expressed the feeling that coordinators jeopardize the doctor–family relationship. Our familiarity with the process by which transplant organizations work helps to avoid these annoyances.

For many of us, neurologists and internists alike, a body with preserved organs but a nonfunctioning (and later, liquefied) brain realistically ceases to be a living human organism. But not for everyone; just as there are climate-change deniers and anti-vaxxers, there are brain-death deniers; no evidence is ever enough for someone skeptical. Creating such a false narrative has resulted in a number of new expressions of concern by families, including questions about apnea test safety (which, with good preparation, is truly a nonissue), and a demand that families consent to the apnea test before allowing physicians to complete their examination. I find myself balking at the implicit assumption that these patients are not dead but just chronically comatose. In a growing number of opinion papers, brain-death determination has misleadingly been stated as an incomprehensible ambiguity. It seems, after many decades of carefully structured practice and oversight, that some want to re-evaluate everything. Future physicians performing a clinical brain-death examination in the United States may now have to anticipate objections in some cases. It is disheartening.

In the past, when discussions on brain death were plentiful and contentious, a whole-brain biological standard for determining brain death was the only tenable approach [5]. It is an arguable stance. Over decades, we have come to accept that a dead brainstem is sufficient in and of itself, as irreversible can be, so we can equate it to death – even if parts of the brain may still appear to function marginally [6, 7]. This extraordinary discovery allows us to concentrate on brainstem reflexes and the vital centers in the brainstem controlling blood pressure and breathing.

Once brain death is pronounced, a completely different discussion follows. It differs from questioning the appropriateness of continued care for unconscious patients deemed to have no chance of recovery. Brain death means the patient has died, which goes beyond the loss of personhood in prolonged irreversible coma where vital signs are unsupported. Brain death is an irrefutable diagnosis, one of the neurocritical care fundamentals, and a "go-no-go" for life. It is a process of going through a series of one-way doors [8]. Death by these neurologic criteria is accepted in most countries, although differences in procedural technicalities will always remain despite good intentions to achieve consensus [9]. These include requirements for multiple examinations by different specialists, ancillary tests, and extended periods of observation. Contrary to belief, the establishment of a uniform worldwide protocol seems unrealistic [5, 10]. Japan likes to do it differently than India, and Italy differs from France. Chile deviates markedly from Mexico. Only the United Kingdom stands out with its pragmatism by testing only the bare necessities [10–12]. Closer to home, there are some additional (unusual) restrictions and expansions in brain-death protocols throughout US hospital practices, not all of which make sense given what we already know. It may already be unrealistic to hope for a consensus in all US states, to say nothing of the whole world.

Neurologists and neurosurgeons, as well as physicians with intensive-care training, all have the capability to perform a clinical examination of brain death. It requires no special knowledge because the number of brainstem reflexes can be standardized into a simple check list; false-negative results are not common. But there is a plethora of claims, counterclaims, and misinformation with brain-death determination. Furthermore, mistakes are possible and do occur in the actual performance of the examination. There is no reason for haste; a quick assessment of brain death, even if requested, is poor practice and prone to error.

When to Proceed

Proceed when you can – alone and truthfully confident or with colleagues if they are available. In many US institutions, trauma surgeons and pulmonary or anesthesia-trained intensivists ask to involve a neurointensivist. Clinical examination should proceed only if prerequisites are met. This section offers a step-by-step checklist, thus avoiding pitfalls and traps. One simple piece of advice: use most of your time in checking the prerequisites for both the neurologic examination and the apnea test. The technical skills required to perform the exam are not that difficult; however, to determine if the examination truly reflects what has happened to the patient is never easy.

Examiners may be asked to see patients in other intensive care units, and these patients are less familiar to us than our own patients in neurosciences intensive care unit. The checklist in electronic medical records poses potential problems because a click is done quickly, and the list might not allow entry of subtleties and nuances. Let us consider how we can best pronounce a person dead after following a neurologic and systemic physical examination. (The oft-heard term "death by neurologic criteria" is insufficient – the criteria are much wider.)

How long should one wait? The answer is when ready (i.e., preconditions met). To add arbitrary time intervals (e.g., 12, 24, or 36 hours) is simply foundationless. Once brain dead, patients do not change (or become undead), and waiting only complicates a necessary resolution (and may obstruct or deny successful organ donation).

Conditio Sine Qua Non

The cause of coma must be known and demonstrably irreversible. This seemingly obvious statement is unfortunately overlooked in actual practice. The cause of coma can usually be established by the history, examination, neuroimaging, and laboratory tests. A certain period of time should have passed since the onset of injury. No physician should perform a definitive examination hours after a patient has entered the emergency room or after transfer from another facility. In these situations, the history is often fragmentary, unreliable, and frequently changing. Also, the use of sedative or analgesic drugs is often unknown or unverifiable, at least not in the first hours. Toxicology screens are rarely done in the acute situation and are often deferred if a cause is already obvious (e.g., massive destructive cerebral hemorrhage). A CT-scan abnormality compatible with brain death, however, should not obviate a search for confounders. Drug or alcohol ingestion may have resulted in a fatal brain injury (e.g., subdural hematoma).

What is irreversibility, and how can we be so sure? Physicians need to appreciate that the complete clinical picture (and prognosis) is far from clear when comatose patients are seen soon after the ictus. Aggressive treatment should be undertaken, such as administration of osmotic agents, surgical evacuation of a mass-induced displacement of the brainstem (particularly in the cerebellum), ventriculostomy placement, decompressive craniotomy, or other measures to reduce intracranial pressure. If the patient does not demonstrably breathe, has no brainstem reflexes, and other contributing factors exist beyond the destructive lesion, then these findings define irreversibility. No patient in this situation will ever improve. No intervention – medical or surgical – can reverse this clinical picture. Much longer wait times or repeat examinations will not make any person more irreversibly dead. "Miracles" or "surprises" can, inevitably, be traced back to failure to confirm all prerequisites, rushing to a conclusion and, unfortunately, pure incompetence. The empiric evidence of irreversibility is absolutely sufficient because 50 years of studying brain-death determination throughout the world has not revealed instances of recovery including in the troubling cases with prolonged somatic support. Mimics of brain death, if any, are very uncommon. The most common mimickers (e.g., profound, accidental hypothermia, or major drug intoxication) are not identical to the clinical findings seen in brain death, and the CT scan is often just normal. Some cases are clearly different; even in cases of severe Guillain–Barré syndrome, for example, patients' progression in days is known, and patients always retain a breathing drive, albeit insufficient to breathe unassisted.

Neuroimaging Shows Catastrophic Damage

The results of the computed tomography (CT) scan should not surprise or cause any doubt. In most comatose patients, CT scanning shows a new mass with a profound shift, multiple hemispheric lesions with brain edema, or diffuse, massive brain edema alone. Conversely, although it occurs only rarely, CT scan will appear normal shortly after cardiac or respiratory arrest and in patients with fulminant meningitis or encephalitis. However, by the time brain-death examination is considered, a repeat CT scan must show diffuse cerebral edema leading to effacement of the basal cisterns. Moreover, in circumstances of overwhelming infection, examination of cerebrospinal fluid (CSF) should reveal diagnostic findings for a major CNS infection, such as pleocytosis, an elevated erythrocyte count, or a positive Gram stain. Some viruses, parasites, and bacteria can be detected by polymerase chain reaction (PCR), although not in due time. Again, if brain death is the ultimate outcome, the brain will show massive swelling from infection and cerebral infarction.

Interpreting the CT scan of a suspected brain-dead patient requires knowledge of the characteristic CT patterns compatible with brain death. For example, in cases of traumatic brain injury, multiple contusions or a subdural or epidural hematoma should be present, causing massive displacement. Effacement of sulci and white or gray matter may show as undifferentiated gray hemispheres with no CSF spaces (a "featureless" CT scan) when there is diffuse, profound cerebral edema. When there are still major discrepancies between the clinical examination and the CT scan, a repeat CT study is warranted; often, this subsequent study will show expansion of the mass, increased shift, or expected brain edema. If the CT scan remains incongruent with loss of brain function, other factors, particularly drugs, poisons, or a major endocrine, acid-based, or electrolyte abnormality, may be in play. When asked if a CT is necessary after a prolonged anoxic injury, my answer is a resounding yes. It is foolish to skip prerequisites when we know they were established for a good reason. If the CT scan does not explain the coma, we should not proceed – heightened caution is not enough. But, a primary injury to the brainstem alone is as destructive as a supratentorial lesion secondarily damaging the brainstem [13].

Sedation Is Not a Factor

The presence of a sedative drug effect may be excluded by history, drug screen, and with calculation of clearance using five times the drug's half-life (assuming normal hepatic function, renal function, normothermia, and no prior use of targeted temperature management [TTM]). The legal alcohol limit for driving (blood alcohol content 0.08%) is a practical threshold; a reliable examination can be expected below this level. If possible, drug–plasma levels should be measured. Commonly used drugs include short-acting benzodiazepines and opioids, but the half-life can be prolonged in patients treated with TTM. Doubling the five times half-life rule may be too conservative, and it is a wild guess. In a stable, just-rewarmed patient, several additional days of observation should be allowed before proceeding with a full examination. It is equally important to emphasize that patients with anoxic-ischemic encephalopathy after cardiopulmonary resuscitation often (>90%) do not meet the criteria of brain death; therefore, the a priori probability is very low even in the worst-case scenarios where patients are placed on ECMO [14].

A Paralytic Drug Is Not a Factor

Muscle relaxants are used with intubation with a typically short-lived effect. Due to serious potential side effects, their use in surgical and medical ICUs is infrequent. When the drug is withdrawn, elimination varies. Absent neuromuscular-blocking effect can be confirmed by the presence of four twitches with maximal ulnar nerve stimulation. However, more simply, the simple presence of tendon reflexes excludes an important effect of the neuromuscular blockers [15, 16].

A Severe Acid-Base, Electrolyte, or Endocrine Abnormality Is Absent

This criterion is important because some serious abnormalities can affect the clinical assessment in patients with structural injury. Exclusion is more relevant (1) if CT scan does not clearly show severe abnormalities and (2) if the cause if not known. Frankly, we should not have arrived at this stage of exclusion because other "hard stops" have likely already been present. These criteria have been included because unfortunately they are not commonly considered. No severe electrolyte or endocrine crisis should be evident. A major acid-based disturbance may indicate an ingested compound not found on a drug screen. Metabolic acidosis may be seen with acetaminophen, alcohols, salicylates, isoniazid, cyanide, cocaine, strychnine, papaverine, and toluene. Respiratory acidosis is seen with opiates, ethanol, barbiturates, and other anesthetics. Some patients may have acidosis due to uncoupled oxidative phosphorylation (salicylates), seizures (isoniazid, cocaine), or anaerobic glycolysis (cyanide). Metabolic or respiratory alkalosis is seldom a manifestation of poisoning. Clinical examination should not proceed if there is evidence of a severe metabolic acidosis after cardiopulmonary resuscitation.

Secure a Normal Core Temperature

The diagnosis of brain death cannot proceed in a markedly hypothermic patient. Brainstem reflexes are generally resistant to hypothermia unless core temperatures decrease below 27 °C. All brainstem reflexes may be lost in a profoundly hypothermic patient (<20 °C), such as after environmental exposure in patients with a catastrophic brain injury. It is usually quite simple to correct hypothermia with a warming blanket or direct-contact pads in order to raise the core bladder temperature to ≥36 °C. In addition, to avoid delaying an increase in partial arterial pressure of carbon dioxide ($PaCO_2$; a result of metabolism) during the apnea test, a normal or near-normal core temperature is preferred.

Secure a Systolic Blood Pressure of at Least 100 mm Hg

A normal blood pressure (systolic blood pressure [SBP] >100 mm Hg) should be achieved using vasoconstrictive agents such as phenylephrine or vasopressin. Hypotension from loss of peripheral vascular tone or hypovolemia (diabetes insipidus) is common. The neurologic examination is usually reliable with an SBP ≥100 mm Hg.

Document the Absence of Respirations

The ventilator should indicate no patient-initiated respirations. Observation is generally unreliable, and perceived triggering may actually be "ventilator autocycling." [17] This false triggering may be more common than generally appreciated, and we have observed ventilatory autocycling in many brain-death examinations. In a pressure-triggered ventilator mode, a decrease in airway pressure from an inspiratory effort by the patient triggers the ventilator to provide a breath. In a flow-triggered setting, gas flows continuously within the ventilator circuit (flow-by). The patient's effort to obtain a breath is much less, making a flow-triggered mode the preferred setting. However, the flow-triggered mode is very susceptible to noise. The ventilator may sense a change in flow in the circuit and provides a mechanical breath. Auto-triggering of the ventilator may occur from leaks in the circuit or flow fluctuations from condensed water in the circuit. Leaks in the endotracheal tube or ventilator tubing or, simply, the presence of chest tubes are common triggers. These changes usually trigger if the sensitivity setting is low, and decreasing the trigger-sensitivity level will cause these breaths to disappear. Changing to pressure triggering is the simplest solution, but with added pressure support and a low pressure-sensitivity setting, ventilator auto-cycling may still occur. Precardiac movements also provoke ventilator auto-cycling as a result of a decrease in airway pressure synchronous with the cardiac heartbeat. Therefore, changing from a flow-triggered ventilator setting to a pressure setting may not always solve the problem. Not uncommonly, another apnea test with disconnection from the ventilator may be needed to differentiate between two possibilities. There is a real concern that some patients with a "retained breathing drive" are excluded from formal testing, or worse, that prolonged waiting for the respiratory drive to disappear may lead to premature cardiac arrest in a potential organ donor.

The Clinical Determination of Brain Death

Next is the actual examination – all steps need to be checked off while proceeding. Brain-death examination is more cerebral than manual and more elementary than digital (Figs. 8.1 and 8.2). The determination of brain death is based on a comprehensive clinical assessment, which requires 25 assessments. This protocol has not resulted in any cases of recovery even if subtle. Ancillary tests are helpful when they do what they are supposed to do: confirm the clinical diagnosis of brain death.

The Pupils Are Nonreactive to Bright Light

The response to bright light should be absent in both eyes. Round, oval, or irregularly shaped pupils are compatible with brain death. The term "fixed and dilated" is not specific enough; in fact half wrong, virtually all pupils in brain death are in the mid-position (4–6 mm). The initial widening of the pupils in patients with an expanding mass is a dysfunction of parasympathetic fibers running with the third nerve and compressed against the uncus or due to a dysfunction of the nucleus. Larger pupils are compatible with brain death because intact sympathetic cervical-spine pathways connected to the radially arranged fibers of the dilator muscle may remain intact.

Many drugs can influence pupil size, but the light response remains mostly intact. High doses of ketamine or propofol may narrow the pupil size, making it difficult to assess the response accurately. In conventional doses, atropine given intravenously has no marked influence on the pupillary response. Short-term neuromuscular-blocking drugs do not noticeably influence pupil size, but a report on escalating doses of atracurium and vecuronium documented nonreactive light responses and mydriasis [15]. Topical ocular instillation of drugs and trauma to the cornea or bulbus oculi may cause abnormalities in pupil size and can produce nonreactive pupils. Preexisting anatomic abnormalities of the iris or effects of previous cataract surgery should be excluded.

The Corneal Reflexes Are Absent

Absent corneal reflexes should be confirmed by squirting water on the cornea or touching it with a cotton swab. Pressure with a cotton-tipped applicator on the cornea is the most sensitive technique to obtain the reflex. Blinking requires intact brainstem-reflex pathways and is not compatible with brain death. Severe facial and ocular trauma with swelling of the eyelids may limit or eliminate interpretation of corneal reflexes. A recent survey showed that a large proportion of (neuro-) critical care physicians do not know how to do a proper test [18], although surveys may easily fool survey takers.

25 ASSESSMENTS TO DECLARE A PATIENT BRAIN DEAD

PREREQUISITES (ALL MUST BE CHECKED)

1. ☐ Coma, irreversible and cause known
2. ☐ Neuroimaging explains coma
3. ☐ Sedative drug effect absent
   (if indicated, order a toxicology screen)
4. ☐ No residual effect or paralytic drug
   (if indicated, use peripheral nerve stimulator)
5. ☐ Absence of severe acid-base, electrolyte, or endocrine abnormality
6. ☐ Normal or near normal temperature
   (core temperature ≥ 36°C)
7. ☐ Systolic blood pressure ≥ 100mmHg
8. ☐ No spontaneous respirations

EXAMINATION (ALL MUST BE CHECKED)

9. ☐ Pupils non-reactive to bright light
   (typically mid-position at 5-7 mm)
10. ☐ Corneal reflexes absent
    (use both saline jet and tissue touch)
11. ☐ Eyes immobile, oculocephalic reflexes absent
    (tested only if C-spine integrity ensured)
12. ☐ Oculovestibular reflexes absent
    (50 cc of ice water in each ear sequentially)
13. ☐ No facial movement to noxious stimuli at supraorbital nerve or temporomandibular joint compression
    (absent snout and rooting reflexes in neonates)
14. ☐ Gag reflex absent
    (gloved index finger to posterior pharynx)
15. ☐ Cough reflex absent to tracheal suctioning
    (at least 2 passes)
16. ☐ No motor response to noxious stimuli in all 4 limbs
    (triple flexion response is most common spinal-mediated reflex)

APNEA TESTING (ALL MUST BE CHECKED)

17. ☐ Patient is hemodynamically stable
    (systolic blood pressure ≥ 100mmHg)
18. ☐ Ventilator adjusted to normocapnia
    ($PaCO_2$ 35-45mmHg)
19. ☐ Patient pre-oxygenated with 100% oxygen for 10 minutes *(PaO2 ≥ 200mmHg)*
20. ☐ Patient maintains oxygenation with a PEEP of 5cm H_2O *(if not, consider recruitment maneuver)*
21. ☐ Disconnect ventilator
22. ☐ Provide oxygen via an insufflation catheter to the level of the carina at 6 liters/min or attach T-piece with CPAP valve @ 10-20 cm H_2O and resuscitation bag
23. ☐ Spontaneous respirations absent
24. ☐ Arterial blood gas drawn at 8-10 minutes, patient reconnected to ventilator
25. ☐ *$PaCO_2$* ≤ 60mmHg, or 20mmHg rise from normal baseline value or
    Apnea test aborted and confirmatory ancillary test
    (EEG or cerebral blood flow study)

DOCUMENTATION

- Time of death *(use time of final blood gas result or use time of completion of ancillary test)*

DISCLAIMERS

- A guideline from a professional organization is an educational tool not a mandate.
- US state laws may have additional requirements (type of specialties, need to repeat the examination by a separate examiner).
- Major differences exist throughout the world.
- Religious and cultural objections may exist.

Fig. 8.1 Tests needed to declare a person brain dead

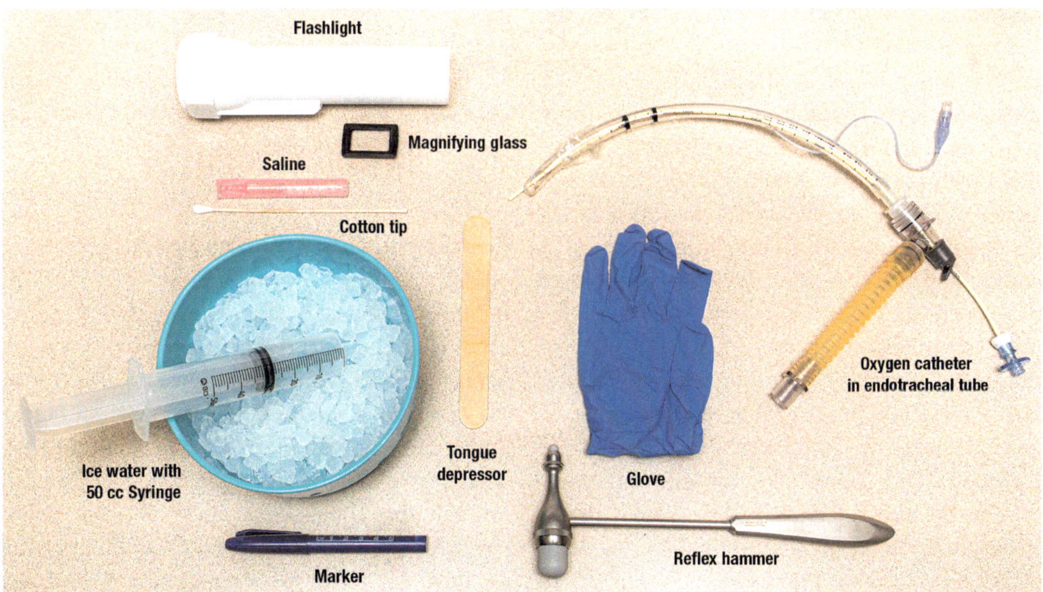

Fig. 8.2 Tools needed to declare a person brain dead

The Oculocephalic Reflexes Are Absent

The eyes are immobile, although they may assume a slightly skewed position. Forced eye deviation also implies stimulation of either the frontal or pontine eye field. Mostly, eyes seem frozen in their sockets; thus, spontaneous movements, including nystagmus beats, do indicate intact brainstem function. The oculocephalic reflex is elicited by grasping the head with two hands while simultaneously keeping the eyelids open with thumb or index finger. The reflex movement (opposite to head movement) is elicited by fast turning of the head from the middle position to 90 degrees on both sides. The oculocephalic reflex cannot be tested in a patient with traumatic brain injury and potential cervical spine injury.

The Oculovestibular Reflexes Are Absent

Ocular movements should also be absent after caloric testing with ice water. Caloric testing should preferably be done with the head elevated to 30 degrees during irrigation of the tympanum on each side. With 30 degrees of elevation, the horizontal canal becomes vertical. Irrigation of the tympanum is best accomplished by inserting a small suction catheter into the external auditory canal and connecting it to a 50-mL syringe filled with ice water. A cold stimulus results in sedimentation of the endolymph and stimulation of the hair cells. The normal response in a comatose patient is a slow deviation of the eyes directed toward the cold caloric stimulus. This response is absent in brain death. Absent eye movement may be very difficult to appreciate, and a reference may be helpful (placement of pen marks on the lower eyelid at the level of the pupil). One should allow up to 1 minute of observation after injection, and the time between stimulation on each side should be at least 5 minutes to reduce a possible overriding effect from the opposite irrigated ear.

Clotted blood or cerumen in the ear may diminish the caloric response, and repeat testing is required. It is prudent to inspect the tympanum directly and to document free access to the cold-water injection. The presence of a ruptured eardrum will enhance the caloric response, but such a maneuver can only be allowed when brain death is highly probable. Theoretically, prior exposure to toxic levels of certain drugs may have diminished or abolished the caloric response. Typical examples are aminoglycosides, tricyclic antidepressants, anticholinergics, antiepileptic drugs, and chemotherapeutic agents, among others, but such confounders are rarely considered in practice. More commonly, as previously mentioned, eyelid edema and chemosis of the conjunctiva may limit movement of the globes. Basal fracture of the petrous bone abolishes the caloric response only unilaterally and may be identified by the presence of an ecchymotic mastoid process (Battle's sign). It should not be an exclusion criterion.

Facial Movement to a Noxious Stimuli Is Absent

Absent grimacing to pain can be documented by applying deep pressure with a blunt object on the nail beds, pressure on the supraorbital nerve, or deep pressure on both condyles at the level of the temporomandibular joint. The jaw reflex should also be absent. Rooting and sucking reflexes should be absent in neonates.

Cough and Gag Reflexes Are Absent

In orally intubated patients, the gag response using a tongue depressor may be difficult to interpret, but a gloved finger inserted deep into the oral cavity should be an adequate stimulus. Lack of a cough response to bronchial suctioning should be demonstrated by passing a catheter through the endotracheal tube at least twice and providing suctioning pressure for several seconds. Although not a required nor validated test, the failure of atropine, 2 mg, to increase the heart rate will corroborate destruction of the central parasympathetic pathways.

Motor Response to Noxious Stimuli Is Absent in Both Arms

The depth of coma can be assessed by examining eye and motor responses with the use of standard painful stimuli, such as pressure on the supraorbital nerve, nail bed pressure, and temporomandibular joint compression. Other painful stimuli, such as sternal rubbing, rubbing the knuckles against the ribs in the axilla, twisting the skin of the arms or shoulder may be equally effective but have not been accepted as the norm. Motor responses to pain in all four limbs should be absent on repeated tests.

Noxious stimuli should produce no motor response other than spinally mediated reflexes. Motor responses may be absent due to a severed cervical cord, which may be suggested on a plain X-ray [19]. The clinical differentiation of spinal responses from retained motor responses associated with brain activity requires expertise. Motor responses may occur spontaneously, after painful stimulation, and during apnea testing, particularly when hypoxemia or hypotension intervenes. These spinal responses include brief, slow movements in the upper limbs, flexion in the fingers, fine finger tremors, or arm lifting. They do not become integrated into truly coordinated decerebrate or decorticate responses [20–23] and diminish with repeated stimulation. Ocular "micro tremors" and eyelid opening have been incidentally noted. Slow head-turning to one side has been observed but, again, is extremely rare.
Vermicular twitching resembling fasciculations has been noted on pectoralis, arm, and abdominal muscles and is likely generated from ischemic anterior horns cells in the spinal cord. Plantar reflexes are usually absent, but some toe flexion may occur, often in combination with a triple-flexion response. An upward toe is seldom found, but a Babinski reflex, assuming a spinal cord origin, is very compatible with brain death.

Prepare for the Apnea Test

Breath is the ultimate metaphor for life, and the absence of breathing under these circumstances confirms cessation of brain function. Apneic oxygenation-diffusion is the most commonly used technique to demonstrate the lack of ventilatory drive. This procedure involves placing a source of 100% oxygen in the trachea, which, through convection, results in a flow of oxygen into the lungs. It follows preoxygenation, which eliminates the nitrogen stores in the respiratory tract and facilitates oxygen transport. On the basis of animal experiments and clinical observations, a target $PaCO_2$ of 60 mm Hg has been proposed as the level at which the medullary respiratory centers are maximally stimulated. This has been based on the assumption that malfunction from brainstem injury causes the respiratory centers to reset to higher levels. The target is actually much lower; the few reported patients who started to breathe after disconnection of the ventilator had $PaCO_2$ levels in the 30–40-mm Hg range.

Breathing can be easily detected by chest expansion, clavicle elevation, and abdominal excursions, but it may also manifest as a single brief gasp that may or may not return during this short testing period.
The increase in $PaCO_2$ is biphasic, with a steep increase in the first minutes due to equilibration of arterial carbon dioxide with mixed central venous carbon dioxide. In the tracheobronchial phase of the apnea test, oxygen flow ensures oxygen uptake in pulmonary capillaries, but carbon-dioxide exhalation does not take place, and therefore, $PaCO_2$ rapidly (3-6 mm Hg per minute) rises due to metabolic production of carbon dioxide [16, 24–26]. Increased $PaCO_2$ causes a decrease in CSF pH, sensed by the respiratory centers in the medulla oblongata, which, when they function, results in a respiratory drive. $PaCO_2$ of 60 mm Hg or 20 mm Hg above normal baseline value maximally stimulates these centers because CSF is unable to buffer acidosis rapidly with blood bicarbonate owing to its slower diffusion than carbon dioxide. However, there are also $PaCO_2$ receptors in the brainstem, and stimulus is not solely through acidosis.

How to Do a Safe Apnea Test

Neurointensivists and neurosurgeons prefer to disconnect the ventilator and allow oxygen to flow through an endotracheally placed catheter [27, 28]. In addition to monitoring oxygen saturation, pulse, and blood pressure, the examiner should observe thoracic and abdominal movement. This is the only way to ensure apnea, particularly when there is a possibility of artifactual reading of a breathing effort on the ventilator display. Complications are minor with good preparation [29–31] and major with bad preparation [32–35] or unusual methods [36, 37].

Switching the ventilator to a CPAP mode and monitoring for apnea is a less-than-optimal approach [38]. Most modern ICU ventilators have a 60-second maximum apnea time before the ventilator goes into a backup-volume or pressure-controlled mode of ventilation, which cannot be turned off. Moreover, interpretation of flow and pressure-wave forms on the ventilator display remains difficult, and distinguishing false CPAP breathing from true breathing requires a particular expertise.

Another group has suggested a possibly attractive method of carbon-dioxide augmentation to reduce observation time, but overshooting to potentially dangerous hypercarbia, acidosis and cardiac arrhythmias is a concern. Monitoring carbon-dioxide levels with a transcutaneous device may reduce carbon-dioxide target overshoot, but discrepancies between transcutaneous and arterial carbon dioxide may be substantial.

Alternatively, manipulating $PaCO_2$ upward by hypoventilation with end-tidal, carbon-dioxide-monitoring devices has been suggested. It can be cumbersome, not only because of the inability to predict $PaCO_2$ but also because the method may lead to gradual CSF buffering and, thus, failure to produce acute acidosis in the CSF compartment. The apnea test may be very difficult to perform (e.g., because of failure to assure adequate oxygenation) in patients with marginal lung function due to contusion or pulmonary edema.

The apnea test is generally safe with careful precautions. First, hypothermia should be corrected from 32 °C (prerequisite for determination of brain death) to normothermia (36–37 °C). In hypothermia, carbon-dioxide production may be delayed because of decreased metabolism. Moreover, in hypothermia, the oxyhemoglobin-dissociation curve shifts to the left, resulting in decreased oxygen release. Second, persistent hypotension should be corrected; this often requires a bolus of 5% albumin or increased administration of intravenous dopamine. It is prudent (but seldom needed) to have 200 μg phenylephrine available when blood pressure declines. A systolic blood pressure of 90–100 mm Hg is acceptable before an apnea test. Before disconnecting the ventilator, reduce the positive end-expiratory pressure (PEEP) requirement to 5 cm water. If this does not lead to oxygen desaturation, maintenance of oxygenation during oxygen insufflation can be expected. If one adheres to these strict guidelines, the apnea test is generally safe.

At the start of the apnea test, a $PaCO_2$ within the normal range (35–45 mm Hg) is preferred. One can expect the $PaCO_2$ to increase 3–6 mm Hg per minute. Concerns about the safety of the apnea test have remained. Premature discontinuation of the apnea test may be related to the method used, susceptibility of the patient, and (most often) pretest cardiovascular and respiratory status. All clinically used apnea tests induce hypercapnia, resulting in concomitant respiratory acidosis, and this acidosis causes problems later in the test.

The five identified factors that predict a possibly problematic apnea test are (1) insufficient preoxygenation, (2) high A-a gradient (>300), (3) T-piece oxygen administration (as opposed to catheter at the level of the carina), (4) hypotension (systolic blood pressure < 90 mm Hg), and (5) baseline acidosis (arterial pH < 7.30). Pretest hypoxemia was a major risk factor in two studies, and in one study, the risk remained despite administration of 100% oxygen. However, preoxygenation – providing an oxygen reservoir and removing alveolar nitrogen – resulted in safe completion of most of our apnea tests. Another study showed that pretest acidemia may predispose to hypotension, but the prevalence of premature discontinuation of the apnea test remains low (<5%). How acidosis causes worsening hypotension is not exactly known. (A transesophageal echocardiogram during the apnea test – with an end pH around 7.0 – did not show any evidence of left ventricular dysfunction.) The apnea test remains a very tricky procedure in young patients after polytrauma and with chest tubes in place. These patients, in an unstable hemodynamic condition from polytrauma, may have lost all brain function, but the apnea test is virtually impossible to perform.

Although it occurs rarely, apnea tests may be discontinued as a result of the sudden development of pneumothorax. High oxygen flows (>10 L/min) or large insufflation catheters may trap gas; moreover, insertion of a catheter with a sharp, pointed tip may damage the bronchial wall.

Of major concern is the apnea test performed without adequate preparation or clinical assessment of the potential risks again. Patients should be normovolemic, normotensive, and normocapnic; preoxygenation should have resulted in a PaO_2 of at least 200 mm Hg. Ventilator requirements should be evaluated, and specifically, adequate oxygenation with low PEEP (5 cm H_2O) requirement should be demonstrated. The apnea test is rarely aborted when these preconditions are met.

Apnea test terminations are usually due to failure to maintain adequate oxygenation and blood pressure after disconnection from the ventilator. Usually, the first minutes of the apnea test will make it clear whether the test can be continued. Oxygen desaturation generally occurs rather quickly after disconnection of the ventilator, leading the S_PO_2 reading to descend below 85%. The apnea test is also aborted if systolic blood pressure descends below 90 mm Hg. However, hypotension may be brief and may respond well to a 200-μg bolus of intravenous phenylephrine. When blood pressures remain marginal, 100 μg from phenylephrine sticks can be administered every 2 minutes (higher frequency than normal due to acidosis, which reduces vasoresponsiveness). Caution remains necessary because systolic blood pressure may decrease quickly and irreversibly, resulting in cardiac asystole.

Cardiac arrest following the sudden appearance of hypotension is very uncommon, with no such instance in our Mayo Clinic series. Cardiovascular collapse occurs if the apnea test is not terminated in time and a rapidly developing hypoxemia or hypotension is underappreciated by the physician. Cardiovascular collapse may also occur with pneumothorax, particularly if not recognized and treated [39].

Another trial of apnea testing can be considered but with the added measures of a T-piece, a CPAP valve of 10 cm H_2O, and 100% oxygen at a rate of 12 L/min. Some evidence indicates that this technique may better secure oxygenation [40, 41]. We have been successful in completing initially aborted tests using a recruitment maneuver (i.e., increasing PEEP levels at 5 cm H_2O increments every three breaths to a maximum of 25 cm H_2O) [40].

Point-of-care i-STAT® testing following the recruitment maneuver should demonstrate adequate oxygenation before proceeding. We then disconnect the ventilator, provide oxygen at a flow of 6 L/min via an oxygen-insufflation catheter, and connect a 20-cm H_2O CPAP valve to the end of the endotracheal tube.

An oxygen-insufflation catheter is preferable to a suction catheter connected to an oxygen source. With an oxygen-insufflation catheter, the position can be more exactly determined. (Usually, it is placed 1 cm beyond the tip of the endotracheal tube.) In the first minutes after insertion, oxygenation should be watched closely before proceeding. Higher flow rates may cause the catheter to wiggle and cause possible trauma, even a pneumothorax. Some oxygenation problems remain difficult to overcome. These patients probably cannot be declared brain dead and could, therefore, become candidates for a DCD protocol following the decision to withdraw support.

When the $PaCO_2$ is in the normal range, 8 minutes of disconnection should be sufficient to reach the target level of 60 mm Hg or to produce an increase of 20 mm Hg [42]. Apnea is determined by visual inspection only. If breathing occurs and is constant, the patient should be reconnected to the ventilator to measure the tidal volume. Visual inspection of the rib cage and abdomen typically may show minimal movement synchronous with the heartbeat; less frequently, there is some intercostal retraction in the upper thorax.

In our experience with the oxygen-diffusion technique, the apnea test was safe and was aborted in only 3% of 212 tests [43]. Hypotension is the most common complication of the apnea test, and the patient should be reconnected to the ventilator when systolic blood pressure drops precipitously to 70 mm Hg. Failure to preoxygenate remains an important cause of hypotension during apnea testing. However, the induced acidosis may reduce myocardial contractility, usually when pH reaches 7.2. Moderate hypercapnia associated with respiratory acidosis does not induce ventricular dysfunction as measured by transesophageal echocardiography. Hypotension without hypoxemia during the apnea test often indicates that the $PaCO_2$ has markedly exceeded 60 mm Hg. Cardiac arrhythmias generally occur in patients in whom hypoxemia is not corrected with oxygen supplementation during the apnea test. Patients who have had cardiac arrhythmias during the evolution of the neurologic catastrophe may not have them during the apnea test, and their presence should not preclude apnea testing.

The Apnea Test in Patients on ECMO

This issue is relatively new and quite pertinent because the use of extracorporeal membrane oxygenation (ECMO) is increasing. According to the Extracorporeal Life Support Organization (ELSO) ECLS Registry report with data from 2015 to 2019, the prevalence of brain death combined with overall neurologic complications in patients receiving ECMO ranges from around 1% and 5% in respiratory failure to about 2% and 7% in cardiogenic shock to 6% and 16% after CPR. This is surprisingly low, considering that this treatment is reserved for the sickest patients, to avoid a nearly certain demise. It has been recently noted [44] that a successful apnea test is rarely achieved and ancillary tests are far from helpful (failure in nearly 1 in 5 tested patients).

What is ECMO? The blending of carbon dioxide (CO_2) into the gas phase of an extracorporeal oxygenator has a long-proven history of safe practice. A gas mixture called carbogen is any mixture of CO_2 and oxygen (O_2). Mixtures enable a normal level of CO_2 in early oxygenators.

Blending CO_2 into the extracorporeal oxygenator is still commonly performed today, especially when implementing deep hypothermic circulatory arrest with pH stat. With a pH-stat protocol, blood-gas values are corrected to the patient's core body temperature, and the oxygenator carbon-dioxide flow is manipulated to maintain a normal "corrected" pH and arterial carbon-dioxide tension during hypothermia. Blood management using a pH-stat technique markedly increases arterial carbon-dioxide tension, which results in cerebral vasodilatation, increased cerebral blood flow (CBF), and changes in regional CBF distribution.

It is easily demonstrated that circulating a crystalloid prime through an extracorporeal circuit with an oxygenator and a sweep gas of 6% CO_2 volume produces a CO_2 around of ~45–50 mm Hg. Likewise, an 8% volume of CO_2 produces a pCO_2 of 65–70 mm Hg, and a 10% volume of CO_2 (as used in pH-stat management) produces a CO_2 of 85–90. If CO_2 is not available for administration into an extracorporeal circuit, the CO_2 can only be increased by diminishing the sweep gas to very low levels. In a case of profound desaturation or critically low hemoglobin, this technique harbors a risk of hypoxia. Adding CO_2 to the oxygenator safely alters CO_2 levels without altering oxygen output of the oxygenator.

The pO_2 of the blood will diminish slightly with the elevation of pCO_2 because of displacement principles. The drop is not significant, however, and the assurance of adequate oxygen levels remains guaranteed. An 8% CO_2 volume will yield similar CO_2 levels independent of the patient blood flow, ECLS circuit volume, or oxygenator size. This technique works without changing any ventilator settings. It also promises swift results and eliminates the "guesswork" of where the sweep gas needs to be, which usually results in multiple blood gases being drawn. By blending CO_2 into the extracorporeal oxygenator, the degree and rate of increase in $PaCO_2$ is very predictable (Fig. 8.3).

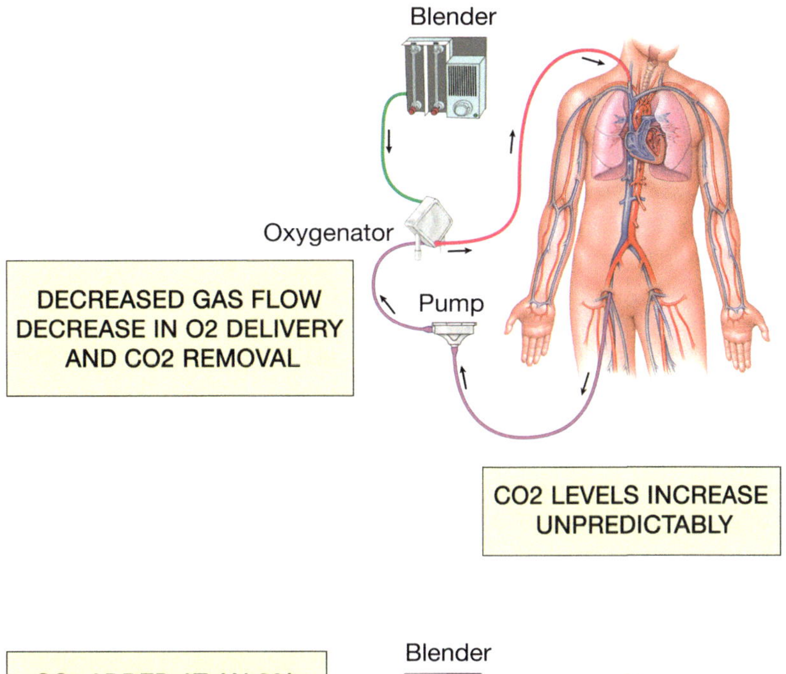

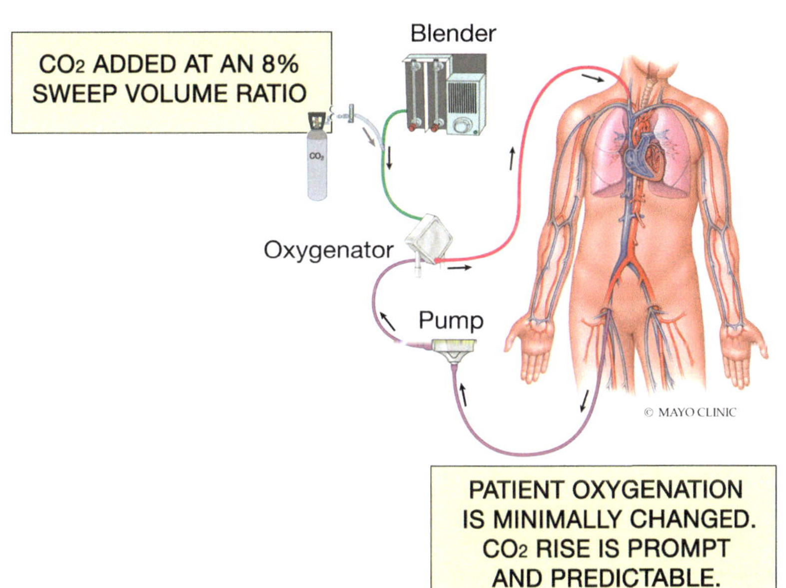

Fig. 8.3 The apnea test in ECMO. Changing sweep vs CO blending technique

Neonates and Children

Brain death determination in neonates is extremely rare with, at most, one or two annual instances in major medical centers. Brain death determination is also very uncommon in younger children, but the incidence in older teenagers (related to motor vehicle accidents) approaches that of adults. Most pediatricians would feel uncertain to declare brain death in a premature baby, and experience is largely absent. Gestational time of 37 weeks is a reasonable cutoff. Otherwise, age (after the first year) should not be a factor in brain death determination, and pediatricians can follow the adult guidelines. Traumatic brain injury and anoxic-ischemic injury from asphyxia are the main causes of brain death in children. The SCCM/AAP/CNS guidelines for determination of brain death in children are similar to guidelines for adults (e.g., neurologic examination beyond 2 months) but also differ in some aspects (e.g., time of observation and number of examinations).

The neurologic examination of neonates and children is, in principle, similar to that of adults but with notable differences. Any pediatrician, pediatric neurologist, or neurosurgeon appreciates the difficulty of visualizing pupil responses (smaller size in newborns), corneal reflexes (less responsive due to dehydration), and caloric testing (narrow ear canal). Generally, corneal reflexes can be obtained, but a small palpebral fissure can pose difficulties in some neonates. Facial trauma or swelling may be substantial, virtually prohibiting accurate assessment of these reflexes. Testing of brainstem reflexes also includes assessment of sucking and rooting reflexes, which are present in normal neonates. The rooting reflex consists of turning the head to bring the mouth toward the finger, alternately placed at both corners of the mouth and at the top and bottom lips. Normal infants typically demonstrate a traction response after grasping the examiner's fingers, but none is present in brain death (and may already be absent in coma). The brain-dead infant should not respond to the Moro reflex (arm extension to a loud noise or a blow to the surface where he/she lies). The tonic-neck reflex, which requires intact labyrinth control, is absent. The apnea test is similar to that in adults, and the $PaCO_2$ targets are comparable. Hypotension and diabetes insipidus are not very different from those in adults.

The criteria for the determination of pediatric brain death diverge from adult guidelines in the greater number of examinations to be performed and the use of confirmatory tests. For years, physicians have been wary of even diagnosing brain death in neonates and children. The concern was simple; it involved uncertainty by the physician about the plasticity of the young child's brain and the difficulty of accepting the irreversibility of a combination of certain signs after a major insult. Similarly, families may be much more reluctant emotionally to accept the death of a child compared to an adult, and some may not understand the clinical situation at all. Involvement of multiple physicians is inadvisable and confusing for parents.

Missing or Unreliable Clinical Tests

Ultimately, the physician who determines brain death will have to rely on his own best judgment. Whether one absent brainstem reflex should prompt an ancillary test is questionable, but loss of medulla oblongata function (hypotension, apnea, gag, and cough reflex) is most essential because (1) it is the last part of the brainstem to go and (2) it defines irreversibility when gone. Focus must be on the functionality of the lower brainstem because there is a vertical loss over time, starting with loss of pupil reflexes, loss of corneal reflexes, oculocephalic reflexes, and, finally, loss of cough, breathing drive, and vascular tone. An absent motor response due to associated traumatic spinal cord severance will also truncate the examination, but attention should be directed to the functioning of the brainstem. It would be very concerning if any single missing test would lead to an ambiguous ancillary test. Clinicians will have to decide in any patient which missing clinical test can be allowed before proceeding with ancillary tests, but they could definitively decide to call it off entirely and to proceed with organ donation using cardiac/circulatory death (DCD) protocol rather than donation after brain death (DBD).

Misinterpretation of Ancillary Tests

Ancillary tests can be divided into those that test the electrical function of the brain and those that test cerebral blood flow [45]. If ancillary tests are used to confirm the clinical examination, the tests should be identical in reliability. Thus, an isoelectric EEG with the appropriate recording settings, no response to any form of stimulation, or a cerebral blood-flow study with absent intracranial blood flow should always coincide with apnea and the absence of all brainstem reflexes, from which no patient can recover. This is, however, not the reality, and persistent blood flow and EEG activity have been found in patients meeting the clinical criteria of brain death. Furthermore, lack of improvement must be proven with a clinical examination – not with laboratory tests (e.g., there are multiple examples in the literature of patients with isoelectric EEGs who improved clinically). One should be concerned about using any surrogate technical test to determine brain death definitively. Patients with high ICPs may have residual breathing drive. Increased ICP does not always correspond to absent intracranial blood flow; often, it has to be massively increased (more than 70 mm Hg). Using a cerebral blood-flow test to diagnose brain death in a patient with a major confounder is bound to cause errors. No one would want to experience reappearance of clinical signs in patients with confounded clinical examination and a false-positive blood-flow test. There are simply too many pitfalls with ancillary tests.

Besides their use as a diagnostic safeguard, the most common indication for ancillary tests for adults with suspected brain death is failure to complete the apnea test. In the United States, ancillary laboratory testing is not mandatory for adults or children. In most other countries, a comprehensive evaluation with a single neurologic examination is adequate. Continued serious consideration of ancillary tests in brain death, rather than acceptance of these tests as relics from the past, is in my opinion unfathomable. Only when an apnea test cannot be performed (which, in my experience, is on the order of 1 in 50 cases), should an ancillary test be undertaken. But, generally, an ancillary test cannot replace an insufficient clinical examination.

More Reflections

Clinical examination of brain death requires training and, eventually, skill. Training programs have been suggested, but there is no evidence to support the validity of any of these. Indeed, it seems self-serving to create online training sites (and charge fees). Simulation of a brain-death determination in a simulation center is inadequate. Although currently used mannequins can demonstrate abnormal vital signs (and they *are* abnormal, with no breathing and hypotension and polyuria), most brainstem reflexes have not been built into the devices, and thus, examination of these reflexes cannot be performed. Pupils have a fixed size and are not mid position as they should be, and that is where it stops. It also feels quite awkward to do pain stimuli in a mannequin and declare it absent. Most importantly, clinical scenarios of patients who do not yet fulfill the criteria of brain death – those with some reflexes spared – cannot be simulated (with the notable exception of a mannequin with only a preserved breathing drive). But, we can simulate the systemic manifestations of brain death (i.e., diabetes insipidus, hypothermia, and hypotension). We and others have found it feasible to develop a skill set required to determine brain death through simulation but only to show learners what comprises a thorough diagnostic evaluation, how to recognize common pitfalls, or how to communicate brain death to family members [46–49]. Through a pitfall-laden experience, these learners may have a new appreciation for the orderliness and complexity of evaluating a patient. Learners in simulation centers can demonstrate understanding of the physiology of the oxygen-diffusion method. Simulation scenarios also can include discussions with family members and the logistics of involving an organ-donation agency after the declaration of brain death.

We decided that the following missteps constitute a failed simulation test result and requires remediation: (1) failure to inquire about alcohol or drugs, (2) failure to recognize and correct hypothermia or hypotension, (3) performance of exam before drug clearance, (4) failure to properly prepare for the apnea test, or (5) improperly performing the apnea test. But despite good efforts, we have no established and accepted way to prove procedural competence (a major understudied area in all of critical care medicine). When we look at who fails and why, it is almost always a failure to inquire about drug or alcohol effects.

In a few patients, the clinical examination cannot be completed, and to pronounce brain death becomes potentially problematic (Table 8.1). Again, central to the determination is the absence of brainstem reflexes and apnea after CO_2 challenge and evidence of pharmacologically supported blood pressure. Equally central to the examination is knowing that loss of brainstem function is in a downward direction – mesencephalon (pupils) first and medulla oblongata (apnea and loss of vascular tone) last. Brainstem loss follows axiomatically after catastrophic hemisphere injury. Thus, inability to examine reliably the function of the dorsal medulla oblongata is far more important than inability to test the mesencephalon and should lead to an ancillary test or, more logically, a decision not to declare brain death. Virtual all apnea tests show apnea. Notably, breathing during the apnea test is so rare that it would need three decades of frequent brain-death determinations to get a handful of patients. One could argue that once the patient is disconnected at a normal or elevated PCO_2 and does not breathe, he or she would be unlikely to do so later. Subjecting the patient to possible hypotension, cardiac arrhythmia (from acidosis), and hypoxemia (from abnormal diffusion due to lung edema) is an unnecessary risk, albeit a small one with adequate precautionary measures. However, several minutes of apnea are far more diagnostic than a quick disconnection or switch to a spontaneous mode. A CO_2 and acidosis stimulus to the CO_2 sensors and H+ respiratory sensors in the medulla oblongata and pons makes good physiologic sense.

Table 8.1 Causes of concern with clinical examination

Inability to check pupils due to trauma
Inability to check corneal reflexes due to massive facial swelling
Inability to check oculocephalic responses due to cervical collar placement
Inability to check oculovestibular responses due to blood clots
Inability to maintain stable blood pressures
Inability to obtain adequate starting arterial blood gas after preoxygenation
Inability to judge motor response due to a severed spinal cord

At the very least, we trust the examiner knows what he or she is doing and defers a declaration of brain death if uncertainties remain. An examiner who has done only a few should not feel overconfident and should ask for confirmation. An examiner without neurologic or neurosurgical expertise has some limitations and would need to have performed a significant (in the teens) number of tests under guidance. These caveats will reduce errors. The brain-death examination should be a multidisciplinary examination with many involved and many observers (Fig. 8.4). Among those present, we should expect a respiratory therapist, a laboratory technician to draw a blood gas at 8- and 10-minute intervals, and a point-of-care blood-gas analyzer. Neurocritical care or critical care fellows may direct the examination under direct supervision of the consultant. Nursing staff and any other trainees can be present, and we will explain what we are doing, why we do a certain test, and what the result (i.e., the absence of a reflex) means. We will explain spinal motor response (mostly a triple-flexion response). We will explain all steps of the apnea test and what we can expect in the first minutes of disconnection (i.e., dropping blood pressure and oxygen saturation). At their request, families can be present but only if they have been told what we will do (e.g., applying noxious stimuli, ice-water insertion, and removing the patient from the ventilator for some time while providing oxygen). In my experience, most families decide to leave the room, although occasionally, some family members felt it helpful to observe a complete brain-death examination. It had little impact on decisions to proceed with organ donation nor did it in any way convince families their loved one was gone when they were unsure. However, one single comprehensive examination should suffice, and frankly two examinations have a

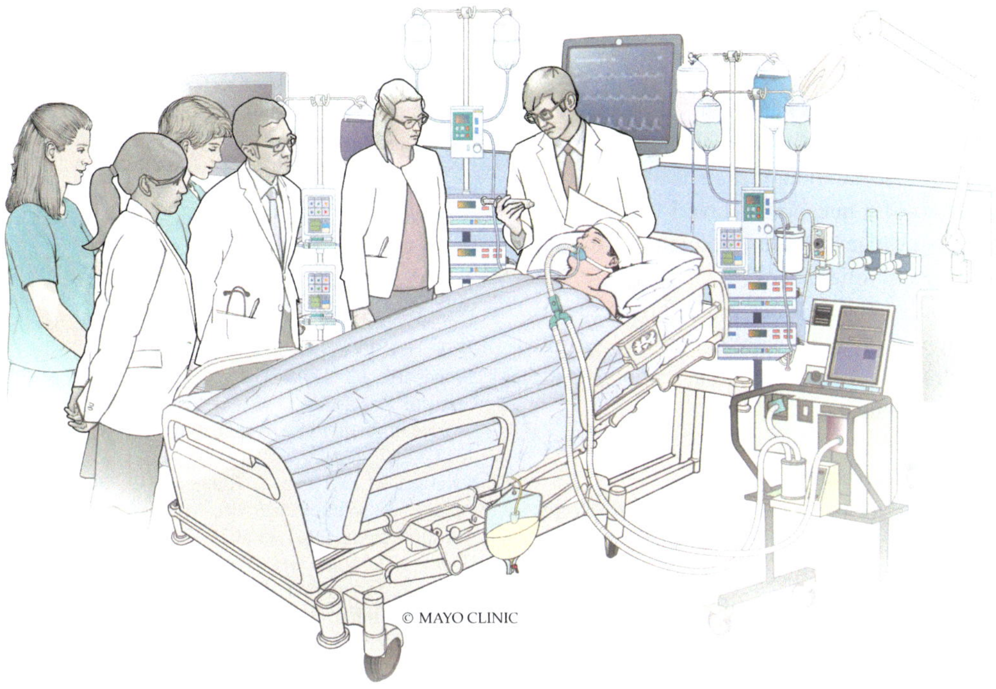

Fig. 8.4 Brain-death testing at the bedside

major negative impact on organ procurement. Delays in performing the second examination reduce the number of transplanted organs as a result of ongoing instability of vital signs [50].

Finally, it is poor practice to resolve clinical uncertainties with an ancillary test, which is actually no solution at all. Every medical institution (in the world frankly) should begin a conversation on its willingness to let go of all the accessories in brain-death determination in favor of the organizational principle of determining futility or irreversibility by carefully examining brainstem function. Ancillary ("confirmatory") tests remain mandated in a minority of countries as a safeguard or when unable to complete the apnea test. These ancillary tests may be throwbacks from a time when brain death as a new neurologic state was not so clear to everyone. To proceed with a technical test was, for many practitioners, the only definitive proof that the brain was "really" dead. Others felt no need to cast particular doubt on the clinical examination. Moreover, studies of ancillary tests have always lacked appropriate controls; comparisons between tests show major discrepancies and technical problems (or even unavailability).

In several countries, repeated comprehensive evaluations are required. A time interval is mandated with great variations from same day to several days. No literature-based evidence supports a second examination contradicting the first. There is no reliable evidence that longer wait times may detect recovery. Rather, longer waiting times may result in loss of donors (from cardiac arrest) or a decrease of suitable organs (from prolonged ischemia or other insults). However, when surveyed, a third of physicians who routinely perform brain-death examinations in practice still order ancillary testing as a part of their standard evaluation [51]. There are multiple reasons for ordering ancillary testing. Some are required by their local hospital policy, while others use them to feel more "comfortable" with the diagnosis or to avoid liability. Ancillary tests are seldom requested by family members.

Pointers and Takeaways

- Considerable experience is needed to proceed with declaration of brain death.
- Only a comprehensive neurologic examination allows declaration of brain death.
- Ask yourself 3 questions: Have I tried everything to change the clinical picture? Can I proceed? Can I be fooled?
- Examination concentrates on testing of seven brainstem reflexes with circuitry going through the mesencephalon, pons, and medulla oblongata.
- Apnea tests are very safe if patients are selected and prepared carefully.
- Apnea tests are quite easy in ECMO (despite initial concerns).
- Ancillary tests can never replace a clinical examination.
- Errors relate to disregarding contributing confounders.

References

1. Wijdicks EFM, Varelas PN, Gronseth GS, Greer DM, American Academy of N. Evidence-based guideline update: determining brain death in adults: report of the Quality Standards Subcommittee of the American Academy of Neurology. Neurology. 2010;74:1911–8.
2. Daneshmand A, Rabinstein AA, Wijdicks EFM. The apnea test in brain death determination using oxygen diffusion method remains safe. Neurology. 2019;92:386–7.
3. Wijdicks EFM. Historical awareness of the brainstem: from a subsidiary structure to a vital center. Neurology. 2020;95:484–8.
4. Marcellino C, Braksick SA, Wijdicks EFM. How does the brain die after a massive posterior fossa lesion? Neurocrit Care. 2021;34:686–90.
5. Bernat JL. Comment: is international consensus on brain death achievable? Neurology. 2015;84:1878.
6. Wijdicks EFM. The transatlantic divide over brain death determination and the debate. Brain. 2012;135:1321–31.
7. Wijdicks EFM, Varelas PN, Gronseth GS, Greer DM. There is no reversible brain death. Crit Care Med. 2011;39:2204–5.
8. Wijdicks EFM. Deliberating death in the summer of 1968. N Engl J Med. 2018;379:412–5.
9. Greer DM, Shemie SD, Lewis A, et al. Determination of brain death/death by neurologic criteria: the world brain death project. JAMA. 2020;324:1078–97.
10. Wahlster S, Wijdicks EFM, Patel PV, et al. Brain death declaration: practices and perceptions worldwide. Neurology. 2015;84:1870–9.
11. Lewis A, Bakkar A, Kreiger-Benson E, et al. Determination of death by neurologic criteria around the world. Neurology. 2020;95:e299–309.
12. Lewis A, Kreiger-Benson E, Kumpfbeck A, et al. Determination of death by neurologic criteria in Latin American and Caribbean countries. Clin Neurol Neurosurg. 2020;197:105953.
13. Manara A, Varelas P, Wijdicks EF. Brain death in patients with "isolated" brainstem lesions: a case against controversy. J Neurosurg Anesthesiol. 2019;31:171–3.
14. Bronchard R, Durand L, Legeai C, Cohen J, Guerrini P, Bastien O. Brain-dead donors on extracorporeal membrane oxygenation. Crit Care Med. 2017;45:1734–41.
15. Kainuma M, Miyake T, Kanno T. Extremely prolonged vecuronium clearance in a brain death case. Anesthesiology. 2001;95:1023–4.
16. Hanks EC, Ngai SH, Fink BR. The respiratory threshold for carbon dioxide in anesthetized man. Determination of carbon dioxide threshold during halothane anesthesia. Anesthesiology. 1961;22:393–7.
17. Wijdicks EFM, Manno EM, Holets SR. Ventilator self-cycling may falsely suggest patient effort during brain death determination. Neurology. 2005;65:774.
18. Maciel CB, Youn TS, Barden MM, et al. Corneal reflex testing in the evaluation of a comatose patient: an ode to precise semiology and examination skills. Neurocrit Care. 2020;33:399–404.
19. Waters CE, French G, Burt M. Difficulty in brainstem death testing in the presence of high spinal cord injury. Br J Anaesth. 2004;92:760–4.
20. Conci F, Procaccio F, Arosio M, Boselli L. Viscero-somatic and viscero-visceral reflexes in brain death. J Neurol Neurosurg Psychiatry. 1986;49:695–8.
21. Marti-Fabregas J, Lopez-Navidad A, Caballero F, Otermin P. Decerebrate-like posturing with mechanical ventilation in brain death. Neurology. 2000;54:224–7.
22. Ropper AH. Unusual spontaneous movements in brain-dead patients. Neurology. 1984;34:1089–92.
23. Saposnik G, Bueri JA, Maurino J, Saizar R, Garretto NS. Spontaneous and reflex movements in brain death. Neurology. 2000;54:221–3.
24. Dominguez-Roldan JM, Barrera-Chacon JM, Murillo-Cabezas F, Santamaria-Mifsut JL, Rivera-Fernandez V. Clinical factors influencing the increment of blood carbon dioxide during the apnea test for the diagnosis of brain death. Transplant Proc. 1999;31:2599–600.
25. Engel GL, Ferris EB, Webb JP, Stevens CD. Voluntary breathholding. II. The relation of the maximum time of breathholding to the oxygen tension of the inspired air. J Clin Invest. 1946;25:729–33.
26. Ferris EB, Engel GL, Stevens CD, Webb J. Voluntary breathholding. III. The relation of the maximum time of breathholding to the oxygen and carbon dioxide tensions of arterial blood, with a note on its

clinical and physiological significance. J Clin Invest. 1946;25:734–43.
27. Rohling R, Wagner W, Muhlberg J, Link J, Scholle J, Rosenow D. Apnea test: pitfalls and correct handling. Transplant Proc. 1986;3:388–90.
28. Ropper AH, Kennedy SK, Russell L. Apnea testing in the diagnosis of brain death. Clinical and physiological observations. J Neurosurg. 1981;55:942–6.
29. Goudreau JL, Wijdicks EF, Emery SF. Complications during apnea testing in the determination of brain death: predisposing factors. Neurology. 2000;55:1045–8.
30. Pitts LH, Kaktis J, Caronna J, Jennett S, Hoff JT. Brain death, apneic diffusion oxygenation, and organ transplantation. J Trauma. 1978;18:180–3.
31. Yee AH, Mandrekar J, Rabinstein AA, Wijdicks EFM. Predictors of apnea test failure during brain death determination. Neurocrit Care. 2010;12:352–5.
32. Ebata T, Watanabe Y, Amaha K, Hosaka Y, Takagi S. Haemodynamic changes during the apnoea test for diagnosis of brain death. Can J Anaesth. 1991;38:436–40.
33. Jeret JS, Benjamin JL. Risk of hypotension during apnea testing. Arch Neurol. 1994;51:595–9.
34. Saposnik G, Rizzo G, Vega A, Sabbatiello R, Deluca JL. Problems associated with the apnea test in the diagnosis of brain death. Neurol India. 2004;52:342–5.
35. Wu XL, Fang Q, Li L, Qiu YQ, Luo BY. Complications associated with the apnea test in the determination of the brain death. Chin Med J (Engl). 2008;121:1169–72.
36. Lang CJ, Heckmann JG. Apnea testing for the diagnosis of brain death. Acta Neurol Scand. 2005;112:358–69.
37. Perel A, Berger M, Cotev S. The use of continuous flow of oxygen and PEEP during apnea in the diagnosis of brain death. Intensive Care Med. 1983;9:25–7.
38. Busl KM, Lewis A, Varelas PN. Apnea testing for the determination of brain death: a systematic scoping review. Neurocrit Care. 2020; https://doi.org/10.1007/s12028-020-01015-0.
39. Gorton LE, Dhar R, Woodworth L, et al. Pneumothorax as a complication of apnea testing for brain death. Neurocrit Care. 2016;25:282–7.
40. Hocker S, Whalen F, Wijdicks EFM. Apnea testing for brain death in severe acute respiratory distress syndrome: a possible solution. Neurocrit Care. 2014;20:298–300.
41. Levesque S, Lessard MR, Nicole PC, et al. Efficacy of a T-piece system and a continuous positive airway pressure system for apnea testing in the diagnosis of brain death. Crit Care Med. 2006;34:2213–6.
42. Marks SJ, Zisfein J. Apneic oxygenation in apnea tests for brain death. A controlled trial. Arch Neurol. 1990;47:1066–8.
43. Wijdicks EFM, Rabinstein AA, Manno EM, Atkinson JD. Pronouncing brain death: contemporary practice and safety of the apnea test. Neurology. 2008;71:1240–4.
44. Migdady I, Stephens RS, Price C, Geocadin RG, Whitman G, Cho SM. The use of apnea test and brain death determination in patients on extracorporeal membrane oxygenation: a systematic review. J Thorac Cardiovasc Surg. 2020; S0022-5223(20)30640-1.
45. Kramer AA, Wijdicks EFM, Snavely VL, et al. A multicenter prospective study of interobserver agreement using the Full Outline of Unresponsiveness score coma scale in the intensive care unit. Crit Care Med. 2012;40:2671–6.
46. Douglas P, Goldschmidt C, McCoyd M, Schneck M. Simulation-based training in brain death determination incorporating family discussion. J Grad Med Educ. 2018;10:553–8.
47. Hocker S, Schumacher D, Mandrekar J, Wijdicks EFM. Testing confounders in brain death determination: a new simulation model. Neurocrit Care. 2015;23:401–8.
48. Hocker S, Wijdicks EFM. Simulation training in brain death determination. Semin Neurol. 2015;35:180–7.
49. MacDougall BJ, Robinson JD, Kappus L, Sudikoff SN, Greer DM. Simulation-based training in brain death determination. Neurocrit Care. 2014;21:383–91.
50. Varelas PN, Rehman M, Mehta C, et al. Comparison of 1 vs 2 brain death examinations on time to death pronouncement and organ donation: a 12-year single center experience. Neurology. 2021;96:e1453–61.
51. Braksick SA, Robinson CP, Gronseth GS, Hocker S, Wijdicks EFM, Rabinstein AA. Variability in reported physician practices for brain death determination. Neurology. 2019;92:e888–94.

9 Recognizing Acute Spinal Cord Injury

Clinical recognition of an acute brain injury requires skill; recognizing an acute spinal cord injury requires skill and aptitude. Acute myelopathy (and cauda equina compression) presenting with early signs are clinical situations that are not always recognized as pressing. Some of these presentations can be quite subtle until exacerbations occur, and failure to recognize these conditions in the earliest of stages, as with so many clinically puzzling cases, remains a trigger for litigation. The time to diagnosis for acute spinal cord injury may not always be critical, but failure to consider the entity at all is problematic. Most notorious is the delay in diagnosing a spinal epidural abscess, but in fairness, physicians may not have reasonably suspected the diagnosis in a patient actively treated for septic shock. Nearly two-thirds of these cases present with loss of bowel or bladder control rather than an initial motor or sensory deficit [1]. Moreover, a delay of more than 48 hours in time to surgery increases the risk of an unfavorable verdict for the responsible surgeon. Internists, hospitalists, and emergency physicians have also faced legal challenges evolving from these situations [2]. Why spinal cord injury is often singled out is unknown, but new, potentially preventable immobility is a major disability. Thus, understanding the clinical manifestations of acute spinal cord injury is paramount because it indicates which segments are involved, and this, in turn, leads to the appropriate MRI scan. It is oddly unsettling to find MRI imaging of the spine occurring outside the area of a possible, anticipated lesion. Examination of the spine is not arduous, but neither should it be perfunctory, certainly not with incipient presentations. There may be an attitude of indifference caused by the perception that the examination of spinal cord is nearly impossible to get right. There is no substitute for an early and full neurologic examination, but awareness of sensory or motor deficit by the patient may be compromised by alcohol, illicit drug use, or impaired consciousness when spinal cord injury is part of polytrauma. It is particularly difficult in emergently intubated patients who cannot express a new deficit.

Circumstances suggesting the possibility of an acute spinal cord injury are severe trauma or a major aortic aneurysm repair. Spinal cord injury is most commonly due to trauma and most commonly in young (late 20s to early 30s) men. Spinal cord compression from cancer is another unfortunate common cause of injury. For a good understanding of spinal cord injury, we need to understand the organization of its motor and sensory tracts first and, second, how combinations of deficits can be explained by certain types of injury. Then, we can make good use of a clinical neurologic examination and compare examination results over time in patients with exacerbating signs.

Another major component of assessment of acute spine injury is the expected presence of mechanical respiratory failure. Catastrophic

E. F. M. Wijdicks, *Examining Neurocritical Patients*, https://doi.org/10.1007/978-3-030-69452-4_9

loss of airway control is a major cause of concern in acute spinal cord lesions affecting the mid- and high-cervical regions (C3–C5) and above the level of the phrenic motor neurons causing complete paralysis of the muscles of both inhalation and exhalation and immediate dependence on mechanical ventilation. Lesions below the level of C5 spare the nerve connections to the diaphragm, but expiratory effort is markedly reduced due to additional involvement of the abdominal and intercostal muscles. Moreover, the impact of acute dysautonomia on respiratory function has only been recently recognized. In the acute phase, sympathetic nerves are interrupted, and the vagus nerve predominates, resulting in increased, potentially airway-blocking, tracheobronchial secretions. Furthermore, unopposed parasympathetic input may facilitate airway narrowing. Thus, a patient with acute spinal cord injury requires a complex systematic clinical approach, and many of these patients are potentially very unstable on arrival in the intensive care unit. It is crucial to stage early and respond accordingly.

A Course in Spinal Cord Anatomy

The spinal cord is basically a cylinder with exiting nerve roots organized in segments. When injury occurs, it is often at a single level that corresponds to a certain segment of spinal cord. When pathologists examine the cord, they view white matter (dark areas) and gray matter (light areas), which have a butterfly shape. At its simplest, white matter in the cord has all the tracts for motor function (going down) and sensory function (going up), and gray matter contains the connecting motor units exiting the cord as spinal nerves. These roots have separate dorsal and ventral parts that merge into a spinal nerve. There are 31 paired bilateral ventral and dorsal nerve rootlets—eight cervical (8), twelve thoracic (12), five lumbar (5), and six sacral (6). Sensory nerve roots enter the dorsal region, while motor nerve roots exit from the ventral region of the spinal cord. The spinal cord is surrounded by the thecal sac, anteriorly by the vertebral body and intervertebral discs (nucleus pulposus and annulus fibrosis), posteriorly by the posterior spinal processes, and laterally by the pedicles and lamina. Between the bony thecal sac and outer layer of the cord (the dura) lies the epidural space, which contains fat and venous plexuses. The thoracolumbar region is the widest epidural space and contains the most fat; thus, it is the most common site of epidural abscesses. The subdural space is a potential space between dura and arachnoid. Between the arachnoid and pia is the subarachnoid space, which contains cerebrospinal fluid (CSF). The white-matter tracts are grouped into three different regions called the dorsal, lateral, and ventral funiculi or columns. The major tracts are the posterior columns (responsible for proprioception, vibration, and ability to discriminate touch), the spinothalamic tract (pain and temperature), and the corticospinal tract (motor). The posterior (dorsal) columns ascend and synapse at the nucleus fasciculus gracilis and nucleus fasciculus cuneatus and eventually project fibers to form the medial lemniscal system in the brainstem. The spinothalamic tract is located in the lateral column (Fig. 9.1). The spinothalamic tract receives input from the dorsal root ganglion cells, continues in the spinal cord, and crosses after ascending one or more levels. There is greater organization within the spinothalamic bundle, and sensory fibers from the arms are more centrally located than the fibers from the leg. Pain and temperature fibers are more dorsal than touch fibers.

Cord location explains processes. Anterior lesions cause motor loss; dorsal and lateral lesions cause sensory loss. Top down are the motor tracts. The corticospinal tract from several cortical regions descends and crosses at the level of the medulla oblongata and continues in the lateral spinal cord, but a minority descends without decussating to form the ventral corticospinal tract. This higher level of crossing means that a lesion in the spinal corticospinal tract can cause weakness on the same side and a pyramidal distribution in arms (muscle weakness in the deltoid and triceps, wrist finger extensors) and legs (hip flexion, knee flexion, and foot dorsiflexion). The stacked structure of the spinal

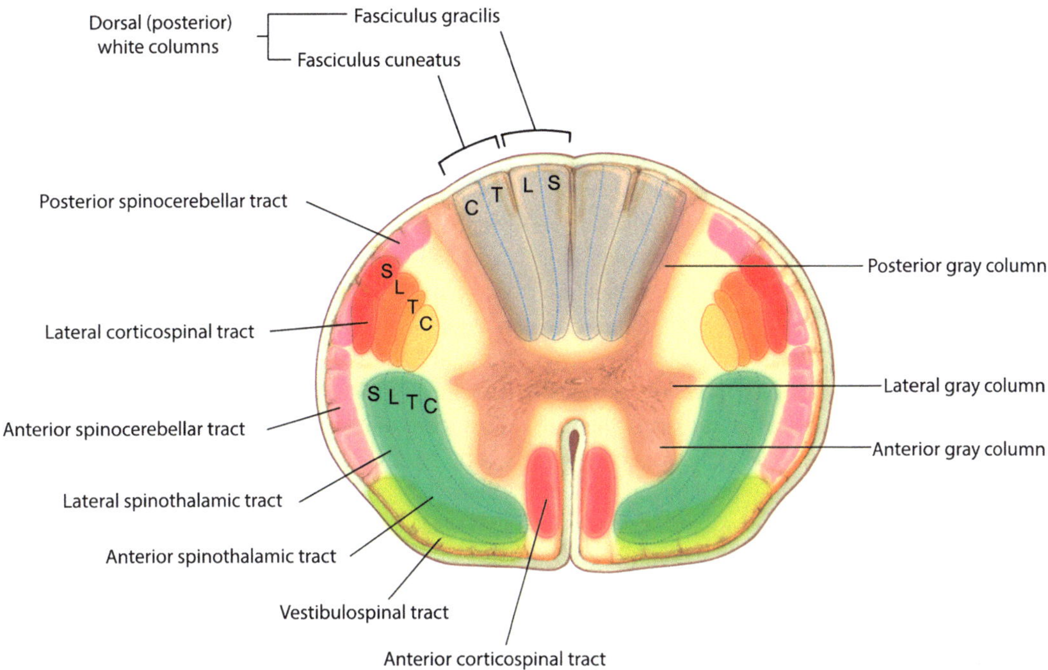

Fig. 9.1 Sectional organization of the spinal cord (*C* cervical, *T* Thoracic, *L* Lumbar, *S* Sacral)

cord explains that eccentrically located lesions in the cervical spinal cord can cause symptoms in the thoracic segments.

The spinal cord blood supply is complicated. In the longitudinal plane, a fairly consistent pattern of vascularization has been recognized. Intercostal or lumbar arterial branches, which give rise to anterior and posterior radicular arteries, supply blood to the spinal cord. These arteries branch to three longitudinal arterial trunks. The anterior spinal artery and both posterior spinal arteries regulate the blood supply to the entire spinal cord. The anterior spinal artery, which supplies approximately 75% of the blood to the cord, is located in the midline on the ventral surface of the spinal cord and is formed superiorly by union of bilateral branches from vertebral arteries in the cervical region. The paired posterior spinal arteries supply major parts of the posterior columns of the white matter. The segmental radicular arteries come from the aorta and enter through the intervertebral foramen [3]. These arteries provide additional blood flow. The largest of these, the artery of Adamkiewicz, supplies blood between the T9 and T12 segments. The lateral spinothalamic tracts receive their vascular supply through the coronal arteries and are, therefore, involved in a typical anterior spinal artery syndrome. Three major segments of arterial organization have been defined in the spinal cord, but anatomical variations at various levels are the rule. The cervicothoracic territory includes the cervical cord and the first two or three thoracic segments. The anterior spinal artery and posterior spinal arteries are branches of the vertebral arteries and costocervical trunk. Below the T3 level, the intercostal arteries from the aorta supply the thoracic segments and are highly variable in number. Although the lengthwise division in the vascularization of the spinal cord may imply the presence of a vulnerable watershed area in the mid-thoracic region (T4–T8), any level in the thoracic region may become affected.

The Clinical Spinal Cord Syndromes

How do acute spinal cord lesions present themselves? An astute physician observes weakness in both legs, an abnormal sensation in the buttock area, and abnormal sphincter function to arrive at the diagnosis. Weakness is front and center, but difficulties with urination and defecation have a more gradual onset. Many patients present with difficulty in initiating micturition, a stream lacking force, and, possibly, progressive difficulty to move the bowels.

Localization principles should be followed, and there is a reasonable consistency to organization of long tracts. Several immediate clues indicate a compression of the spinal cord from a nearby mass. In central cord compression lesions, weakness is more profound in the arms than in the legs. Generally, patients have weakness below the level of the lesion, often with upper and lower motor neuron signs, abnormal sphincter function, and sensory loss involving all modalities. The sacral dermatomes are involved because the spinothalamic tract fibers lie close to the surface of the cord and are the first to compress.

Motor weakness is a better indicator of level in the spinal cord than sensory loss. The weakness of certain muscle groups localizes the lesion to the spinal cord. It is good to remember that full tetraplegia occurs at C4–C5 level and paraparesis with retained hand function occurs at C6–C7 level. Most importantly, the diaphragm is innervated by the phrenic nerve, which can be damaged by high cervical lesions because it originates from the C3–C5 spinal nerves. Moreover, for cervical cord lesions, we start by noting leg paralysis and then categorize which key arm muscles are weak. If there is abduction and flexion using the deltoid and biceps muscles, we are at level C5. With preserved extensor muscles function testing at the brachioradialis and extensor carpi radialis longus, we are at C6 and, with a normal triceps, at C7. Only weakness in the hands with leg paralysis places the lesion one or two levels lower, at C8 or T1, when flexor muscles in the wrist and fingers and intrinsic muscles of the hand are weak.

Sensory levels usually are found at upper chest area for C3 and C4, the nipple at T4, and navel at T10. When a sensory level exists, it may feel like a tightly buckled belt but with tingling below that level. Sensory loss involving temperature may be experienced as a lack of sensation when in contact with water (shower or bath). A sensory level may indicate the level of the compression; however, it may also involve several segments above the lesion. This is important because MRIs of the spine typically involve a narrow spectrum of segments of the spinal cord, and other segments can be easily missed if not fully imaged. (For example, lumbar MRI scans will not image the vast majority of the thoracic spine.) Any patient with spinal cord compression, therefore, should have an MRI of the entire spine to avoid missing a surgically treatable lesion. This is one of the most commonly missed opportunities; that is, failing to image the lesion when guided by the sensory level alone.

Several cord syndromes have been described. This includes a *central cord syndrome*, in which patients have damage to the central gray matter but are spared long tracts. These patients have more conspicuous lower motor neuron signs in the arms without long tract features. These are typically seen in intramedullary spinal tumors or after traumatic spinal injury. *Anterior cord syndrome*, in which the anterior spinal artery fails, results in paraplegia or tetraplegia, no pain, and loss of temperature sensation below the lesion, but it spares the dorsal column function with intact vibration and position sense. A *hemi cord syndrome* diagnosed with corticospinal tract and dorsal column function on one side but pain and temperature sensation on the opposite side is indicative of trauma or a compressive cord lesion. This so-called *Brown-Séquard syndrome* may occur in radiation myelopathy, trauma with locked facets, but most often in penetrating stab wounds. Peacock and colleagues published a large series of stab wounds and reported that the majority of persons with incomplete SCIs have primarily Brown-Séquard syndrome. Although rare (certainly in its original form), Brown-Séquard syndrome remains the most common neurological presentation seen with penetrating SCIs and in almost two-thirds of cases [4, 5]. Its clinical hallmark is loss of pain and temperature (involvement of the crossed spinothalamic tract), sensation opposite the lesion, with loss of position and vibration (involvement of the ascending posterior column tracts), and more prominent leg weakness (corticospinal tract) at the level of the lesion. (The patient may be puzzled by numbness in one leg and weakness in the other.) Variants have been noted including a midline-cord syndrome. In a well-described case, a large nail bisected the spinal canal at the thoracic level into two halves. The midline structures at possible risk were the fasciculus gracilis, the anterior and posterior gray commissures, anterior white commissure, and anterior spinal artery; however, the patient described displayed only sensory deficits [6].

The most commonly missed diagnosis is a *cauda equina syndrome*, in which the weakness of the lower limbs is mild but there is profound sphincter abnormality and saddle (S3–S5) anesthesia [7, 8]. There are some obvious rules of thumb. First, there is absence of any neurologic symptoms in the arms. Second, there is no sensory level. Third, there are diminished or absent lower-limb reflexes with flexor plantar responses and a fairly typical sensory loss in the proximal lower limbs and perineal or saddle region if examined well. There is largely proximal lower-limb weakness with initially good foot function that later becomes flaccid. Bladder impairment in cauda equina syndrome manifests as incomplete emptying and, if progressive, leads to urinary retention with overflow incontinence. An extradural tumor or central disk prolapse is the usual culprit that compresses the cord and causes cauda equine symptomatology. Bilateral leg pain is also an important clue. The outcome is largely determined by whether the patient can walk at presentation, although long-term improvement has been described in patients with severe spinal cord involvement.

Neurologic examinations should focus on several principles. If the motor fibers in the corticospinal tract are involved, expect flaccidity and, later, increased tone and brisk tendon reflexes to clonus. As previously mentioned, evolving weakness has a classic upper motor–neuron (pyramidal) distribution affecting the flexor muscles of the legs and extensor muscles of the arms. Increased tone, increased muscle-stretch reflexes up to clonus, and Babinski signs are present in more chronic presentations. Involvement of the corticospinal, pyramidal, and spinothalamic tracts and the anterior horns results in complete loss of muscle activity in the legs. Many patients have fine fasciculations in the lower limbs that usually resolve within a couple of days. Decreased reactions to pinprick and to hot and cold sensation are present, but light touch and position sense are preserved.

In sensory examinations, sensation to pinprick should be recognized. Failure to react to pinprick should be considered abnormal (comparison with facial pinprick is very useful). As alluded to earlier, there are easy-to-remember landmarks: the clavicle at C3, the nipple at T4, navel at T10, and midway arm to chest (the C4–T2 border). Because T3 and C4 dermatome may vary and overlap, mistakes are made here. Moreover, high cervical C2 lesions cause a level at the trigeminal border both in front and back of the head. This unusual level may not be easily found or appreciated. Sensory examinations of the genital and rectal area are exceedingly important but (understandably) not often performed. Important tests should include the anal-wink reflex (a safety pin poking resulting in anal contraction). The sensory dermatomes are shown in Fig. 9.2, and known abnormal patterns are shown in Fig. 9.3.

Motor examination can be further differentiated by testing 20 muscles (the so-called "key muscles") and should differentiate between voluntary and involuntary movements (spasms may be common). Palpation of muscles may be needed to detect contraction. Several muscles can indicate the region of abnormalities: deltoid and biceps for C5, brachioradialis and extensor carpi radialis longus for C6, triceps for C7, wrist and finger flexors for C8, intrinsic muscle of the hand for T1, quadriceps for L3, quadriceps and tibialis interior for T4, extensor hallucis longus for L5, and gastrocnemius for S1.

Combinations of these abnormalities also help to determine the abnormality of the lesion. Another sign associated with incomplete myelopathy is the *Beevor* sign. The umbilicus moves upward when a patient attempts to sits up or raise the head from a reclining position (relative weakness of the lower abdominal muscles compared to the upper abdominal muscles) [9, 10]. An autonomic

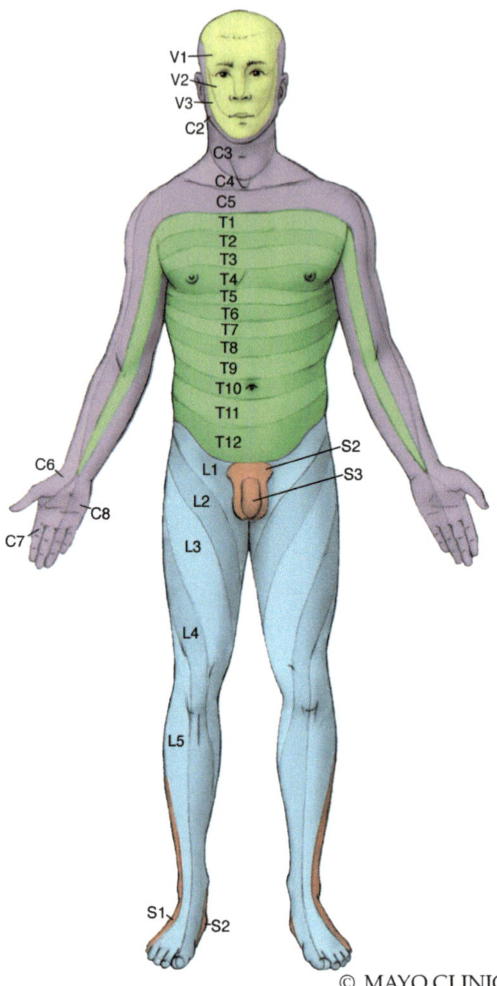

Fig. 9.2 Sensory dermatomes. *V* trigeminal nerve divisions, *C* cervical, *T* thoracic, *L* lumbar, *S* sacral

level can occasionally be demonstrated in the back by scratching the skin longitudinally in the paraspinal region to test for the wheal-and-flare reaction. In patients with myelopathy, the flare reaction can be missing below the level of spinal cord injury, suggestive of autonomic dysfunction below that level. Digital rectal examination may not be the first go-to test in any patient, but it must be assessed in some way. Rectal sensation is far more important than the assessment of rectal tone, which is too variable to be clinically useful unless fully flaccid. However, in a retrospective study of MRI-confirmed cauda equina syndrome, rectal tone findings did not correlate with the presence of a cauda equina syndrome [11]. (This study of digital rectal examination to assess rectal tone suggested very limited physician accuracy in determining reduced tone on an artificial anal canal using sphygmomanometry.)

The abnormalities are best categorized into complete or incomplete status and summarized in an ASIA scale (Table 9.1) [12, 13]. ASIA grade A in traumatic injury indicates a low probability of recovery with less than 5% improving to functional strength [14]. Incomplete paraplegia levels increase the change in mobility greatly (up to approximately two-thirds of the patients). As a general rule, spared function, either returning in a number of days or present at onset, increases the chance of later improvement. The delicate autonomic balance of sympathetic and parasympathetic output is skewed toward vagal output in lesions involving the low cervical to mid-thoracic cord (C1–T6). This leads to hypotension (reduced sympathetic arteriolar tone) and bradycardia (vagal innervation unopposed). Tonic and reflex control of sympathetic and sacral parasympathetic function in cord lesions above T5 leads not only to hypotension with head-up tilt but also to frequent bradycardia. Many patients are poikilothermic (temperature fluctuations from ambient temperature). Hypothermia may occur from lack of shivering (loss of sympathetic tone) and profound vasodilatation, and some patients may not notice decreases in core temperature to 33 °C. Lack of sweating from loss of sympathetic control of the apocrine glands may produce hyperthermia. However, sweating above the lesion often prevents hyperthermia.

Respiratory Failure in Acute Spinal Cord Injury

Although respiratory difficulties may not be immediately obvious, they soon will be. Again, the sensory level can inform the clinician of what to expect of diaphragm function. A simple rule of thumb is the presence or absence of sensory function at the level of the acromioclavicular joint. The presence of pinprick sensation strongly

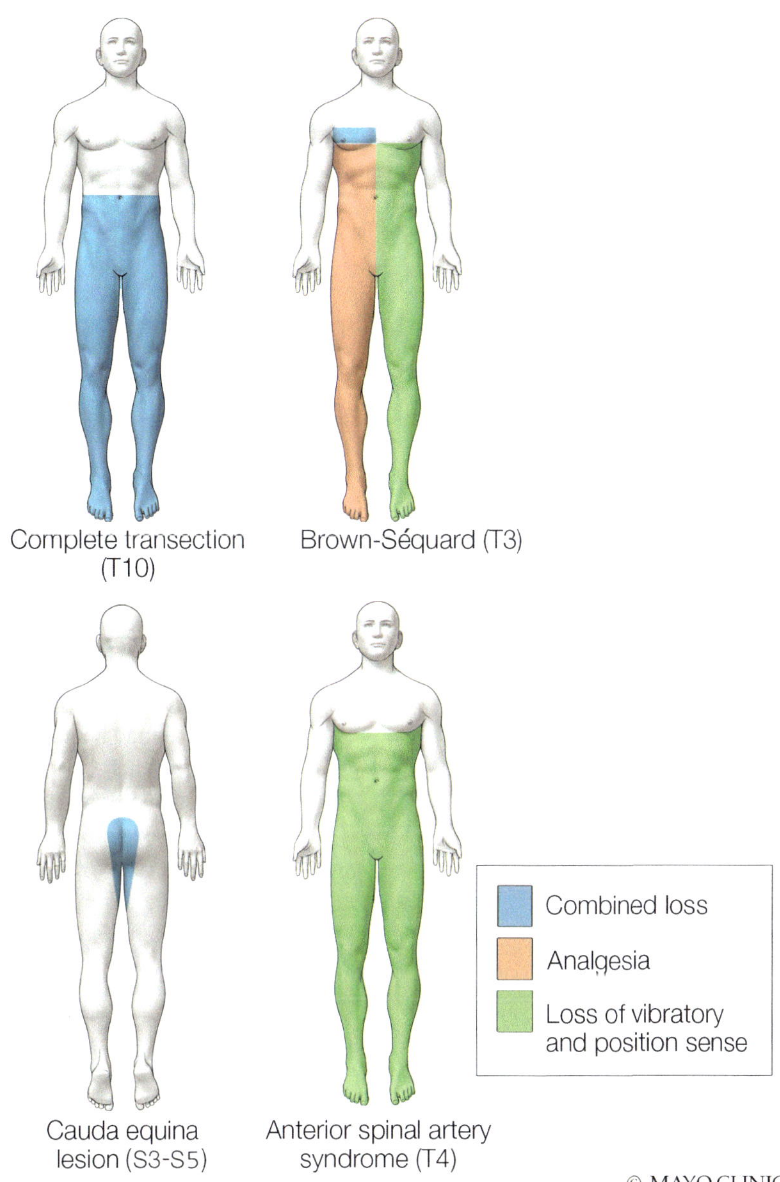

Fig. 9.3 Patterns of sensory loss

suggests that the diaphragm is still innervated and will become functional even if it does not yet seem that way.

Bach [15–17] has described three categories of ventilatory failure complicating spinal injuries. The first category comprised patients with high spinal cord (C1–C4) injuries who are apneic at the time of injury. If these patients survive the initial injury, they require long-term ventilation. About 60% of patients in this category can be eventually weaned off, but it may take up to 5 years. Patients in the second category have autonomous breathing at the time of presentation to the hospital regardless of the level of injury. These patients are at high risk of developing respiratory failure, which may occur 12 hours or more after injury. More than 50% of patients with cervical spine injuries are intubated. Ventilatory failure among these patients may last up to 5 weeks, and most can be weaned off. The third category comprises patients who did not have ventilatory failure immediately after the injury or during the hospital admission but may have pulmonary complications, including respiratory failure, many years later. Noninvasive ventilation may be helpful after acute spinal cord injury, even for patients with sleep-disordered breathing.

Respiratory complications are very prevalent in at least one of five patients [18, 19]. Moreover, acute traumatic spinal cord is most often complicated by traumatic brain injury (and marked decrease in alertness reducing respiratory drive), coexisting chest trauma, and adult respiratory distress syndrome, but also, in less severe cases, profound anxiety exacerbated by inadequate ventilator assistance. Pain often leads to high-dose opioids, all of which can render noninvasive management ineffective.

Table 9.1 ASIA impairment for spinal cord injury

A	Complete	No motor or sensory function preserved in S4–S5 segments
B	Incomplete	Sensory only preserved below level including S4–S5 levels
C	Incomplete	Motor function preserved below level (more than half of key muscles and >MRC 3)
E	Incomplete	Motor and sensory function normal
Proceed as follows: 1. Determine sensory level (most caudal). 2. Determine motor level (lowest key muscle of MRC ≥3). 3. Determine complete or incomplete.		

The breathing pattern in tetraplegia is typically small tidal volumes and increased respiratory rate but with no change in minute ventilation (Fig. 9.4). With complete cervical lesions above the C2 level, the vital capacity is immeasurable, and ultrasound of the diaphragm will show no excursions. Patients with complete lesions at C5–C6, on the other hand, are typically left with approximately 30% reduction of the vital capacity in the sitting position and more when supine. The vital capacity decreases in the sitting position because abdominal contents sag and the diaphragm has fewer excursions to generate negative pressures for air to enter the lungs. On the other hand, patients with a failing diaphragm but relatively intact accessory muscle function can increase tidal volumes when sitting.

There may be abnormalities in chemoreceptor function that could blunt the response to hypercapnia. Early respiratory complications are rapidly developing atelectasis and pneumonia, occurring in over 50% of patients, and the prevalence increases with the time in the ICU. When patients undergo anterior cervical

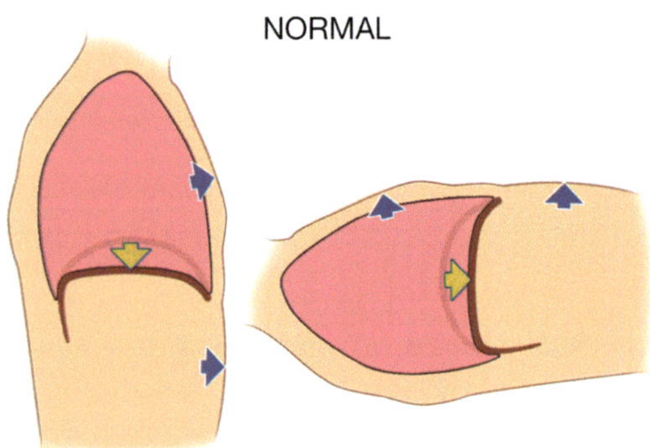

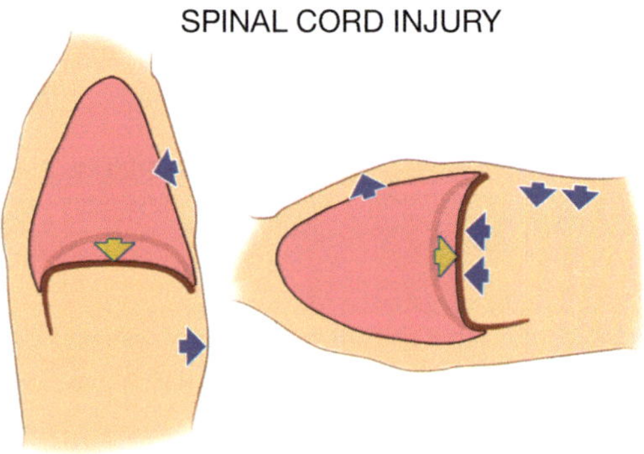

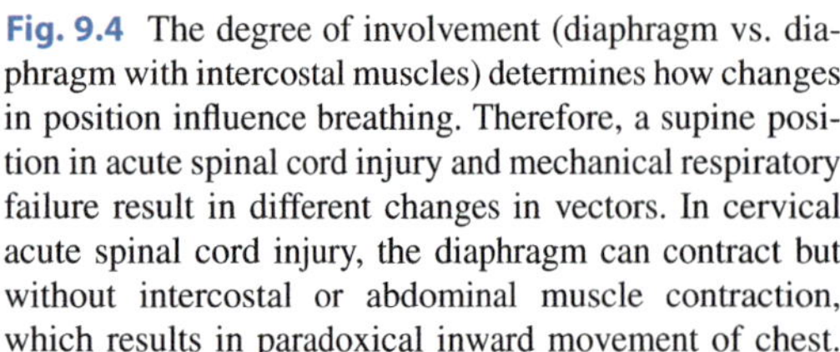

Fig. 9.4 The degree of involvement (diaphragm vs. diaphragm with intercostal muscles) determines how changes in position influence breathing. Therefore, a supine position in acute spinal cord injury and mechanical respiratory failure result in different changes in vectors. In cervical acute spinal cord injury, the diaphragm can contract but without intercostal or abdominal muscle contraction, which results in paradoxical inward movement of chest. This improves with supine position, when abdominal viscera push up the diaphragm. The vital capacity is greater in the supine position for patients with SCI, as abdominal contents displace the diaphragm cephalad, lengthening it to a more favorable length-tension position. If the diaphragm is very weak or completely paralyzed, the accessory muscles can ventilate the lungs but barely so (note the arrows show direction)

spine surgery, regional edema may considerably compromise airway; in particular, if three or four cervical levels are exposed. Many patients may need prolonged intubation (1–2 days) after surgery. Most patients with C2- to C3-level injury will recover diaphragmatic function, and weaning can be successful eventually.

To summarize: the ability to drive respiration can be very seriously affected when the lesion involves the C3 and C4 cervical cord segments. Patients often have shortness of breath with increased effort required to breathe, as well as a sensation of air hunger and chest tightness. When lesions reach C3, diaphragmatic expansion and abdominal muscle function are absent, requiring rapid intubation and full mechanical ventilation. Innervation of these muscles is organized as follows: diaphragm (phrenic nerve, C3–C5), intercostal muscles (T1–T12), and abdominal muscles (T7–L1) [20]. Therefore, effective coughing is determined by the level of injury.

A Major Clinical Urgency: Compression of the Spinal Cord

Next, causes need to be considered. It is practically useful to separate acute spinal cord compression into traumatic versus nontraumatic and to separate nontraumatic into cancer-related and other causes. Selective posterior column dysfunction occurs in syphilitic spinal cord disease (tabes dorsalis), posterior spinal artery infarction, multiple sclerosis, trauma, and after platinum-based chemotherapy. In syphilitic disease, involvement of the dorsal root ganglion also occurs, likely accounting for the shooting pain and diminished reflexes distally. A separate syndrome is the combination of dorsal column and corticospinal tract dysfunction generating the so-called posterolateral column syndrome. This clinical syndrome is marked by bilateral corticospinal tract-pattern weakness below the level of the lesion with decreased vibratory and position sense but spared pain/temperature sensation. This clinical pattern is commonly found in vitamin B12 deficiency, copper deficiency, human immunodeficiency virus (HIV)-associated vacuolar myelopathy, adrenomyeloneuropathy, HTLV-I-associated myelopathy, methotrexate toxicity, and spinal dural arteriovenous fistula (Table 9.2).

Acute spinal cord injury is frequently due to spinal cord compression. Signs may be easily identifiable (inability to stand or walk) or very difficult to detect (fever and back pain without obvious initial clinical signs). The diagnosis of spinal cord compression has been greatly (almost revolutionarily) advanced with the availability of spinal MRI, and true enough, clinical examination has not always perfectly predicted what to expect. Spinal MRI scans can not only detect compression but can also identify

Table 9.2 Common diagnostic considerations in acute spinal cord disease

Disorder	Diagnostic test
Myelopathy	
Compressive myelopathy	MRI of spine
Acute necrotic myelopathy	MRI of spine, biopsy
Vacuolar myelopathy	CSF (PMN), HIV-1
Anterior spinal artery occlusion	MRI of spine, RF, SLE, ANA
Foix-Alajouanine syndrome	MRI of spine, spinal angiogram
Radiation myelopathy	Radiation field, irradiation dose
Paraneoplastic myelopathy	CT scan of chest–abdomen, bone marrow, thyroid scan
Myelitis	
Acute disseminated encephalomyelitis	CSF (MN), MRI of brain
Postinfectious myelitis	Echovirus, coxsackievirus
Demyelinating myelitis	CSF (protein, IgG, oligoclonal bands)
Neuromyelitis optica	VEP, MRI of spine, CSF protein
Viral myelitis	Herpes zoster, CSF (PCR), HTLV-1
Bacterial myelitis	FTA-ABS, CSF (cells, protein)
Tropical myelitis	Circulating antigen, stools, (schistosomiasis, trichinosis)

ANA antinuclear antibody, *CSF* cerebrospinal fluid, *CT* computed tomography, *ELISA* enzyme-linked immunosorbent assay, *FTA-ABS* fluorescent treponemal antibody absorption test, *HIV* human immunodeficiency virus, *PMN* polymononuclear, *VEP* visual evoked potential intrinsic cord lesions and abnormal vasculature. The reality, however, is that the availability of MRI is limited—particularly, after hours. However, it is still more common that physicians miss the connection of signs and symptoms with acute spinal cord injury. Neurologists and neurosurgeons should be able to recognize spinal cord compression urgently, and implicit in this is a good understanding of neuroanatomy and specific syndromes. Compression of the spinal cord remains one of the most important neurologic emergencies because of the very real and significant potential for irreversible injury with the passage of time. Acute spinal cord compression is usually seen under three circumstances—compression from cancer, blood, or abscess. Any patient with new leg weakness and prior metastasis of a known cancer should be screened for it. In cases of abscess, plain X-rays can be helpful and often diagnostic, leading to a definitive MRI study. But in other situations, the presentation can be more difficult to recognize. Astute clinicians may suspect the development of an acute epidural hematoma in anticoagulated patients with significant thoracic or lumbar pain, but the most difficult-to-recognize spinal cord compression is in the patient who presents mainly with sepsis and progressive hypotension. In those patients, managing septic shock, including resuscitative fluid and vasopressor management, takes precedence; recognizing significant limb weakness depends on the physician's level of experience, vigilance, and dissatisfaction with an unresolved clinical problem. Such a presentation may catch an inexperienced physician completely off guard. Spinal cord compression is serious, and the spinal cord cannot sustain many hours of acute compression. Spinal epidural abscess may present distinctly from other causes of spinal cord compression. Some patients are clearly at risk, such as those with a recent spinal surgical procedure, indwelling spinal hardware, diabetes mellitus, intravenous drug use, chronic liver or kidney disease, and other known sites of infection. Not uncommonly, vertebral osteomyelitis is seen in these patients. Epidural abscesses are more frequent in immunocompromised patients.

Back pain has been identified in approximately 90% of patients, but many may not mention it when they are sick. When we ask, we will hear that symptoms have occurred fairly rapidly, within a matter of days to a week, with back pain, fever, and then, often, a phase with radiating pain and tingling. Fever is present in approximately two-thirds of the patients but less often in immunocompromised patients and patients with prior IV drug use.

Clinical features of epidural spinal abscess correspond to its localization. Most often, limbs are initially flaccid and areflexic. There may also be relative hypotension that can be less attributed to spinal shock (only when suddenly severely tetraplegic) and more often to emerging sepsis syndrome. In many instances, conus and cauda equina symptomatology occurs, typically showing paraparesis, loss of sphincter function, and loss of sensation in several lumbar dermatomes with a sensory level at the waist. Cauda equina also shows reduced sensation from the sacral saddle region and further in the groin area. Sacral sparing of the sensory symptoms is an important sign because it implies a centrally located intramedullary lesion. (The representation of the sacral fibers is very peripheral in the cord; thus, pinprick and temperature sensation may be spared in acute central cord lesions.)

Mostly seen in ICU consultations in patients with recent aortic aneurysm repairs (albeit infrequently) is an acute spinal cord infarct. Its presentation is acute, but as mentioned in Chap. 1, initial symptoms may occur gradually ("stuttering") before a very sudden loss of motor function, which usually occurs up to 12 hours after initial onset. A common problem is its recognition in sedated and intubated ICU patients and determining the onset of paraplegia. We often find that MRIs already show early longitudinal hyperintensities indicating ongoing ischemia. Patterns of acute spinal cord have been identified, but most involve the anterior two-thirds of the cord with paraplegia, dissociative sensory loss, and some autonomic symptoms. (It is important to recognize that hypotension may be a clinical sign and requires vasopressors.) Involvement may be strictly localized at the level of the anterior horns, and therefore, sensation and sphincter function are intact. However, a more central spinal infarct occasionally occurs after cardiac arrest or prolonged hypotension, and because it is a watershed area, its clinical presentation includes bilateral spinothalamic sensory deficit with sparing of the posterior columns. Motor deficit and sphincter dysfunction may be absent [21, 22]. Unfortunately, many patients with severe prolonged hypotension, who are also predisposed from prior recognized (or unknown) severe aorta atherosclerotic disease, will demonstrate more typical signs with paraplegia and a sensory level.

Finally, and seen with a regular frequency in large referral centers, is functional weakness. It may come after unrelated surgery [23] or after a trivial incident or injury. (I have seen sudden functional weakness for days after an injury from slipping on ice.) There is either a full new sudden quadriplegia or tetraplegia with loss of all sensory modalities but often with sensory levels not respecting dermatomas. Vibration splitting can be seen (patient detects vibration on part of a bone and not on another part of the same bone). In less severe manifestations, there is global weakness that equally affects both flexors and extensors and thus is inconsistent with an upper motor–neuron pattern of weakness. Moreover, patients cannot move any limb in the supine and seated positions yet demonstrate muscle activation of the upper limbs while in a tripod position. The patient can sit upright with fair balance due to activation of the abdominal and trunk muscles. In contrast, people with tetraplegia due to a recognized neurological disease cannot support themselves while seated at the edge of bed unless supported by a healthcare worker. The diagnosis of functional weakness cannot be made without carefully excluding injury by MRI because functional overlay may exist in patients with real injury.

More Reflections

Recognizing an acute spinal cord lesion does not require a great clinician, but neither is it an easily discoverable clinical syndrome. Recognition requires a combination of clinical history (acute, trauma, vascular surgery, fever, and sepsis following severe back pain) and the clinical finding that the patient is unable to move arms, legs, or both. With hindsight, we will always find sufficient clues; the question is whether we will be able to identify them in real time. With a few notable exceptions, textbooks seldom cover examination of the spinal cord thoroughly. Most chapters focus on MRI patterns and neurosurgical interventions when we need to be aware of the major pitfalls of an evolving cauda equine syndrome or an evolving epidural hematoma or abscess.

The degree of deficit on clinical examination is also dependent on urgent management of the inciting event. Immediate reversal of anticoagulation and evacuation within 12 hours can restore good ambulatory function and bladder control. In epidural abscess, emergency drainage after surgical exploration followed by an 8-week course of antibiotic therapy has a less successful outcome; delayed recognition may be implicated in some but not all patients. It is a very serious injury with few options other than trying to drain (often multiple) pus pockets. When there is an ischemic spinal cord we increase the MAP and reduce the CSF pressure with a lumbar drain.

In our simulation center, we do anticipate that learners could fixate on neck stiffness in a patient with prior undiagnosed epidural abscess and thus could consider acute bacterial meningitis, leading to a lumbar puncture, which would worsen the clinical features to complete paraplegia. Others would fail to appreciate this as a neurosurgical emergency.

Another common error in spinal cord injury assessment is the lack of anticipation of respiratory events and a tendency to react only after there has been respiratory deterioration [24]. The vast majority of high-level spinal cord injury patients need ventilatory support 12 hours to 6 days post-injury with spinal shock and ascending cord edema transiently extending the level of neurologic impairment. Some neurological deterioration (often of a temporary nature) following spine surgery is well recognized, and extubation on the day of surgery may be ill advised. Extubation can only be entertained when there is a vital capacity of

15 ml/kg. Moreover, it is commonly remembered that supplemental O_2 diminishes ventilatory drive, certainly when combined with opioids. The length of time for paralyzed diaphragms to recover can be a matter of months with prolonged weaning programs only offered in respiratory units. Many units offer a rigid respiratory expansion program, which will maintain the lung capacity and compliance until the diaphragm returns to function. The pulmonary morbidity and mortality for patients with spinal cord injury are caused by chronic aspiration due to invasive airway tubes and inadequately prevented and treated upper respiratory infections. All this causes major mucous congestion, atelectasis, and pneumonia in patients who cannot expel debris due to ineffective cough flows.

Pointers and Takeaways

- Any weakness with a combination of long tract signs leads to diagnosis of a myelopathy.
- Dissociated sensory loss leads to a diagnosis of myelopathy.
- A sensory level is clinically comparatively easy to find and an equivocal clinical sign for a myelopathy but less accurate in topography.
- Testing key muscles can be helpful in the determination of the level involved.
- Specific cord syndromes will further refine localization.
- Neurologic examination is absolutely essential to determine which part of the spine should be imaged by MRI.
- The vast majority of high cervical-level spinal cord injury patients need ventilatory support within 12 hours following the injury.

References

1. French KL, Daniels EW, Ahn UM, Ahn NU. Medicolegal cases for spinal epidural hematoma and spinal epidural abscess. Orthopedics. 2013;36:48–53.
2. DePasse JM, Ruttiman R, Eltorai AEM, Palumbo MA, Daniels AH. Assessment of malpractice claims due to spinal epidural abscess. J Neurosurg Spine. 2017;27:476–80.
3. Carmichael SW, Gloviczki P. Anatomy of the blood supply to the spinal cord: the artery of Adamkiewicz revisited. Perspect Vasc Surg Endovasc Ther. 1999;12:113–22.
4. Thakur RC, Khosla VK, Kak VK. Non-missile penetrating injuries of the spine. Acta Neurochir. 1991;113:144–8.
5. de Villiers JC, Grant AR. Stab wounds at the craniocervical junction. Neurosurgery. 1985;17:930–6.
6. Sarkar B, Ahuja K, Choudhury AK, Jain R. Penetrating spine injury bisecting thoracic spinal canal with no significant neurological deficits-the midline cord syndrome. Spinal Cord Ser Cases. 2018;4:102.
7. Fraser S, Roberts L, Murphy E. Cauda Equina syndrome: a literature review of its definition and clinical presentation. Arch Phys Med Rehabil. 2009;90:1964–8.
8. Korse NS, Pijpers JA, van Zwet E, Elzevier HW, Vleggeert-Lankamp CLA. Cauda Equina syndrome: presentation, outcome, and predictors with focus on micturition, defecation, and sexual dysfunction. Eur Spine J. 2017;26:894–904.
9. Pearce JM. Beevor's sign. Eur Neurol. 2005;53:208–9.
10. Mathys J, De Marchis GM. Teaching video neuroimages: Beevor sign: when the umbilicus is pointing to neurologic disease. Neurology. 2013;80:e20.
11. Sherlock KE, Turner W, Elsayed S, et al. The evaluation of digital rectal examination for assessment of anal tone in suspected cauda Equina syndrome. Spine (Phila Pa 1976). 2015;40:1213–8.
12. International Standards for neurological classification of spinal cord injury. 2019. https://asia-spinalinjury.org/wp-content/uploads/2019/10/ASIA-ISCOS-Worksheet_10.2019_PRINT-Page-1-2.pdf. Accessed 4 Nov 2020.
13. Kirshblum SC, Waring W, Biering-Sorensen F, et al. Reference for the 2011 revision of the international standards for neurological classification of spinal cord injury. J Spinal Cord Med. 2011;34:547–54.
14. Metastatic spinal cord compression: diagnosis and management of patients at risk of or with metastatic spinal cord compression. Cardiff (UK): National Institute for Health and Clinical Excellence: Guidance; 2008.
15. Bach JR, Burke L, Chiou M. Conventional respiratory management of spinal cord injury. Phys Med Rehabil Clin N Am. 2020;31:379–95.
16. Bach JR. Prevention of respiratory complications of spinal cord injury: a challenge to "model" spinal cord injury units. J Spinal Cord Med. 2006;29:3–4.
17. Bach JR. Alternative methods of ventilatory support for the patient with ventilatory failure due to spinal cord injury. J Am Paraplegia Soc. 1991;14:158–74.
18. Peterson P, Brooks C, Mellick D, Whiteneck G. Protocol for ventilatory management in high tetraplegia. Top Spinal Cord Inj Rehabil. Aspen Publishers, Inc. 1997;2:101–6.

19. Jackson AB, Groomes TE. Incidence of respiratory complications following spinal cord injury. Arch Phys Med Rehabil. 1994;75:270–5.
20. Martirosyan NL, Feuerstein JS, Theodore N, Cavalcanti DD, Spetzler RF, Preul MC. Blood supply and vascular reactivity of the spinal cord under normal and pathological conditions. J Neurosurg Spine. 2011;15:238–51.
21. Zalewski NL, Rabinstein AA, Krecke KN, et al. Characteristics of spontaneous spinal cord infarction and proposed diagnostic criteria. JAMA Neurol. 2019;76:56–63.
22. Yadav N, Pendharkar H, Kulkarni GB. Spinal cord infarction: clinical and radiological features. J Stroke Cerebrovasc Dis. 2018;27:2810–21.
23. Scheitler KM, Robin CR, Wijdicks EFM. Charcot in the ICU: functional tetraplegia after surgery. Pract Neurol. 2020;20:476–8.
24. Hassid VJ, Schinco MA, Tepas JJ, et al. Definitive establishment of airway control is critical for optimal outcome in lower cervical spinal cord injury. J Trauma. 2008;65:1328–32.

10 Sorting Through Acute Neuromuscular Diseases

There is a justifiable understanding among intensivists that acute neuromuscular disease becomes a critical illness as a result of mechanical respiratory failure. While it is mostly true, acute neuromuscular disease can also target the autonomic nervous system; this, in turn, affects other organ systems. Some muscle disorders affect the heart, which may lead to acute interventions. Some neuropathies affect vasomotor control of blood vessels.

When patients are admitted, we—intensivists and neurointensivists—have a distinct advantage because the neurologic diagnosis has become obviously clear in most patients we receive from the emergency department or the neurology ward. Newly admitted patients with an undiagnosed neurologic disorder could present first with respiratory failure. This often occurs when patients have undiagnosed amyotrophic lateral sclerosis; previously consulted physicians may have been uncertain (and therefore unwilling) to pronounce a devastating diagnosis, or they may simply have failed to connect the dots and misjudged the rapid muscle atrophy attributed to weight loss. Other patients are admitted to intensive care units from the ward following a severe bout of aspiration while undergoing treatment for acute neuromuscular disease; this unfortunately includes patients who were emergently intubated on the ward and, in worse cases, progressed to asystole requiring cardiopulmonary resuscitation [1].

The recognition of acute neuromuscular disease with its diverse presenting clinical syndromes has remained discouragingly elusive for many physicians [2, 3]. It is daunting to see how long it takes to diagnose myasthenia gravis in a patient presenting with weight loss, swallowing difficulties, and drooping head, jaw, and eyelids. (Some have made the rounds and been seen by gastroenterologists, otolaryngologists, and oncologists.) Also disconcerting is how many young patients with early Guillain–Barré syndrome (GBS) have been sent home with "hyperventilation and functional weakness" only to return hours later with more weakness and the earlier overlooked key finding of areflexia on further examination. Precisely because the two situations described above occur relatively often, this chapter tackles two major issues. First, how do we best examine, quantify, and clinically diagnose acute neuromuscular disease? How can we amalgamate all the clinical findings into a working diagnosis? It may seem convenient to diagnose from test results, such as nerve conduction testing (NCV) and electromyography (EMG), but we should acknowledge that the clinical examination, particularly in acute neuromuscular diagnosis, triumphs over any test. This is particularly pertinent in acute neuromuscular disorders, where NCV/EMG may be too complex or confounded by prior illness such as diabetes or prior radiculopathies [4]. (The NCV/EMG is, however, greatly helpful

E. F. M. Wijdicks, *Examining Neurocritical Patients*, https://doi.org/10.1007/978-3-030-69452-4_10

in confirming amyotrophic lateral sclerosis, unlike a nerve or muscle biopsy, which is rarely helpful.) Second, how do we best recognize and quantify neuromuscular respiratory failure at the bedside? How can we perfect our observational skills? The need for ventilatory assistance in acute neuromuscular respiratory failure completely depends on how to settle a presentation at the bedside and does not solely depend on the pulse oximeter, arterial blood gas, or even a chest X-ray.

The reasoning is obvious—move a patient to the intensive care unit if there is any cause for concern and if the respiratory mechanics are failing. Intubate before the situation becomes an emergency.

The Spectrum of Acute Neuromuscular Disease

There are more than a few different stratifications within peripheral neuropathies. The most logical approach for many physicians, including non-neurologists, is to look for certain disorders based on which part of the nerve is involved [5] (Fig. 10.1). At the most basic level, acute neuropathies can be polyneuropathies (bilateral and symmetric disturbance of function), focal neuropathies (a single nerve dysfunction), and multifocal neuropathies (an asymmetric disturbance of function due to a number of isolated nerve involvements) [6]. This clinical classification also helps to identify causes for polyneuropathies commonly due to toxins, nutritional deficiencies, systemic metabolic or endocrine disorders, and hereditary origins. Focal and multifocal neuropathies, on the other hand, are caused by entrapment, mechanical damage, vascular injury, and a number of neoplastic, granulomatous, or infiltrative processes.

Classification also depends on which structure is preferentially involved (axons, myelin sheets, or ganglions). Axonal injury starts at the end of large-diameter fibers and is followed by myelin degeneration and muscle denervation. Myelin injury involves myelin

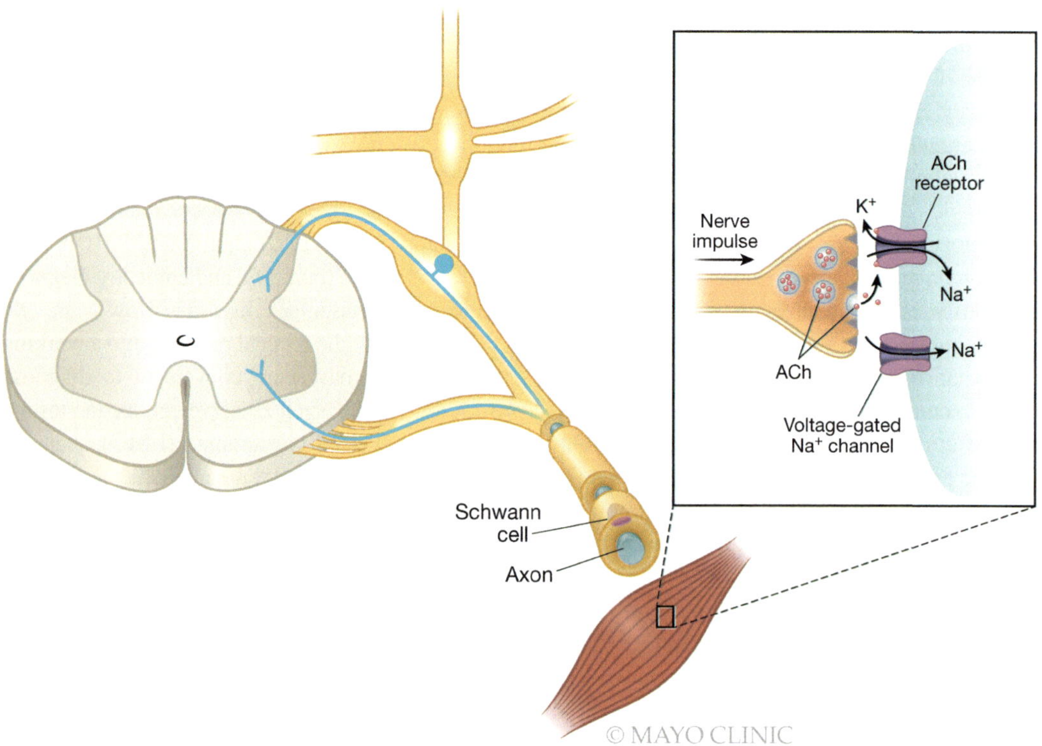

Fig. 10.1 The motor unit and neuromuscular junction

sheets first, then Schwann cells, inter-nodal segments, and eventually axons. Ganglion or neuron cell body neuronopathy is a bit of an outlier type of neuropathy seen with systemic illness (e.g., Sjögren syndrome or systemic lupus erythematosus), HIV infection, or cancer and its treatment—notoriously cisplatin and checkpoint inhibitors. More refined classification is based on size, designating small or large fibers. Large-fiber neuropathy is clinically recognized by loss of position sense, causing pseudoathetosis (slow, writhing movements more obvious with eye closure), loss of touch, and typical lower motor neuron involvement (areflexia, muscle atrophy, fasciculations). Small-fiber neuropathy involves both myelinated and unmyelinated fibers with preserved strength and reflexes but mostly burning and stabbing pain, ataxia, and dysautonomia; these symptoms rarely bring patients into the intensive care unit, although the underlying illness often does. However, in half of these cases, a cause cannot be established, and an autoimmune mechanism is implied. Unfortunately, treating cancer or immunosuppression will not affect ganglion necrosis once it is established, often weeks into the illness [7].

Neuropathies, usually immune-mediated, differ in acute settings. Previously normal nerves become suddenly and acutely abnormal, and the mechanism at play may be acute demyelination or functional blockade of the neuromuscular junction that translates signals to muscular action. Generally, we divide these acute neuropathies into acute demyelinating, immune-mediated neuropathies (classic Guillain–Barré syndrome); acute axonal, immune-mediated (acute motor axonal and acute motor and sensory) neuropathies; and acute neuronopathies (autoimmune ganglionopathies). Patients experiencing any of these three entities may show up in the intensive care unit for management of respiratory weakness, severe dysautonomia, or systemic complications. Chronic myopathies eventually will involve the oropharyngeal musculature and diaphragmatic function; common examples are the major dystrophies (Duchenne, limb-girdle), acid maltase deficiency, and metabolic myopathies such as X-linked myotubular myopathy. Centronuclear myopathy can manifest in adulthood and be accompanied by respiratory failure, ophthalmoparesis, and contractures. In myotonic dystrophy, myotonia, typical facial features with bifacial palsy, temporal baldness, and predominant distal weakness are clinical clues, but respiratory failure remains underrecognized [8, 9]. These patients should be recognizable by clinical, telltale signs of frontal baldness, temporal muscle wasting, bilateral facial palsy causing a sad expression, and myotonia with grasping. Skeletal muscle weakness involves the distal finger flexors, ankle dorsiflexors, and facial muscles, although, over time, the weakness can progress to involve any muscle. The distal finger flexors are often affected first. Along with the progressive muscle weakness, muscle atrophy is prominent. In the facial muscles, patients often present with bifacial weakness and mild ptosis. Patients with myotonic dystrophy describe myotonia as muscle stiffness or delayed muscle relaxation. The myotonia often worsens with cold or stress and may improve with use. It may be seen on physical examination by asking the patient to squeeze his or her hand tightly for 5 seconds and then open quickly, with the clinician monitoring for delayed relaxation. Percussion of the thenar muscle, forearm extensors, or tongue may also produce myotonia.

Mitochondrial encephalomyopathy with lactic acidosis and stroke-like events (or MELAS) may manifest with early respiratory failure. Patients with predominantly diaphragmatic weakness must be investigated for Pompe disease; the blood spot screen is highly specific and based upon a ratio calculated between the creatine (Cre) and creatinine (Crn) and the activity of acid-alpha glucosidase (GAA). The presentation is rarely acute, and gradual oropharyngeal weakness requiring tracheostomy is a more common presentation than abnormal respiratory mechanics. Hereditary myopathies comprise other myopathies of interest. Titin gene mutations are a well-recognized cause of hereditary myopathy with early respiratory failure (HMERF), in which respiratory weakness is the predominant feature and disproportionate to sometimes mild, distal weakness [10–13].

Table 10.1 Uncommon causes for acute weakness and respiratory failure

Tick paralysis (*ptosis, very rapid onset*)
Botulism (*ptosis, ophalmoplegia*)
Organophosphate poisoning (*miosis fasciculations*)
Fish poisoning (tetrodotoxin) (*severe paresthesias*)
Snake bite (*ptosis,oropharyngeal weakness*)
Vasculitis (*skin lesions, more asymmetric*)
Hypophosphatemia (*respiratory failure early, facial tingling*)
Hypokalemia/hyperkalemia (*neck and trunk weakness*)
Hypermagnesemia (*ptosis, dry mouth, facial flushing*)
Acute porphyria (*fever, severe abdominal cramps*)
Acute buckthorn neuropathy (*normal CSF, ingested fruit of plant*)

Additional distinctive signs in italics

Mayo Clinic saw ten patients with chronic myopathies and early respiratory failure in one decade, suggesting it is extremely rare even when taking referral bias into account. None of these patients needed endotracheal intubation or mechanical ventilation and could be managed with BIPAP alone [14].

When diagnosing acute neuromuscular disease—either a new diagnosis or an exacerbation, it becomes clear that infections preceded the neuromuscular manifestations. This is most evident in Guillain–Barré syndrome, which commonly has been associated with respiratory infections and, often, viral infections including the recent Zika and SARS-CoV-2 viruses, although there is no definitive evidence of an appreciable increase in GBS due to COVID-19 (or vaccinations) [15–17]. Other infections can be associated with later GBS including cytomegalovirus, influenza, Mycoplasma pneumonia, dengue, and the alphavirus, chikungunya. Still of all infections gastroenteritis with campylobacter predominates, in a third of cases [17, 18].

A number of other disorders can be considered (Table 10.1). Nevertheless, despite all these rare (and, sometimes, extremely rare) disorders, the most common ICU admissions for patients with acute neuromuscular disease primarily consist of amyotrophic lateral sclerosis, Guillain–Barré disease, or myasthenia gravis, and thus we should look for their characteristics.

Examination of Acute Neuromuscular Disorder

Neurologic examination of the weak patient starts with inspection of the skin and muscle movement underneath it to ascertain whether the muscles look to be of normal bulk. The skin is rarely affected in acute neuromuscular disorders (see Chap. 3 for rash in dermatomyositis), but skin manifestations are more common in the chronic neuropathies of systemic illness such as scleroderma. Some have noted facial flushing in hypomagnesemia (hypomagnesemia from renal disease, excessive magnesium intake, or lithium therapy) causing progressive weakness and areflexia. Arsenic and thallium poisonings show hair loss and transverse white nail bands, but it is fortunately an exceedingly rare differential in the diagnosis of Guillain–Barré. Still, from the descriptions in the rare case reports, it can look pretty similar due to tetraplegia, facial, and oropharyngeal weakness [19]. Attention should be paid to fasciculations in tongue and extremities. They can often be induced by tapping the muscle, but in motor neuron disease (ALS), they are easily found. Fasciculations are random, repetitive twitches, often in regions where you are not looking (like falling stars or fireflies). Fasciculations in the tongue that resemble worms crawling under the surface and are accompanied by wasting and weakness (inability to distend the cheek with the tongue muscle) are highly characteristic of ALS. Myokemias (often in weak facial muscles) also present with repetitive but irregular fasciculations on muscle. Prominence of the hand tendons and tibia (e.g., in amyotrophic lateral sclerosis) indicates a peripheral nerve injury causing profound atrophy.

The tone is examined first and is generally flaccid or normal. Hypotonia is characteristic of neuromuscular disease, but the decreased tone is somewhat subjective and depends on the ability of the patient to relax. (Asking patients to relax usually causes them to do the opposite.) Generally, hypotonia is the absence of resistance felt with passive movements in hip, knee, or elbow. Whether or not a limb falls easily remains a judgment call. (Bend the knee and hold it with one hand while raising the lower leg with the other hand and letting it drop.) One can also try shaking and wiggling the limb to see its increased excursions. Whether it can be considered floppy is, again, in the eye of the beholder. Another trick is suddenly to flex at the thigh, which will not cause the heel to briefly rise from the bed. Bilateral flaccid paralysis of the arms and legs (or of the arms more than the legs) may indicate an acute lesion of the spinal cord, certainly if more cord signs are present such as an absent anal reflex or low body temperature with warm skin, priapism, and dysautonomia.

Muscle weakness can be graded, but strength is often an approximation because of little cooperation from the sick patient. The MRC grading system (see Chap. 3) is used nearly universally, but a number of other scales and scores for specific disorders are better at categorizing which muscles are involved and to what extent. Muscle-strength testing is often performed incorrectly, with more attention to grip and foot strength (the "squeeze my hands, wiggle your feet" perfunctory examination.) rather than focusing on the proximal muscles, which are predominantly involved in acute polyradiculopathies and myopathies. (Remember, distal weakness is common in neuropathies, and proximal weakness is common in acute inflammatory polyneuropathies.) Neck flexion weakness is more common than extensor muscle weakness in myasthenia gravis, and a "dropped-head syndrome" can occur with severe weakness of the cervical paraspinal muscles in 10% of patients, often as a later-onset manifestation [20]. (Dropped-head syndrome is more common with motor neuron disease or inflammatory myopathies [21].) If good strength is found, fatigability should be tested. This can be done by a "pump handle" test or the so-called Jolly test. (Inflate blood pressure cuff and have the patient squeeze the cuff repeatedly. The mercury level on the blood pressure device will decrease after a number of squeezes. It remains useful surrogate but this device may have been replaced by an automated one). Many patients with myasthenia gravis and oropharyngeal manifestations have a flaccid and hypernasal dysarthria. When they read a lengthy passage out loud, a fatigable dysarthria becomes more and more obvious; they become harder to understand over the course of 2 to 3 minutes. Poorly enunciated words are due to a marked degree of nasality; that is, weakness of the soft palate muscles causes air to escape nasally and not orally with speech. While eating, patients may notice that oral muscles weaken with extended chewing, and at times, there is a nasal escape of liquids. The voice may become hoarse; this can be demonstrated by asking the patients to say "eeeee." Jaw weakness may already be obvious if the patient holds the jaw closed with the thumb and with the index finger on the cheek, creating a studious pose that is more socially acceptable than an opened mouth.

Ptosis is common and often asymmetric. It can be brought on with sustained upward gaze for 2 to 3 minutes and improved with an ice pack or washing the face in cold water. Ophthalmoparesis is present. Often the weakness of the medial rectus muscle, which pulls the eye medially, predominates, but it becomes identifiable as a global weakness rather than a third-nerve cranial-nerve deficit because pupils are spared and the eye does not have a down-and-out position (Fig. 10.2). It should be noted that patients with acute inflammatory demyelinating polyneuropathies progressing to neuromuscular respiratory weakness will have bifacial palsy. Its presence increases the probability of oropharyngeal weakness and a future need for mechanical ventilation.

Neuropathies lead to loss of sensation usually in all known modalities. Sensory loss may be suggested by the failure of the patient to grimace or retract, and careful evaluation of sensory level is not always reliable. This applies to alternate diagnoses such as a spinal cord lesion, where the patient needs to display adequate attention for the physician to search for possible thoracic-level pinprick analgesia. Sensation tests several modalities: fine touch, pinprick, vibration, and position sense, with the last two indicating the function of the posterior columns in the spine. Sensation can be totally lost or differentially lost. In polyneuropathies, the sensory loss is mostly distal (the classic glove-and-sock distribution).

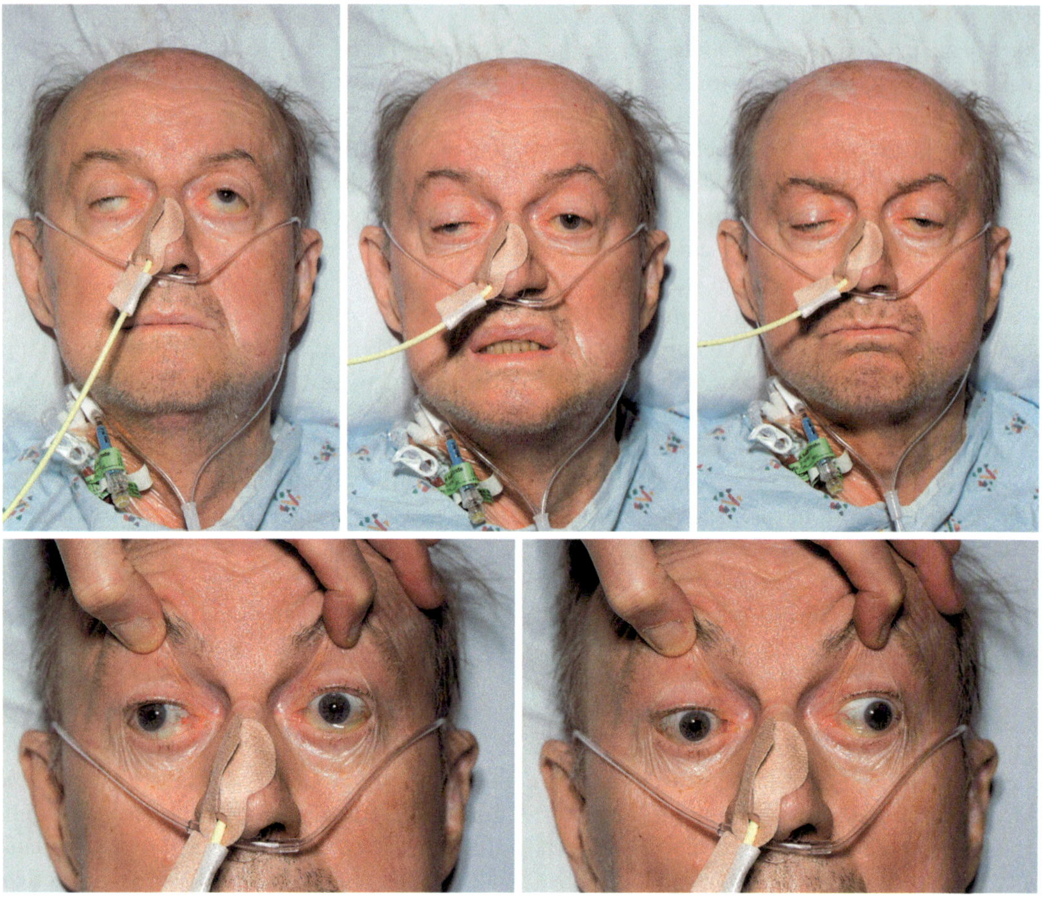

Fig. 10.2 Myasthenic crisis with a number of key findings. Bilateral (often asymmetric) ptosis worsening with prolonged upgaze; bifacial palsy, snarl with "show me your teeth," inability to blow up cheeks, and pseudointernuclear ophtalmoparesis

Progression of sensory loss is more proximal (to knee and elbow), but patches of sensory loss are often found on the abdomen and, more commonly, the vertex. It may involve the full body to the nipples and may mimic a sensory level as seen with a spinal cord lesion. Some hereditary neuropathies have a characteristic sparing of sensory loss in the paraspinal region. Tendon reflexes are helpful and, when they are present, will exclude GBS and a number of very unusual entities (Fig. 10.3).

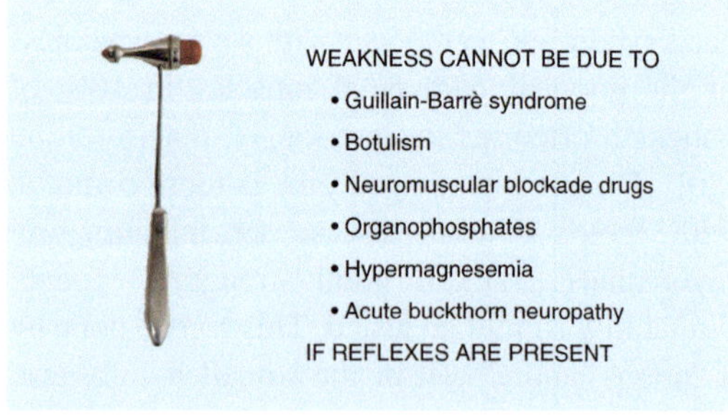

Fig. 10.3 Usefulness of tendon reflexes in excluding disorders

Neurologists also see a good number of neuromuscular disorders when we are asked to consult in other ICUs. In medical or surgical intensive care units, it can take some time to identify a neuromuscular disorder in a critically ill patient, but once discovered, it triggers a neurologic consultation. The ICU nurses may have already noted the patient's failure to move limbs spontaneously or against minimal resistance during daily hygiene routines. Other patients cannot tolerate weaning from the mechanical ventilator, and this failure should alert attending intensivists to a previously undiagnosed neurologic disorder. Rapidly developing hypercapnia often indicates diaphragmatic weakness. Critical illness myopathy (and neuropathy) is common in sepsis, but the systematic review found no association between glucocorticoids or neuromuscular-blocking agents and the development of myopathy or neuropathy [22]. Prednisone for treatment of acute respiratory distress syndrome seemingly does not offer protection either [23].

Unfortunately, many patients with undiagnosed neuromuscular respiratory failure, including signs of upper and lower motor neuron involvement, such as widespread fasciculations, tongue fibrillations and atrophy, prominent atrophy of the interossei, and brisk reflexes, will ultimately be diagnosed with amyotrophic lateral sclerosis (ALS). In my experience, sadly, ALS appears more frequently than critical illness polyneuropathy or myopathy.

Once these findings have been quantified and tabulated—patterns of muscle weakness, sensory domains, and reflex patterns—we can come to a diagnosis: acute (polyradiculo) neuropathy, myopathy, or a neuromuscular-junction disorder.

Respiratory failure will be most often seen with GBS and myasthenia gravis and less commonly in ALS. Most ALS patients already have their diagnosis established and often refuse admission to intensive care units [24]. Respiratory difficulties in myasthenic gravis are more difficult to recognize. Patients may have been overdosed with cholinesterase inhibitors and may have excessive salivation and sweating, abdominal cramps, and urinary urgency. Myasthenia gravis associated with muscle-specific tyrosine kinase (MuSK) antibodies have more prominent oculobulbar weakness, although they eventually develop more generalized weakness. Patients with MuSK antibody-positive myasthenia gravis have a much less successful response to acetylcholine esterase inhibitors [25, 26].

It is rare to see chronic myopathies in the intensive care unit because cramps, myalgia, and stiffness are not life-threatening symptoms. However, in later stages, the condition will eventually involve the oropharyngeal musculature and diaphragmatic function. Many myopathies have a number of associated clinical findings that "look different" (e.g., dysmorphic features, cardiac dysfunction, connective tissue-disease findings, skeletal contractures, and skeletal deformities including Paget disease).

A recently discovered, very serious problem is checkpoint inhibitor-induced myositis, and severe forms cause respiratory failure [27–30].

Acute Neuromuscular Respiratory Disease

We need to acknowledge that mechanical respiratory failure may be central or peripheral. Figure 10.4 shows the anatomy of central and peripheral compartments of respiratory mechanics and how they can be affected by the disease of the higher cervical regions, phrenic nerve, neuromuscular junction, and diaphragm.

Upfront there are difficulties and limitations with recognition of acute neuromuscular respiratory failure. This is how I understand it. The patient cannot catch a breath, needs several pillows at night, and gasps for breath when walking uphill, up a slope, or even up a flight of stairs. Patients cannot talk without pausing in a sentence. (Try speaking after you have speed-walked up 10 flights of stairs, and you will feel the same thing.) If the respiratory weakness continues and $PaCO_2$ rises, patients may then experience "air hunger," which is associated with an increased respiratory drive. Eventually, the patient becomes markedly drowsy and gasps for air—the agonal "fish out of water"—as a result of marked hypercarbia ("CO_2 narcosis"). This is a late-stage neuromuscular respiratory failure that strongly indicates earlier clinical cues have been missed.

There are difficult decisions to be made, including assessment of the severity of respiratory failure, triage of the patient, how to instruct the nursing staff to recognize worsening, and which set of tests to order, which should not only

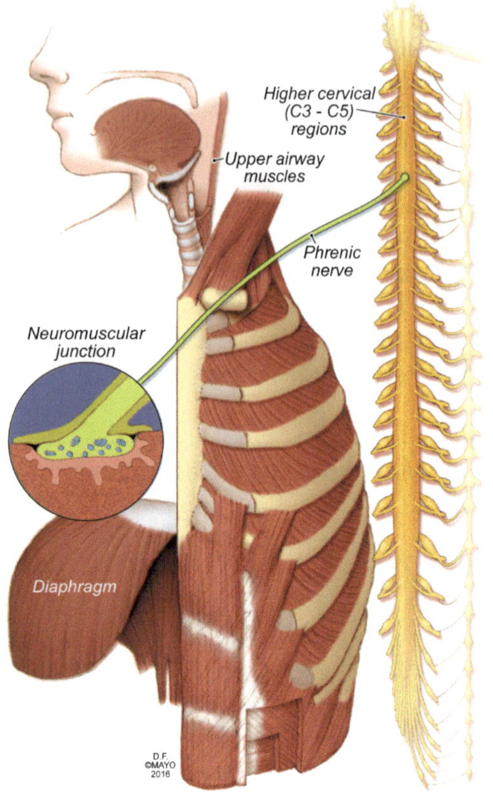

Fig. 10.4 The neuromuscular unit of respiration

assist in the diagnosis but also anticipate serious problems down the line.

So let's look more closely look at the mechanics of respiration and what is involved with moving air in and moving the chest out for lung deployment. During respiration, the lungs can expand and recoil in two ways: by downward and upward movement of the diaphragm that lengthens and shortens the chest cavity and by elevation and depression of the ribs to increase and decrease the anteroposterior diameter of the chest [31–33]. Normal, quiet breathing is largely accomplished by contraction of the diaphragm. In general, the diaphragm is responsible for approximately 2/3 of the ventilatory effort to generate inspiration. It may be supplemented by accessory inspiratory muscles including the external intercostal, scalene, and sternocleidomastoid muscles. Expiration is mostly due to recoil of the thoracic cage, but abdominal-wall muscles are also necessary to generate forceful expiration and are responsible for effective cough [34–36].

Although the upper airway muscles do not contribute directly to chest expansion, they are essential for keeping the airways open during respiration. They play an important role in preventing the collapse of the pharynx during inspiration and preventing aspiration during swallowing.

With the exception of laryngeal muscles, the oropharyngeal muscles have a higher proportion of fast fibers. Their weakness can therefore be seen early on in acute neuromuscular disorders because slow fibers have a higher fatigue resistance than fast fibers due to their highly oxidative metabolism. The diaphragm, on the other hand, has an equal proportion of slow and fast muscle fibers, which, in association with small fiber size, high aerobic oxidative enzyme activity, and large numbers of capillaries, make it more resistant to fatigue [37].

Another major component of breathing relates to the chest-wall mechanics. A large muscle plate such as the diaphragm is necessary to cause a significant change in lung volume. During contraction, there is tilting and flattening of the diaphragm in the anterior/posterior direction. Function of the diaphragm is also affected by intra-abdominal pressure (e.g., in a traumatic abdominal compartment syndrome) and by chest wall elasticity (e.g., advanced Parkinson disease). Most commonly, marked obesity (more specifically, a "potbelly") may result in intra-abdominal hypertension in a supine position creating a cranial shift of the diaphragm and increased pleural pressure. It will increase the work of breathing.

Moving air into the lungs is dependent on respiratory load (the sum of resistance of inspiratory flow, the resistance of the chest wall and lungs, and the positive pressure at peak expiration). When inspiratory muscles contract, a negative force overcomes this respiratory load resulting in inward movement of air. With weakness, the respiratory load can only be partly overcome, leading to less airflow and collapse of lung areas. Compliance, the willingness of the lung to distend, is determined by the change in volume divided by change in pressure ($C = \Delta V/\Delta P$). The inverse of compliance is elastance, defined as the willingness of the lung to return to a resting position. Neuromuscular weakness is basically a decreased inter-thoracic compliance, and maximal inspiratory flow is limited by muscle strength and the poor compliance of the lung and chest wall.

Neuromuscular respiratory failure thus follows a predictable pattern: failure of diaphragm and intercostal muscles followed by compensatory use of accessory muscles but eventually resulting in hypoventilation and atelectasis, further leading to shunting and hypoxia. Respiratory muscle weakness leads to low tidal-volume ("shallow") breathing and poor gas exchange, leading to tachypnea and later hypercapnia. These patients also have increased dead-space ventilation next to their elevated respiratory drive. The rapid breathing is the result of signals to the respiratory center from the abnormal, weak respiratory muscle. Usually, arterial PCO_2 will decrease due to this rapid breathing; however, when respiratory muscle strength is more than 25% of normal, arterial PCO_2 will increase.

Thus, the first indication of diaphragmatic weakness is alveolar hypoventilation and impaired CO_2 exchange. Alveolar hypoventilation may also become more apparent with sleep or when the patient performs a simple exercise. These changes are followed by an increased respiratory rate as a compensatory mechanism to maintain minute ventilation. Later, the assessory muscles of the neck are recruited in response to increased ventilatory demand. Paradoxical breathing, also known as thoraco-abdominal asynchrony, occurs with severe respiratory weakness. Normally, the abdomen and chest expand and contract in a synchronized fashion. During inspiration, downward movement of the diaphragm pushes abdominal contents down and out as the rib margins are lifted and moved out, causing both the chest and abdomen to lift. With diaphragmatic weakness or paralysis, the diaphragm moves up rather than down during inspiration and the abdomen moves in, contracting during chest rise. Clinically, the patient will pause frequently during the speech. Typically, breathlessness improves in the upright position. How the mechanics change in acute neuromuscular respiratory failure is shown in Fig. 10.5.

Oropharyngeal muscle weakness threatens the collapse of the upper-airway musculature. Furthermore, coughing, deep inspiration, closure of glottis, and contraction of abdominal muscles are understandably weak and could eventually lead to aspiration, further worsening the hypoxia.

A single breath-count test is reasonably reliable, although it has never been compared to vital capacity. The relationship is monotonic (changes in the same direction but not at the same rate) and nonlinear (same rate and straight line) but

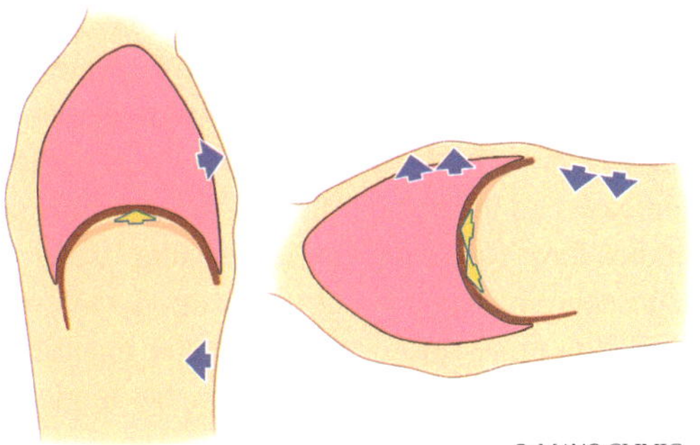

Fig. 10.5 Changes in pulmonary mechanisms with diaphragmatic paralysis. Normally, chest and abdomen move out with inspiration, and the contracting diaphragm pushes abdominal contents down. No measurable change with sitting or supine positions. In diaphragmatic paralysis, the diaphragm is immobile or moves slightly upward due to accessory muscle contraction. This paradoxical breathing (chest out, abdomen in) worsens in a supine position (note arrows for direction of movements)

possibly exponential (curved up). A correlation of single-breath count test and neck flexor-muscle strength has been found in myasthenia gravis [38].

At the bedside, pulmonary function tests are an important way of assessing diaphragmatic function and respiratory muscle strength. These usually involve vital capacity, maximal inspiratory pressure (MIP), and maximal expiratory pressure (MEP) [39]. Pulmonary studies typically reveal a pure ventilatory defect with otherwise normal pulmonary parenchyma. Ventilatory defect may worsen with the development of atelectasis [40]. As expected, neuromuscular weakness of the respiratory muscles is characterized by the inability to generate or maintain normal respiratory pressures.

Respiratory muscle-force examination depends on several factors. The first is the position, particularly in acute neuromuscular respiratory failure; patients who lean forward may have higher inspiratory pressures, and patients in recumbent position will have lower inspiratory pressures. Leaks are common when there is pressure generation in the muscles of the cheek and buccal muscles. This can be prevented by the placement of a mask device.

Generally, several repetitions are necessary to obtain a valid and reliable measurement. All these measurements are volitional and therefore may be effort dependent. If all three test results are similarly normal or decreased, they more likely represent the maximal effort of the patient [41]. Usually, the MEP is twice the MIP, and when the MEP is less than the MIP, it may be due to a leak. Questionable efforts could be followed up by a sniff nasal-pressure maneuver. The sniff maneuver is more useful because it is more natural and easier to understand for the patient. Several studies have found the test to be quite useful in acute neuromuscular disease, but it may be inferior to MIP in more severe neuromuscular disorders. These volitional tests might also be hindered by insufficient nasal passages in some patients [42–45]. A high MIP (>80 cmH_2O), particularly in combination with vital capacity, makes a neuromuscular respiratory failure unlikely [46]. In more slowly progressive neurologic disorders, pulmonary function tests are also clinically important because, for example, a normal MIP or MEP in ALS means that the patient is spared mechanical ventilation for 6 months. MEP >60 cmH_2O may also predict the ability to cough in patients with neuromuscular disease, and in one study, a MEP >70 cmH_2O also correlated with a >50% predictive value of tracheostomy-free survival. This, again, emphasizes the importance of expiratory pressure; effective coughing reduces the chance of mucus plugging and, as a result, pneumonia.

Pulse oximeter is important in any patient with neuromuscular respiratory disease, but obviously it does not identify CO_2 retention. Rapid, shallow breathing leads to chronic hypercapnia in patients with neuromuscular disease. With rapid, shallow breathing, the total volume is markedly decreased, inspiratory time is shortened, and vital capacity is truncated resulting in hypercarbia.

Arterial blood-gas measurements can be informative or not. They may show a hypoxemic–hypercapnic respiratory failure in a patient in obvious respiratory distress. Normal arterial blood gas may be seen in markedly fatigued patients. Normally one would expect a reduced $PaCO_2$ in a tachypneic patient. A normal arterial $PaCO_2$ in a tachypneic patient, therefore, is a sign of pending fatigue because the patient cannot "blow-off CO_2" as a result of mechanical failure. Thus, hypercapnia is a late feature in acute neuromuscular failure. In other words, poor bellows lead to poor ventilation due to alveolar collapse resulting in hypoxemia. There is a "normal" $PaCO_2$ only to rise when the system completely fails. The relationship between $PaCO_2$ and alveolar ventilation is hyperbolic. As ventilation decreases below 4–6 l/min, $PaCO_2$ rises precipitously.

Problems Related to a Specific Disorder

A common problem in evaluating a critically ill patient for neurologic disease is when weaning from the ventilator is not only prolonged but appears to be impossible. The challenge here is to distinguish whether respiratory muscle failure due to critical illness is from acute or subacute neurologic disease. Mechanical ventilation can cause substantial damage, and diaphragmatic weakness may already occur by the end of the first week of ventilation. Diaphragm dysfunction (and evolving atrophy) may occur in approximately 30% of critically ill patients when assessed with ultrasonography [47].

Respiratory failure and GBS can be assessed clinically as with any other case of acute neuromuscular respiratory failure. Others have found that the time between the onset of weakness and hospital admission, the presence of facial weakness or oropharyngeal dysfunction, and the severity of limb weakness assessed by the medical research council score predicted respiratory failure and intubation. This confirms the clinical impression that difficulty clearing secretions in patients with a rapid weakness, defined as between 3 days after onset, may indicate a high risk of respiratory failure [2, 3, 31, 33, 48–55].

Again, imminent neuromuscular respiratory failure in GBS is recognized by restlessness, tachycardia with a rate of >100 bpm, tachypnea, respiratory rate of >20 per min, use of sternocleidomastoid or scalene muscles, hesitant or constantly interrupting speech, asynchronous and sometimes paradoxical breathing, and the presence of forehead sweating [2, 3, 31, 33, 48–55]. It should be pointed out that patients will continue to have a sensation of breathlessness even in the presence of normal blood gas. When $PaCO_2$ rises, patients will then experience "air hunger," which is the result of an increased respiratory drive and of a hypercapnic stimulation of chemoreceptors. The sensation of hypoxemia, however, can be different, and patients will develop a sensation of "rapid breathing." Patients often have an unpleasant and frightening struggle to breathe, which is an indicator that intubation is necessary [56, 57].

Noninvasive ventilation is usually a first option in patients presenting with respiratory failure from ALS. The indications are a combination of clinical findings (orthopnea, nocturnal awakening) and pulmonary pressures (usually less than 50% of predicted). Noninvasive mechanical ventilation reduces the work of breathing and may reverse hypercapnia and atelectasis. Many patients with bulbar dysfunction will need a tracheotomy, a practice considered acceptable by quite a few patients with ALS.

Weaning of the Ventilator

Any patient on ventilation for a prolonged time period will need a plan for weaning from the ventilator. Generally, weaning from mechanical ventilation should be guided by improvement in strength and normalization of values on serial pulmonary function tests. Several conditions need to be considered before even attempting to wean the patient from the ventilator.

Bronchial function is likely impaired in Guillain–Barré syndrome, because bronchoconstriction and bronchodilatation are under the control of vagal and sympathetic innervation. There is some evidence that impaired bronchoconstriction and dilation due to abnormal innervation of bronchial smooth muscle can lead to profoundly impaired ability to clear already increased secretions and, in turn, lead to atelectasis of large lung segments.

In GBS, diaphragmatic weakness may reverse before extremity weakness. Thus, the timing of weaning should not be gauged solely by the recovery of extremity muscle strength. Weaning from mechanical ventilation should be undertaken as early as possible because of the many significant complications related to prolonged intubation. After intubation, however, respiratory function parameters often continue to fall. Reducing intermittent mandatory ventilation rates or reducing pressure-support levels can be used as weaning approaches, at the discretion of the treating physician. However, one should anticipate weeks on the ventilator. The weaning process can be initiated once VC reaches 25 mL/kg and spontaneous tidal volumes of 10–12 mL/kg are attained. PImax exceeding −50 cm H_2O and VC improvement by 4 mL/kg from pre-intubation to pre-extubation are associated with successful extubation.

In myasthenia gravis, an important priority for weaning is the satisfactory treatment of the myasthenic symptoms. In addition, the patient should not have a major pulmonary problem, atelectasis, pleural effusion, or marked difficulty handling secretions. Several factors, including secretion volume, patient's comfort level with a T-piece trial, and a normal chest X-ray, are good predictors of successful extubation in a patient with acute neuromuscular respiratory failure. An optimal dose of pyridostigmine needs to be found, because patients cannot be liberated from the ventilator without adequate treatment despite multiple IVIG courses or plasma-exchange courses.

The weaning process in patients with MG often is challenging because of the fluctuating nature of the disease. Reintubation is not uncommon [58]. In selected patients, noninvasive ventilation can be used for bridging during the weaning process to prevent reintubation. Older age, pneumonia, and atelectasis are major risk factors for poor outcome.

It is important to reintroduce cholinesterase inhibitors before initiating extubation trials. The timing for reintroducing pyridostigmine is variable among neurologists. Some start pyridostigmine in a lower dose around the time of extubation to optimize oropharyngeal and diaphragmatic strength. Careful monitoring for increased secretions after reintroduction of pyridostigmine is imperative. Weaning methods may vary. Patients can be switched to continuous positive airway pressure (CPAP) with pressure-support ventilation (PSV) and the level decreased 1–3 cm H_2O each day. Decreased tidal volume and increased respiratory and heart rates are indicators of fatigue. Once the patient demonstrates good endurance at low-pressure support (5 cm H_2O), usually for more than 2 hours, extubation can be accomplished, but often several T-piece trials are needed first. After extubation, incentive spirometry may reduce the risk of atelectasis and re-intubation.

Acute Neuromuscular Cardiac and Circulatory Disease

There are other major clinical manifestations with acute neuromuscular disease. Involvement of the autonomic nervous system can cause organ injury. The overriding mechanism of organ involvement is uncontrolled hypertension and cardiac arrhythmias. There is a complex interplay between blood pressure swings and pulse, and these hemodynamic changes will (in the more severe cases) create an additional concern for the neurointensivist (Fig. 10.6). As part of the screening for dysautonomia, patients should also be carefully examined for the development of adynamic ileus. This occurs in about 1 in 10 patients with severe GBS and is recognized by absent abdominal sounds, expansion of the abdominal girth, and enlarged colonic loops on abdominal X-ray. Perforation of the colon is a major complication that can substantially change the outcome of a recoverable neurologic illness. As with central causes (where it is much more common), the neuromuscular disease can cause stress cardiomyopathy. In GBS, it is the extreme hypertension in the setting of dysautonomia, where there might be acute ventricular strain; in myasthenia gravis, the cause is largely unknown. It has been noted after plasmapheresis for MG crisis, but more likely hyperadrenergic drive and consequent cardiomyopathy are due to patients' stress when they feel they may not be able to breathe freely. Stress cardiomyopathy is diagnosed on the basis of echocardiography with markedly reduced ejection fraction. When the regional wall-motion abnormalities are outside single epicardiac vascular distribution, it can result in troponin delta and ST-segment elevations very closely mimicking an acute coronary artery occlusion, but the coronary angiogram is normal. Clinically, we may see "unexplained reduced systolic blood pressures" or pulmonary edema with hypotension. Patients with hypotension should undergo echocardiography to look for stress cardiomyopathy [59, 60].

Dysautonomia is common in the severe forms of GBS. Dysautonomia in GBS is recognized by blood-pressure fluctuations, exaggerated drug responses, cardiac arrhythmias, hypersecretions, gastrointestinal dysfunction, and bladder dysfunction. Curiously, profound flushing and sweating may cause some clinicians to consider a coexisting pheochromocytoma, and of course, many patients have increased urinary catecholamines. Paroxysmal or sustained

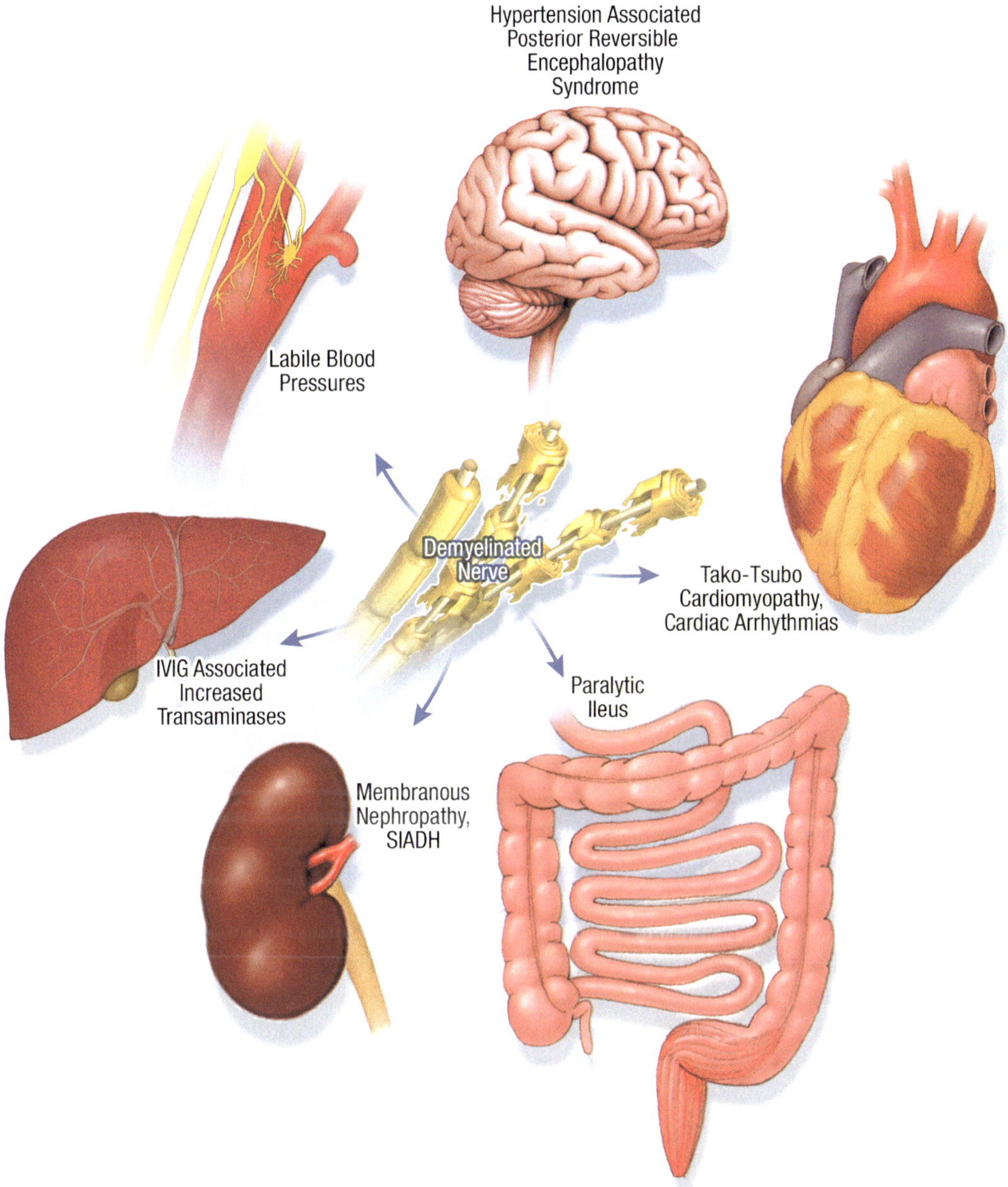

Fig. 10.6 Dysautonomia and its potential effect on organ systems and the brain

hypertension is seen in nearly one in four patients with GBS but not always the more severely affected. Substantially elevated systolic blood pressures can reach values that not only cause acute left ventricle strain but can even predispose the patient to posterior reversible encephalopathy syndrome. Because encephalopathy with new-onset seizures or visual disturbances is unexpected in GBS, MRI of the brain is essential to find the characteristic vasogenic edema in the posterior parieto-occipital white-matter regions. The cause of these extreme blood-pressure fluctuations is not entirely known, but a cardiac baroreflex abnormality is the best explanation. Baroreceptor sensitivity might be altered as a result of vagal nerve demyelination; because sympathetic nerves have less myelin, altered sensitivity results in a sympathetic overdrive. Dysfunction of afferent input from atrial-stretch receptors could also affect blood-pressure swings [61, 62].

These blood pressure elevations require treatment, but treatment might lead to marked hypotension due to exaggerated drug sensitivity.

The whole gamut of cardiac arrhythmias can be seen in GBS including complete heart block [2, 63, 64]. Sinus tachycardia (Fig. 10.7) and so-called vagal bradycardia spells are most frequent in patients with GBS. Persistent sinus tachycardia may appear at any time during the illness and is not generally associated with hypotension or chest pain, but the slowing of rate is indicated with signs of myocardial ischemia on EKG. Vagal spells are brief episodes of bradycardia or sinus arrest, and nursing staff know that tracheal suctioning is a trigger. Vagal spells are usually seen in the worsening and plateau phases but may also extend into the recovery phase. These bradycardic spells may be severe enough to cause a brief pause. In some patients, atrioventricular block or other more benign arrhythmias (e.g., bigeminy) become apparent. A pacemaker may be considered if these episodes are symptomatic and recurrent [62].

The mechanism of stress cardiomyopathy in GBS could be explained by sympathetic overdrive resulting in markedly reduced ventricular ejection from a sudden ventricular strain with hypertension. Morphologic EKG abnormalities are uncommon and nonspecific in GBS, but ST-segment abnormalities are frequent when they are present. It is uncertain whether they represent myocardial damage. Myocarditis has been found in fatal cases that went to autopsy, but this entity remains poorly understood. It might be difficult to distinguish it from a co-existing viral infection affecting the heart. Bradycardia, including extreme decreases in pulse rate, may occur spontaneously, but asystole remains unlikely [63].

A few words on more chronic conditions that eventually become acute. The heart is involved in most genetic myopathies and requires comprehensive cardiology care. Cardiac phenotypes (cardiomyopathies and rhythms disorders) can be the first or predominant manifestations. The best characterized is myotonic dystrophy. The disease causes cardiac conduction abnormalities, most often progressive atrioventricular block, and may be fatal if unaddressed. Other cardiac arrhythmias include sinus node dysfunction, atrial or ventricular fibrillation, and atrial flutter. Diastolic heart failure has also emerged as a common manifestation with myotonic dystrophy.

The spectrum of cardiac manifestations in neuromuscular diseases is wide and quite complex, but we need to be aware of these associations during a clinical examination [65]. Cardiac phenotypes may be dilated, hypertrophic, or restrictive, with potential overlap. Rhythm disorders occur in cardiomyopathies or present as isolated manifestations, especially in myotonic

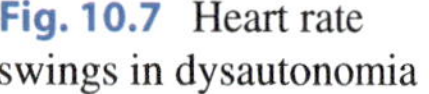

Fig. 10.7 Heart rate swings in dysautonomia

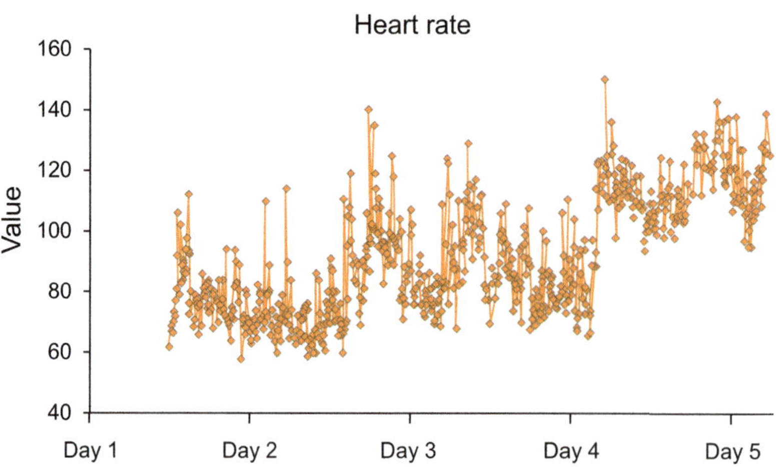

dystrophies and muscle channelopathies. The most common inheritable muscle diseases with dilated and hypokinetic cardiac phenotypes include the dystrophinopathies, limb girdle muscular dystrophies (LGMD), and Emery–Dreifuss muscular dystrophies (EDMD). The prevalence of dilated cardiomyopathy in Duchenne is high, and nearly all young adult patients have cardiac complications. Early resting electrocardiography (ECG) changes include right axis deviation, Q waves in the left precordial leads, and conduction defects. Dilated cardiomyopathy is also prevalent in limb girdle muscular dystrophy. Hypertrophic cardiomyopathies are seen in genetically heterogeneous metabolic disorders, in which the mechanisms causing left ventricular hypertrophy differ from the classic (sarcomeric) hypertrophic forms. Friedreich ataxia and mitochondrial myopathies are characterized by impaired synthesis and use of energy substrates, with the proliferation of abnormal organelles. In glycogen-storage diseases, the use of energy substrates is impaired, leading to intracellular accumulation. Adult-onset or late-onset Pompe disease (LOPD) usually appears between the third and the fourth decades of life. Fasting may trigger manifestations of illness, which include exercise intolerance, fatigue, myalgia, cramps, and stiffness in the absence of contractures. Patients with LOPD are referred to neurologists because of the prevalent skeletal muscle-related symptoms and post-exercise rhabdomyolysis with episodes of pigmenturia. Mild, nonspecific cardiac abnormalities are detectable only in a small proportion of patients with LOPD. The disease has a slowly progressive course, especially in patients treated with enzyme replacement treatment.

More Reflections

With acute neuromuscular disorders, even the astute intensivist has difficulty assessing the severity of mechanical failure and the decision to intubate the patient. We may not fully grasp the degree of unpleasantness and distress. Usually, it is a patient visibly struggling, but it is purely subjective. Struggling patients do not need to have facial expression of distress, tachypnea, or tachycardia, and there is a marked variation in how patients respond to this challenge. Once hypoxemia occurs it is recognized by the carotid bodies, which send signals to the medulla oblongata causing increased minute ventilation and dyspnea (this feeling of distress is likely due to cortico-medulla connections). But these responses to hypoxia are muted by low $PaCO_2$. We may also underestimate the patient's failure to manage secretions, which, as simple as it is, may be the most common reason for acute intubation. We may not fully appreciate that the degree of limb weakness does not always correlate with diaphragm involvement [66]. Emergency intubation for acute respiratory failure is often a result of escalating wrongs rather than something that came out of the blue.

Clinically, it is not difficult to diagnose acute neuromuscular disease if we (1) determine which part of the motor unit is involved, (2) assess the vital signs, and (3) anticipate problems with both the pulmonary and cardiac pump. I have seen impressive manifestations and mostly spontaneous, wild blood-pressure swings—one moment in frank shock, then markedly hypertensive after placing in Trendelenburg. Although I have personally not encountered dysautonomia-associated fatality or cardiac arrest or a patient who needed a pacemaker, I still feel uneasy keeping patients with GBS and dysautonomia on the ward. Similar issues arise with getting patients off the ventilator. The paradox of GBS is that we do not need to see improved strength for weaning to occur. In myasthenia gravis, one important priority is to achieve satisfactory treatment of the myasthenic symptoms before getting the patient off the ventilator. In addition, the patient should have no major pulmonary problem and no evidence of clinically significant atelectasis, pleural effusion, or marked difficulty in handling secretions. Secretion volume, the patient's comfort level with a T-piece trial, and a completely normal chest X-ray are good predictors of successful extubation in any patient with acute neuromuscular respiratory failure.

Pointers and Takeaways

- The major principles of examination of acute neuropathies are assessment of proximal and distal strength, judgment of tone and muscle bulk, grading reflexes, and looking for fasciculations.
- Patients with acute neuromuscular disorders may need critical care when there is mechanical respiratory failure, cardiac failure, or autonomic function failure.
- Look for repeatedly catching a breath during speech (i.e., "staccato speech") and use of accessory muscles lifting the shoulders.
- Mild tachycardia and mild tachypnea may be due to increased work of breathing.
- It is preferable to intubate preemptively in GBS if oropharyngeal weakness emerges.
- Better to admit to the ICU when arms are weak in a rapidly progressive GBS.

References

1. Alshekhlee A, Miles JD, Katirji B, Preston DC, Kaminski HJ. Incidence and mortality rates of myasthenia gravis and myasthenic crisis in US hospitals. Neurology. 2009;72:1548–54.
2. Griggs RC, Donohoe KM, Utell MJ, Goldblatt D, Moxley RT 3rd. Evaluation of pulmonary function in neuromuscular disease. Arch Neurol. 1981;38:9–12.
3. Hutchinson D, Whyte K. Neuromuscular disease and respiratory failure. Pract Neurol. 2008;8:229–37.
4. Callaghan BC, Kerber KA, Lisabeth LL, et al. Role of neurologists and diagnostic tests on the management of distal symmetric polyneuropathy. JAMA Neurol. 2014;71:1143–9.
5. Khamees D, Meurer W. Approach to acute weakness. Emerg Med Clin North Am. 2021;39:173–80.
6. Herskovitz S, Scelsa S, Schaumburg H. Peripheral neuropathies in clinical practice. 1st ed. New York: Oxford University Press; 2010.
7. Amato AA, Ropper AH. Sensory ganglionopathy. N Engl J Med. 2020;383:1657–62.
8. Mathieu J, Allard P, Potvin L, Prevost C, Begin P. A 10-year study of mortality in a cohort of patients with myotonic dystrophy. Neurology. 1999;52:1658–62.
9. Sansone VA, Gagnon C, participants of the 207th EW. 207th ENMC Workshop on chronic respiratory insufficiency in myotonic dystrophies: management and implications for research, 27–29 June 2014, Naarden, The Netherlands. Neuromuscul Disord. 2015;25:432–42.
10. Newsom-Davis J. The respiratory system in muscular dystrophy. Br Med Bull. 1980;36:135–8.
11. Palmio J, Evila A, Chapon F, et al. Hereditary myopathy with early respiratory failure: occurrence in various populations. J Neurol Neurosurg Psychiatry. 2014;85:345–53.
12. Shahrizaila N, Kinnear WJ, Wills AJ. Respiratory involvement in inherited primary muscle conditions. J Neurol Neurosurg Psychiatry. 2006;77:1108–15.
13. Tasca G, Udd B. Hereditary myopathy with early respiratory failure (HMERF): still rare, but common enough. Neuromuscul Disord. 2018;28:268–76.
14. Naddaf E, Milone M. Hereditary myopathies with early respiratory insufficiency in adults. Muscle Nerve. 2017;56:881–6.
15. Smith DW, Mackenzie J. Zika virus and Guillain-Barre syndrome: another viral cause to add to the list. Lancet. 2016;387:1486–8.
16. Katyal N, Narula N, Acharya S, Govindarajan R. Neuromuscular complications with SARS-COV-2 infection: a review. Front Neurol. 2020;11:1052.
17. Lunn MP, Cornblath DR, Jacobs BC et al. COVID-19 vaccine and Guillain-Barre Syndrome: let's not leap to association. Brain. 2021;144:357–60.
18. Poropatich KO, Walker CL, Black RE. Quantifying the association between Campylobacter infection and Guillain-Barre syndrome: a systematic review. J Health Popul Nutr. 2010;28:545–52.
19. Misra UK, Kalita J, Yadav RK, Ranjan P. Thallium poisoning: emphasis on early diagnosis and response to haemodialysis. Postgrad Med J. 2003;79:103–5.
20. Sih M, Soliven B, Mathenia N, Jacobsen J, Rezania K. Head-drop: a frequent feature of late-onset myasthenia gravis. Muscle Nerve. 2017;56:441–4.
21. Burakgazi AZ, Richardson PK, Abu-Rub M. Dropped head syndrome due to neuromuscular disorders: clinical manifestation and evaluation. Neurol Int. 2019;11:8198.
22. Stevens RD, Dowdy DW, Michaels RK, Mendez-Tellez PA, Pronovost PJ, Needham DM. Neuromuscular dysfunction acquired in critical illness: a systematic review. Intensive Care Med. 2007;33:1876–91.
23. Steinberg KP, Hudson LD, Goodman RB, et al. Efficacy and safety of corticosteroids for persistent acute respiratory distress syndrome. N Engl J Med. 2006;354:1671–84.
24. Burakgazi AZ, Hoke A. Respiratory muscle weakness in peripheral neuropathies. J Peripher Nerv Syst. 2010;15:307–13.
25. Guptill JT, Sanders DB, Evoli A. Anti-MuSK antibody myasthenia gravis: clinical findings and response to treatment in two large cohorts. Muscle Nerve. 2011;44:36–40.
26. Lacomis D. Myasthenic crisis. Neurocrit Care. 2005;3:189–94.
27. Haddox CL, Shenoy N, Shah KK, et al. Pembrolizumab induced bulbar myopathy and respiratory failure with necrotizing myositis of the diaphragm. Ann Oncol. 2017;28:673–5.

28. Shah M, Tayar JH, Abdel-Wahab N, Suarez-Almazor ME. Myositis as an adverse event of immune checkpoint blockade for cancer therapy. Semin Arthritis Rheum. 2019;48:736–40.
29. Kao JC, Brickshawana A, Liewluck T. Neuromuscular complications of programmed cell death-1 (PD-1) inhibitors. Curr Neurol Neurosci Rep. 2018;18:63.
30. Touat M, Maisonobe T, Knauss S, et al. Immune checkpoint inhibitor-related myositis and myocarditis in patients with cancer. Neurology. 2018;91:e985–94.
31. Laghi F, Tobin MJ. Disorders of the respiratory muscles. Am J Respir Crit Care Med. 2003;168:10–48.
32. Mier-Jedrzejowicz A, Brophy C, Moxham J, Green M. Assessment of diaphragm weakness. Am Rev Respir Dis. 1988;137:877–83.
33. Moxham J. Respiratory muscle fatigue: mechanisms, evaluation and therapy. Br J Anaesth. 1990;65:43–53.
34. Derenne JP, Macklem PT, Roussos C. The respiratory muscles: mechanics, control, and pathophysiology. Am Rev Respir Dis. 1978;118:119–33.
35. Derenne JP, Macklem PT, Roussos C. The respiratory muscles: mechanics, control, and pathophysiology. Part 2. Am Rev Respir Dis. 1978;118:373–90.
36. Derenne JP, Macklem PT, Roussos C. The respiratory muscles: mechanics, control, and pathophysiology. Part III. Am Rev Respir Dis. 1978;118:581–601.
37. Polla B, D'Antona G, Bottinelli R, Reggiani C. Respiratory muscle fibres: specialisation and plasticity. Thorax. 2004;59:808–17.
38. Elsheikh B, Arnold WD, Gharibshahi S, Reynolds J, Freimer M, Kissel JT. Correlation of single-breath count test and neck flexor muscle strength with spirometry in myasthenia gravis. Muscle Nerve. 2016;53:134–6.
39. Leech JA, Ghezzo H, Stevens D, Becklake MR. Respiratory pressures and function in young adults. Am Rev Respir Dis. 1983;128:17–23.
40. Estenne M, Gevenois PA, Kinnear W, Soudon P, Heilporn A, De Troyer A. Lung volume restriction in patients with chronic respiratory muscle weakness: the role of microatelectasis. Thorax. 1993;48:698–701.
41. Hamnegard CH, Wragg S, Kyroussis D, Aquilina R, Moxham J, Green M. Portable measurement of maximum mouth pressures. Eur Respir J. 1994;7:398–401.
42. Chaudri MB, Liu C, Watson L, Jefferson D, Kinnear WJ. Sniff nasal inspiratory pressure as a marker of respiratory function in motor neuron disease. Eur Respir J. 2000;15:539–42.
43. Fitting JW, Paillex R, Hirt L, Aebischer P, Schluep M. Sniff nasal pressure: a sensitive respiratory test to assess progression of amyotrophic lateral sclerosis. Ann Neurol. 1999;46:887–93.
44. Hart N, Polkey MI, Sharshar T, et al. Limitations of sniff nasal pressure in patients with severe neuromuscular weakness. J Neurol Neurosurg Psychiatry. 2003;74:1685–7.
45. Heritier F, Rahm F, Pasche P, Fitting JW. Sniff nasal inspiratory pressure. A noninvasive assessment of inspiratory muscle strength. Am J Respir Crit Care Med. 1994;150:1678–83.
46. Polkey MI, Green M, Moxham J. Measurement of respiratory muscle strength. Thorax. 1995;50:1131–5.
47. Doorduin J, van Hees HW, van der Hoeven JG, Heunks LM. Monitoring of the respiratory muscles in the critically ill. Am J Respir Crit Care Med. 2013;187:20–7.
48. Durand MC, Porcher R, Orlikowski D, et al. Clinical and electrophysiological predictors of respiratory failure in Guillain-Barre syndrome: a prospective study. Lancet Neurol. 2006;5:1021–8.
49. Gibson GJ, Pride NB, Davis JN, Loh LC. Pulmonary mechanics in patients with respiratory muscle weakness. Am Rev Respir Dis. 1977;115:389–95.
50. Misuri G, Lanini B, Gigliotti F, et al. Mechanism of CO(2) retention in patients with neuromuscular disease. Chest. 2000;117:447–53.
51. Roussos C. Ventilatory muscle fatigue governs breathing frequency. Bull Eur Physiopathol Respir. 1984;20:445–51.
52. Roussos C. Function and fatigue of respiratory muscles. Chest. 1985;88:124S–32S.
53. Tobin MJ, Chadha TS, Jenouri G, Birch SJ, Gazeroglu HB, Sackner MA. Breathing patterns. 2. Diseased subjects. Chest. 1983;84:286–94.
54. Tobin MJ, Chadha TS, Jenouri G, Birch SJ, Gazeroglu HB, Sackner MA. Breathing patterns. 1. Normal subjects. Chest. 1983;84:202–5.
55. Wijdicks EFM. Short of breath, short of air, short of mechanics. Pract Neurol. 2002;2:208–13.
56. Galtrey CM, Faulkner M, Wren DR. How it feels to experience three different causes of respiratory failure. Pract Neurol. 2012;12:49–54.
57. Simon PM, Schwartzstein RM, Weiss JW, Fencl V, Teghtsoonian M, Weinberger SE. Distinguishable types of dyspnea in patients with shortness of breath. Am Rev Respir Dis. 1990;142:1009–14.
58. Seneviratne J, Mandrekar J, Wijdicks EFM, Rabinstein AA. Predictors of extubation failure in myasthenic crisis. Arch Neurol. 2008;65:929–33.
59. Fugate JE, Wijdicks EFM, Kumar G, Rabinstein AA. One thing leads to another: GBS complicated by PRES and Takotsubo cardiomyopathy. Neurocrit Care. 2009;11:395–7.
60. Lichtenfeld P. Autonomic dysfunction in the Guillain-Barre syndrome. Am J Med. 1971;50:772–80.
61. Flachenecker P, Hartung HP, Reiners K. Power spectrum analysis of heart rate variability in Guillain-Barre syndrome. A longitudinal study. Brain. 1997;120(Pt 10):1885–94.
62. Flachenecker P, Lem K, Mullges W, Reiners K. Detection of serious bradyarrhythmias in Guillain-

Barre syndrome: sensitivity and specificity of the 24-hour heart rate power spectrum. Clin Auton Res. 2000;10:185–91.
63. Emmons PR, Blume WT, DuShane JW. Cardiac monitoring and demand pacemaker in Guillain-Barre syndrome. Arch Neurol. 1975;32:59–61.
64. Greenland P, Griggs RC. Arrhythmic complications in the Guillain-Barre syndrome. Arch Intern Med. 1980;140:1053–5.
65. Arbustini E, Di Toro A, Giuliani L, Favalli V, Narula N, Grasso M. Cardiac phenotypes in hereditary muscle disorders: JACC state-of-the-art review. J Am Coll Cardiol. 2018;72:2485–506.
66. Demedts M, Beckers J, Rochette F, Bulcke J. Pulmonary function in moderate neuromuscular disease without respiratory complaints. Eur J Respir Dis. 1982;63:62–7.

11 Clinical Course and Anticipating Outcome

The *Corpus Hippocratum* stated: "I hold that it is an excellent thing for a doctor to practice forecasting. For indeed, if he discover and declare unaided by the side of his patients their present, past, and future circumstances, he will be able to inspire greater confidence that he knows about illness, and thus people will decide to put themselves in his care." From ancient Greece to our times, physicians have been predicting outcome based solely on their clinical impression. Early attempts at prognostication were often crude but definitively decisive; for example, calling patients "moribund" or "terminally injured" and showing much less optimism because targeted interventions and cures were few and far between. When rational medical treatments and successful surgical interventions appeared, prognosis improved for many disorders. This also applies to neurology despite its caricature of a specialty of no effective treatments. Even before the inception of the specialty, treatments have made a major difference (e.g., phenytoin discovery in the late 1930s). Prognostication therefore also had to change and will continue to change as treatments become more successful and care improves. There is one more major caveat. Prognostication figures based on information gathered decades ago may not resonate well or even resemble current practices.

As acute neurology became a more established branch, the need arose to predict how acute brain injury would affect outcome, particularly in patients who were comatose. Whether a structural injury to the brain was permanent became a pressing question. Whether there was a chance of outcome that was "meaningful" (an arguable term) was not an unreasonable question, in particular when care could escalate to high levels with minimal chances of recovery. On the one hand, neurologists tried to allow patients a fighting chance when others had found it very unlikely they would make a significant recovery. When asked to opine on whether interventions were futile, they often predicted an intermediate outcome rather than something gloomy.

In the early 1960s, intensive care units mushroomed in the United States and the rest of the world. As a result, rising medical costs and, in particular, high ICU costs prompted a review of care protocols in patients with hopeless neurologic conditions. Often, these discussions focused on how to approach the long-term care of persistently comatose patients, which soon became a pressing issue. Serious prospective studies on outcome determination in neurology only started in the 1980s. For example, early studies looked at 500 comatose patients in nontraumatic coma at several major centers in the United States and United Kingdom. Patients were identified and studied serially by the investigators while attending neurology rounds and visiting intensive care units and emergency departments daily for over a year [1, 2]. Despite lack of CT scan data, lack of details on level of care, lack of details in specific disease groups, and very wide confidence intervals of

E. F. M. Wijdicks, *Examining Neurocritical Patients*, https://doi.org/10.1007/978-3-030-69452-4_11

the reported percentages (due to small groups of patients in each prognosis category), the so-called Levy algorithms, became a popular predicting tool for neurologists. The study found that in comatose patients not only the absence of pupillary light responses, corneal reflexes, and caloric responses were each associated with a very low probability (<5%) of independent outcome but also that no single sign predicted good outcome. How they have influenced practice is unknown, but pupillary, corneal, and oculocephalic responses now became part of the clinical tools to assess the degree of coma and outcome probability in coma associated with structural damage. These investigators eloquently found out that assessment of injury to the brainstem was crucial in order to prognosticate accurately. As expected, the worst outcomes occurred in structural brain injury and much better outcomes in metabolic derangements causing brain dysfunction. But many patients with a poor prognosis still lived over 6 months, indicating reluctance on the part of physicians and families to withdraw support or withhold new interventions. This observation illustrates that not all predictions of poor outcome had consequences. Feeling less fatalistic than their physicians, families often took a chance despite a poor prognosis (partly driven by denial and partly as a result of major distrust).

Neurologic prognostication for years was based on an early assessment with neurosurgeons questioning the benefit of intervention, families asking to honor the patient's clearly stated prior wish not to be placed on a ventilator, and more generally, to provide comfort to the patient when the acute neurologic illness was the only the most recent in a series of severe complications. Even with further treatment options available, the team reviewed the bigger picture and, of course, the patient's functional state. None of this could be clearly measured, and there was a certain kind of arbitrariness. Medical practice of outcome prediction could not be relegated to statistical odds or "back-of-the-napkin" math.

The neurologic community is reasonably comfortable stating that neurologic examination (and, to a lesser extent, neuroimaging) determines outcome. Part of that is not only what is presented at onset but how clinical signs progress. We know patients initially may have fluctuating signs. Some are a result of interobserver variation – one physician finding different deficits than another physician – and neurologists have earned a reputation for squabbling about asymmetries and interpretation of signs.

With each disease entity, we can expect an initial trajectory to some maximum deficit, which can be sudden in seconds or over hours (Chap. 1). Following stability is improvement (very common) or no change (plateau). Trajectories vary considerably in each patient, and only estimates are possible. Acute brain injury may remain static for months or even years before some level of recovery occurs. Time counts; early improvement often predicts further recovery, and late improvement often predicts disability.

It has been said that physicians do not want to prognosticate because they do not know (and do not want to pretend otherwise). Family members may ask, "How will he/she look a few months (or a year) from now?" and therein lies the problem. Neurologists may not know the future outcome for the patient they see in the midst of an acute brain or spine injury. Really, come to think of it, how many patients have we actually followed for years after the injury? It is difficult to be certain. In other words, precise answers are not possible except in the recognizable extremes of obviously good to truly very bad, where we have much better confidence – all the more reason to be wavering. Most patients who barely survive a major brain injury (often with a neurosurgical intervention) may not thrive for years or at all and stay in a joyless condition. But, we have all seen others improve well beyond expectations [3]. Physicians have, on occasion, even seen patients do reasonably well after being sent to a hospice [4]. We have seen unexpected awakenings, and frankly, family claims of a miracle. The last thing we want families to tell us is, "well, you were completely wrong," but some patients will defy the binary and not fit algorithms. This does not absolve us from a moral obligation to discuss our best assessment of the patient's condition and make decisions about which level of care is appropriate (full resuscitative measures or no further escalation).

Before we prognosticate, we diagnose the condition. Wrong diagnoses lead to incorrect prognoses (but not always). More broadly, much may be said for experience [5]. Indeed, some studies suggest that receiving care from senior staff (the "oldsters" who have "seen a thing or two") boosts patients' chances of survival. Younger staff may be cagey about this responsibility to tell families what to expect and may not have a good answer (if there is one) to prodding questions such as "what is likelihood he/she will never leave a nursing home?" Accurate prognostication must be based on good judgment, and the key to good judgment is good evaluation of all available information and careful assessment of all possible variables. In some overly clear situations, quick but reasoned decisions are valuable; in others, it is better to postpone a final decision. Lest I forget the obvious, before we prognosticate, we must have considered all potential interventions. No neurologist should entertain an outcome prediction if the results of a medical or surgical treatment are not yet known. Prognosis will be poor if the wrong intervention is chosen or the correct one not considered; prognostication in acute neurology should never be done in haste or be based predominantly on neuroimaging. We may all be impressed by the catastrophic injury on a CT scan, but when we assess the impact, the clinical examination plays a very important, if not crucial, role. I can recall several situations when walking up to the bedside with the picture of a major cerebral hemorrhage fresh on my mind only to find that the patient was tolerating the seeming onslaught reasonably well. We must avoid placing too much value in neuroimaging even if it all seems very clear. While the abnormalities may support clinical findings, they may not necessarily tell us how the patient will fare over time.

Generally, we need to remember the three tenets of neurologic prognostication. First, prognostication works best in identifying the extremes of injury. Second, prognostication cannot be an easily refutable judgment call, and third, prognostication misjudgments often relate to failure to identify confounders (Chaps. 3 and 8). The following questions are useful in avoiding pitfalls: "Am I being fooled by the presence of lingering drugs?" and "Are there still considerable metabolic derangements that require better correction?" With the very best efforts, some patients pull through, and some may surprise us. (However, surprise is not really a desirable outcome in any area of medicine because we *should* know what to expect and if we do not know, we should know *not* to speculate or make predictions.)

Acute brain injury has a number of later trajectories, and they have been summarized in Fig. 11.1. These include rapid improvement (hours to weeks), prolonged improvement with recovery, and prolonged recovery without much improvement. Patient may go from an intensive care to a rehabilitation center or nursing home, where they will spend some time before their condition can be upgraded and disposition to home considered. The time they spend in each institution is somewhat indicative of the degree of injury but not a reliable parameter because other immeasurable factors (e.g., finances) may play a role. Outcome categories are often a result of further deterioration, and patients with more brain injury are at higher risk of brain edema, increased intracranial pressure, and systemic complications such as hypoxemia (inability to adequately protect the airway or pulmonary edema) or hypotension (in polytrauma). In some instances, it is highly predictable (e.g., cerebral vasospasm in aneurysmal subarachnoid hemorrhage) or not anticipated (a second stroke shortly after the first one). Recognizing clinical deterioration (Chap. 6) remains a major part of our clinical responsibility. In this chapter, I approach this complex task of determining prognosis by reviewing general principles first, followed by review of the recovery potential of individual neurologic symptoms, and finally, providing prognostic information organized by specific neurologic pathology. Outcome determination is what we do, based on findings on repeat clinical assessment and, thus, deserves a chapter in this book.

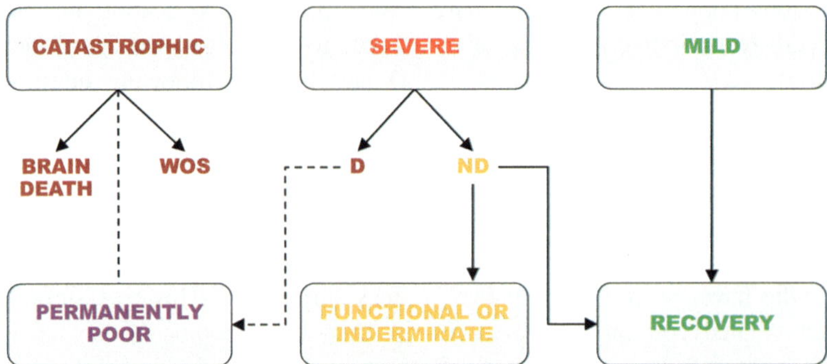

Fig. 11.1 Outcome trajectories. Catastrophic brain injury may lead in 1 in 10 instances to brain death. Failure to improve often leads to withdrawal of support (WOS), and the remaining patient's outcome is permanently poor (vegetative state or a minimally conscious state with some return of communication but fully dependent on close nursing care). Deterioration (D) or not (ND) from any cause will determine outcome in severely injured patients, and in this group there is a potential to lose more than wanted

Absolutely Calamitous

Let there be no doubt. Neurointensivists see a fair share of brain injuries in which the damage is overwhelmingly catastrophic [6]. CT correlates are diffuse brain edema causing significant tightness and obliteration of CSF spaces and, often, radiographic signs of tissue herniation under the falx, into the tentorial opening, and coning of the tonsils by cerebellar lesions. Other CTs may show a catastrophic, large-volume hemorrhage or infarct with mass effect and large horizontal shift obstructing CSF circulation at the foramen of Monro. Again, a clinical correlate is absolutely essential, and all these patients should be comatose with loss of many brainstem reflexes and extensor-posturing motor responses or no responses at all. Secondary involvement of the upper brainstem marks a tipping point. Loss of all brainstem reflexes may be next – but only infrequently (approximately 1 in 10 patients with a catastrophic injury). The word "unsurvivable" is often used, and indeed, it can be strongly argued in some cases but not all of them. For example, a patient with many absent brainstem reflexes and dramatic hydrocephalus can improve after a ventriculostomy. (We have seen this in the dreaded third-ventricle colloid cysts.) Also, removal of a cerebellar hematoma or traumatic epidural hematoma may relieve compression on the brainstem, quickly restoring brainstem reflexes. However, as discussed in Chap. 8, once absent brainstem reflexes are accompanied by demonstrable apnea (and, often, pharmaceutical blood pressure support), damage is irreversible and recovery is simply not possible (Table 11.1) [7].

Table 11.1 Anticipated poor prognosis

Clinically
Prolonged coma with extensor posturing
Any sequential loss of brain stem reflexes with a hemispheric mass (any type)
Failed endovascular retrieval of large vessel occlusion with a severe deficit
No awakening after craniotomy for large subdural hematoma
Imaging
Diffuse cerebral edema with anoxic-ischemic injury
Diffuse axonal injury with hemorrhages in both the corpus callosum and midbrain- pons hemorrhages
Pontine hemorrhage extending into midbrain and thalamus

As always, the clinical examination findings must be related to the neurocritical illness in question. It is important primarily to look the destruction of brain parenchyma, injury location in eloquent areas, and whether secondary effects are contributing to illness. Therefore, a massive destructive thalamic hemorrhage with breakthrough seepage filling and enlarging the ventricles will neither benefit from removal of the clot nor placement of a ventriculostomy.

Is There a Fighting Chance?

Is there are small but real possibility of major improvement? Will the patient awaken? How will the patient do? These questions most often come with a comatose patient. Of course, the ideal starting point is the early recognition of a treatable cause of coma [8]. Outcome is linked to time of intervention. Untreated hydrocephalus, prolonged pressure effect from a mass, untreated large-vessel occlusion, and untreated (or inadequately treated) seizures or infection can all pose impediments to recovery. However, substantial recoveries are possible with timely, aggressive treatment (Table 11.2) [7].

There are less controllable, less predictable factors. Age remains a key factor; an acute brain injury in an older person is very often a major watershed event. ("He was doing great until he fell off the ladder.") Age decreases the chance of withstanding injury; it reduces neuronal regeneration and reemployment of other structures needed for a satisfactory recovery. In the elderly, the outcome of any major traumatic brain injury (TBI) resulting in coma is generally poor. Age and major trauma (brain or polytrauma) remain the most important determinants and most indicative variable [9–11]. It can be quite different in acute stroke. Older patients who receive aggressive management of acute ischemic stroke have nearly as good an outcome after endovascular treatment as their younger counterparts, but this pertains to a small group of patients with a large vessel occlusion. Most neurointensivists and neurosurgeons take age into account, because with age comes the presence of significant comorbidity. Many patients are affected greatly by a stroke, and their lives change abruptly. Motility and adequate communication remain important factors in determining outcome after a stroke. There is little question that poor outcome usually means nursing home placement and full dependence on nursing care. Where it becomes dicey is determining appropriate management for less disabled patients, who "just need help from others." Many patients with acute brain injury will suffer from cognitive and emotional difficulties and are less bothered by loss in physical stamina. These symptoms may last for years. Another issue is the subjective definition of "poor." Poor outcome may not be that "poor" for some family members, while it is absolutely devastating for others.

Table 11.2 Recovery anticipated

Craniotomy and removal of mass (any)
Treated meningitis
Drained acute hydrocephalus
Managed non-convulsive status epilepticus

When "Good" Recovery Is Expected

Some useful observations have been made over a number of years of closely following patients. First, full recovery of cognition and motility after a major initial brain injury does not always protect the patient from later risk of epilepsy, major mood swings (posttraumatic stress syndrome), and depression, often all requiring drug treatment. Difficult-to-measure panic and fear attacks resembling acute stress reaction are common and may limit the survivor's ability to rejoin the workforce. Second, the resilience of individuals younger than 40 is enormous. Although they may need to find significantly less demanding work than they had before the injury (with some struggling to hold a job), recovery can be quite good after traumatic brain injury, meningoencephalitis, and intracranial hemorrhage when the injury does not affect level of consciousness at the nadir of illness.

Resolution of a few other disorders promises full improvement. These include many acute neuromuscular disorders (despite reaching a tetraplegic state), although recovery can be protracted for months and even years. Acute metabolic disorders (such as severe hyponatremia or hyperglycemia) can be corrected, and in each of these disorders, the clinical examination shows fairly rapid or sustained improvement.

When Poor Recovery Is Unanticipated

Supportive care allows recovery from brain damage. Patients admitted to the intensive care unit are at very high risk of complications, both expected and unforeseen. Complications occur more often after a neurosurgical procedure. Some examples are status epilepticus or severe brain edema after debulking of brain tumor, massive hemorrhage associated with diagnostic biopsy of a mass, or reaccumulation after removal of a large subdural hematoma. There are no known explanations for these setbacks, and fortunately, they are rare. In medically managed patients with acute brain injury, failure to respond rapidly to early signs of sepsis may lead to multiorgan failure and subsequent higher mortality. In some patients, one complication leads to another, and this is more often anticipated in patients with significant medical comorbidity where a new acute neurologic problem pushes patients over the edge. Often, these patients have prior (underappreciated) cognitive difficulties, which markedly compromise later functionality. Genetic predisposition may play a role but cannot be used in decision making.

Recovery of Disabling Neurologic Signs

How can physicians reliably document outcome? Although multiple scores and scales have been published, the modified Rankin scale (Table 11.3) has been universally adopted and dichotomized to show poor versus good (or reasonably good) outcome (see also Chap. 3). The scale can be taught, but with any scale assessing functionality without specific tests, the inter- and intravariability are less than perfect. A 90-day modified Rankin scale has been used in many acute stroke trials and is considered a useful metric [12, 13]. The modified Rankin scale, scored on an ordinal scale of 0–6, measures the degree of disability and dependency in daily activities. In each study, it is important to use the Rankin scale appropriately and understand what constitutes a poor or good outcome. The modified Rankin scale has been divided into a Rankin scale of 0–2 and >2, 0–3 and >3, or 0–4 and >4. The scale, however, is not specific to stroke [13].

A widely used metric is the functional independence measure (FIM) score (scale 1–7), which assesses ability to perform several important activities of daily living: eating, grooming, bathing, sphincter control, mobility, locomotion, communication, and social cognition. It is an 8-item instrument graded on a 7-point ordinal scale (Table 11.4). The scale is useful for tabulation, but the most clinically relevant change in the score, which is dependent on score at baseline, remains difficult to find. One study found changes in scores had to be large in

Table 11.3 The modified Rankin scale (mRs)

0	No symptoms
1	No significant disability. Able to carry out all usual activities, despite some symptoms
2	Slight disability. Able to look after own affairs without assistance, but unable to carry out all previous activities
3	Moderate disability. Requires some help, but able to walk unassisted
4	Moderately severe disability. Unable to attend to own bodily needs without assistance, and unable to walk unassisted
5	Severe disability. Requires constant nursing care and attention, bedridden, incontinent
6	Dead

Table 11.4 The FIM rating scale

FIM rating	Definition
1	Maximum dependence – patient performs less than 25% of the task
2	Patient can perform 25–49% of the task, the remaining 50–75% dependent on the caregiver or assistive device
3	Patient can perform 50–74% of the task, the remaining 25–50% being performed by the caregiver or device
4	Patient can perform ≥75% of the task and requires ≤25% from the caregiver
5	Supervision – patient requires verbal cuing or set-up to perform the task; caregiver needs to be on "stand-by" or "contact guard" assist
6	Modified independent – patient can perform with assistive devices, or with increased time; no assistance required from the caregiver
7	Complete independence

order to represent notable improvement: in total FIM score (22 points), motor part of the FIM scores (17 points), and cognitive parts of the FIM score (3 points) [14].

Outcome may be determined not only by the intrinsic recovery potential but also by opportunities for neurorehabilitation. Recovery from major disabling symptoms such as aphasia, hemiplegia, neglect or ophthalmoplegia is variable, as expected, and with very small steps. Not surprisingly, then, looking back 1 week may not detect changes; looking back several months will. One neurologic deficit may define the disability, and resolving it would improve the patient's ability to function. Other patients may have a number of neurologic handicaps such as memory loss or impaired vision or gait; these are seen, for example, in a brainstem stroke from a basilar artery occlusion. Based on these observations of many rehabilitation physicians, what can we expect? Moreover, are we able to generalize? For example, does a hemiplegia or aphasia from traumatic brain injury improve differently than a similar condition due to a stroke, and if so, are there differences between a hemorrhage and infarct? We can only work with what has been published and from personal observations.

Substantial improvement in language performance occurs within the first 2 weeks after stroke and may improve further up to 6 months, after which it may reach a plateau. Semantics (meaning of a sentence) and syntax (structure of sentence) may improve notably in up to 6 weeks; phonology (how speech sounds) and token test (comprehension of instructions) may improve within 3 months. In patients with global aphasia, speech output may improve but with a significant lag in verbal communication. To a certain degree and depending on the severity of aphasia, speech therapy may improve outcome, but no controlled studies have compared natural course with speech therapy interventions. Most of us who have seen speech therapists at work must agree that they are expert at uncovering elements of communication, and practicing their techniques may speed up communication recovery. Currently, "aphasiologists" assume that there is a method to what we perceive as an unpredictable recovery. A patient with small left hemisphere lesion usually recovers well due to restitution of perilesional language networks. With larger lesions, we recruit other areas of the left hemisphere surrounding the lesion. With a severely damaged hemisphere, activation of the opposite right hemisphere comes into play [15, 16]. Among clinical variables, initial aphasia severity seem to be one of the best predictors of aphasia outcome [15, 16]. For instance, it has been demonstrated that the initial Aphasia Quotient is a good predictor of 6- and 12-month aphasia recovery [17]. The right hemisphere language network seems to be important in aphasia recovery after left hemispheric stroke [18]. The size of the long segment of the arcuate fasciculus in the right hemisphere (contralateral to the lesion) is an important predictive factor for language recovery after stroke. The volume of the other segments in the right and left hemisphere is not associated with recovery [18]. MRI studies indicated an important role of the right hemisphere for aphasia recovery after stroke. Two possible mechanisms have been suggested: unmasking of previously ready language capacities or reorganization of right hemisphere language areas. Right hemispheric language functions could be mediated by a specific segment of the arcuate fasciculus.

Hemiplegia recovery may be protracted, with many patients persistently weak and without function 6 months after the event. Arm function recovery is much less expected than leg function recovery. Voluntary ability to extend fingers (with shoulder abduction within 5 days of stroke) strongly predicts further recovery [19]. If both movements are absent, only one in ten recovered some (mostly nonfunctional) dexterity. Increasing tone in the leg will allow stance. Walking is very difficult to predict at the outset, but there is reason for optimism if a stance with little (e.g., supporting-arm) assistance is achieved.

Body awareness after stroke may improve, but approximately one third of the patients will still have clear signs of spatial disorientation a full year after the stroke, which interferes with rehabilitation of a paretic limb and transfers. Several disturbances of body awareness occur after right hemisphere stroke including asomatognosia (i.e., feelings of nonbelonging or nonrecognition of the limb) and somatoparaphrenia (also known as "delusional ideas of disownership") [20]. These disturbances reveal that our seemingly effortless sense of body ownership (i.e., the sense that "my" body belongs to "me") is actually quite fragile.

Dysphagia is also markedly worse in patients with persistent loss of awareness (asomatognosia) with dribbling and potential for choking and aspiration [21]. The incidence of dysphagia after stroke ranges widely from 2% to 33% after 1 month and from 0.4% to 50% after 6 months. A bedside test for swallowing dysfunction is important, and aspiration to thin liquid resulting in spontaneous cough during test swallows increases the risk of aspiration greatly. Further evaluation of swallowing is necessary, but enteral nutrition may have to be started in patients with little reserve. Outcome studies in patients with severe strokes and head injury suggest that early nutritional support reduces mortality and incidence of nosocomial infections. The main goal of nutrition should be to preserve muscle mass and to provide adequate fluids, minerals, and fats [22, 23]. One study found that almost two third of people with an initially severe dysphagic stroke do not recover functional oral intake within 7 days and therefore benefit from NGT feeding. The Predictive Swallowing Score (PRESS) is available, predicting need for PEG placement in the coming weeks and months. This robust prognostic model was based on prospective cohorts and predicts the recovery of oral intake and return to prestroke diet on days 7 and 30. Five factors, including age ≥70 years, NIHSS at admission, lesion of the frontal operculum, initial risk of aspiration, and initial score of functional oral intake scale (Table 11.5) [24], were selected to develop the prognostic score system, which was also externally validated in a multicenter approach [25].

Table 11.5 Functional oral intake scale (FOIS)

Level 1	Nothing by mouth
Level 2	Tube dependent with minimal attempts of food or liquid
Level 3	Tube dependent with consistent oral intake of food or liquid
Level 4	Total oral diet of a single consistency
Level 5	Total oral diet with multiple consistencies, but requiring special preparation or compensations
Level 6	Total oral diet with multiple consistencies without special preparation, but with specific food limitations
Level 7	Total oral diet with no restrictions

Crary et al. [24]. Reprinted with permission

Specific Neurocritical Disorders

Every neurocritical illness has its own trajectory, and its path determines outcome. Outcome is determined by clinical findings assessed *after* interventions and *after* stabilization or "neuroresuscitation." In other words, looking poor initially does not mean a later poor outcome. Moreover, outcome models or datasets become rapidly outdated if a new therapy comes along, and certainly, if it approaches a cure. This may be dramatic as with retrieval of clot from a large cerebral artery (recovery on the angiogram table) or not so dramatic as in therapeutic hypothermia in comatose patients after cardiopulmonary resuscitation (improvement as measured in days in the ICU or the time needed to return to a new but functional baseline). The major disease entities are reviewed in this section.

Traumatic Brain Injury

Both the IMPACT (Fig. 11.2) and CRASH (Fig. 11.3) databases have found several variables that consistently predict death or poor outcome; strong clinical predictors for outcome are motor response and pupil reactivity [9, 11]. Patients with "fixed and dilated" (often as alluded to in Chap. 8), round or pear-shaped pupils with spontaneous

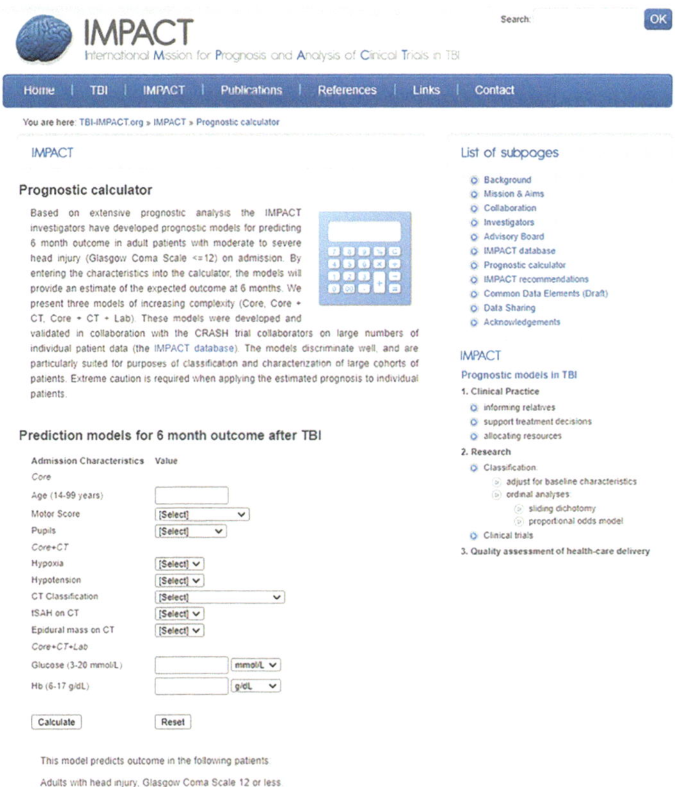

Fig. 11.2 IMPACT score calculator

extensor posturing may have a severe diffuse axonal injury and additional CT or MRI proof of white-matter shear lesions and contusions, primary brainstem injury, subarachnoid and subdural hemorrhage—never to improve or worsen from cerebral edema. The paucity of reliance on clinical exam findings in these databases is very concerning. The rehabilitation trajectory after TBI is hard to predict and, often, is surprisingly good in younger individuals. Only a relative minority are cared for in nursing homes. Patients who become severely disabled after severe head injury may be able to return to work in sheltered workplaces, usually with responsibilities far less demanding than those before the injury. Visual difficulties due to cranial nerve involvement, central causes of vertigo, and also more difficult-to-measure panic attacks in closed spaces are common and may limit the survivor's ability to rejoin the workforce. Cognitive dysfunction can be substantial after TBI and is the source of less functional independence. The same applies to aggressive agitation, which is common in 70% of patients admitted to rehabilitation centers. Depression and apathy may occur in combination and may respond to dopaminergic medication such as selegiline, bromocriptine, and amantadine. Many rehabilitation physicians prescribe psychostimulants. Posttraumatic stress disorder with major emotional and behavior abnormalities may occur with or without objective findings of brain injury, and cognitive rehabilitation is currently an active field of research in combat veterans within the Department of Veterans Affairs.

Traumatic brain injury databases may create very large datasets and are very helpful as screening tools but should never be used as metrics in clinical practice, and when communicating these numbers to families, it is

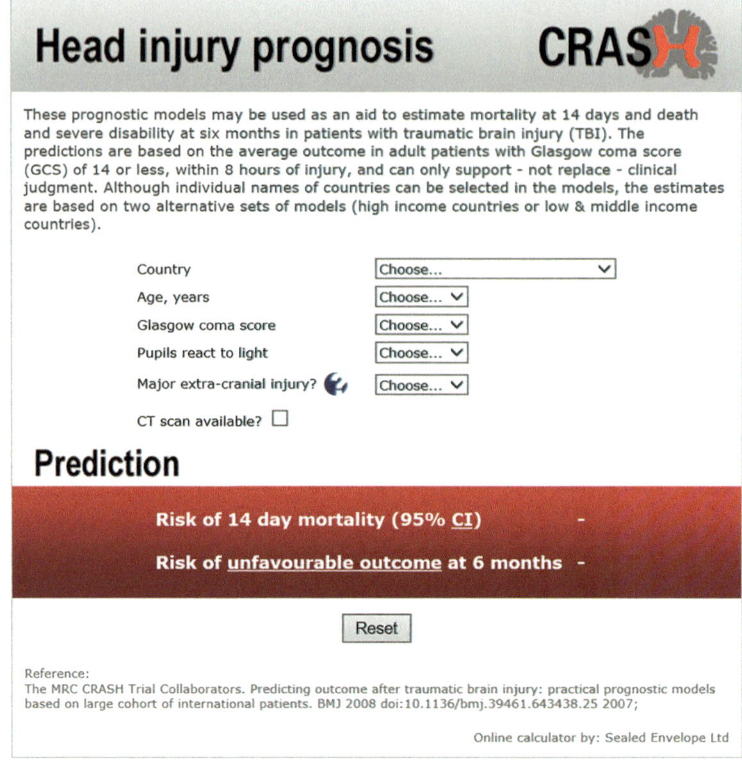

Fig. 11.3 CRASH score calculator

important to "individualize" them to the specific patient by including underlying comorbidities, age, and hospital course. We assume that these sets of big data (even if less fine grained) must be superior to any individual physician's cumulative experience as they encompass thousands of patients. Still, extreme caution is advised when applying these prediction models to individual patients, as they represent probabilities only and cannot provide any level of certainty in an individual patient's outcome.

Here are some other facts. Improvement in traumatic head injury in hospitalized patients differs from improvement of patients in specialized rehabilitation centers. Length of coma is an unreliable predictor, because serious and deep sedation may have been used to control agitation, ICP, or cooperation with the mechanical ventilator. Once patients awaken, simple motor skills improve over time. Postcraniotomy patients may have major mood disturbances, headache, and long-standing fatigue. They may develop a full-blown trephined syndrome with delayed contralateral hemiparesis that improves with cranioplasty.

So how do patients fare after a severe traumatic head injury? What can we (and their families) expect? It only takes a minute to find "miracle" patients on the web, patients purportedly beyond hope but with an "unexpected" recovery or awakening from a supposedly "vegetative state." What is the reality? Persistent vegetative state can be diagnosed 1 month after TBI, when patients open their eyes and develop sleep–wake cycles. It may take up to a full year to declare it permanent with confidence. Recovery from a prolonged PVS to a more responsive, minimally conscious state has been documented in several cases but, invariably, in younger individuals. PVS mostly occurs in polytraumatized, critically ill patients, in which other systemic injuries (prolonged shock or refractory hypoxemia) have contributed to the initial injury. The opposite condition (hyperoxemia) in the first day does not affect outcome [26]. A far more subtle prognosticator is the neurosurgeon's prognostic beliefs of

6-month functional recovery and the decision to proceed with surgery. One study [27] found that prognostic predictions among neurosurgeons were highly variable and only modestly influenced by evidence-based risk predictions. The majority of neurosurgeons recommended craniotomy for hematoma evacuation, but not when they "believed" the prognosis was very poor.

Stroke Syndromes

Because stroke is so variable in cause and presentation, large databases are not able to find key predictors for functional outcome after ischemic and hemorrhagic stroke. Certain generalities have emerged, and they are age, prior stroke, abnormal consciousness, and disorientation at onset. Obviously, which territory is involved matters most. Infarcts in posterior circulation are mostly benign, and small brainstem strokes (with their many clinical variants) have a good functional outcome. However, a cerebellar infarct, infarcts in the thalami (particularly bilateral), or infarcts in the occipital lobe can be quite disabling because they impair stance, motility, vision, and speech. An embolus in the posterior circulation in the basilar artery that causes coma or locked-in syndrome is associated with high mortality and little functional recovery. In general, long-term care of patients with massive pontine infarcts does not lead to a good functional outcome despite all goodwill and rehabilitation. The majority of patients in locked-in syndrome will remain in a wheel chair, head control is poor, bladder and bowel control will be absent, but swallowing may improve with later removal of gastrostomy months later. Bilateral PCA infarcts can also be quite disabling, causing a Balint syndrome (inability to perceive the entire visual field as a whole) or cortical blindness with very protracted, often only moderate, recovery.

The prognosis for recovery from acute intracerebral hemorrhage (ICH) is strongly related to size, location, and extent of blood seepage into the ventricles. Infratentorial location of ICH (due to its proximity to the brainstem and hemorrhage in a tight compartment) also predicts a higher likelihood of death or dependency. Location of the cerebral hemorrhage is quite important. A large study of 2066 patients with ICH from the INTERACT2 trial (Intensive Blood Pressure Reduction in Acute Cerebral Hemorrhage Trial 2) found thalamic location more likely to be fatal and associated with poor stroke outcome [28]. A strong association between posterior limb internal capsule and globus pallidus/putamen caused not only worse stroke outcomes, including early mortality, but also later disability. Caudate ICH location was not associated with significant effect [29] on functional outcomes after adjustment for ICH and IVH volumes. Moreover, ICH involving the posterior limb of the internal capsule, thalamus, and infratentorial sites was each associated with poor quality-of-life indices.

The cause of cerebral hemorrhage does not affect outcome when it pertains to deep-seated hemorrhage because most are related to hypertension. Aggressive blood pressure control (i.e., <130 mm Hg) after these cerebral hemorrhages is important to prevent recurrence, but unfortunately the general population once recovered and followed up for risk factor modification, shows that good control of blood pressure is not frequently achieved with 40–50% control at best [30, 31]. Blood pressure control is even more important in patients with cerebral amyloid angiopathy to reduce recurrence [32]. Only when lobar hematomas are caused by treatable vascular anomalies is long-term outcome improved.

Age remains a major factor [33]. Patients aged 85 years or older showed a poor prognosis as shown by the higher mortality rate, longer hospital stays, and higher rates of complications, particularly respiratory events, compared to younger patients. Elderly patients are more prone to aspiration due to increased immobility-related secretions, predisposing this age subgroup to respiratory infection. It is unclear whether associated comorbidity or de-escalating care level solely determined by age is a factor [33, 34].

Aneurysmal Subarachnoid Hemorrhage

The initial presentation of a poor-grade subarachnoid hemorrhage defined as comatose does not necessarily imply a poor outcome; one of five patients may still have a good outcome without major disabling cognitive defects [35, 36]. Approximately 20% of post-SAH patients will remain markedly impaired (rated as a modified Rankin scale of 4–5).

Long-term outcome may change. Of patients discharged to a nursing home in a poor condition, one in three will improve and recover to independent functioning within the first 2 years of admission. Patients who survive aneurysmal subarachnoid hemorrhage regain functional independence, although approximately half complain of cognitive impairment and dissatisfaction with their quality of life [35]. Approximately half the patients have memory deficits persisting beyond a year after aneurysmal subarachnoid hemorrhage. In the second year after aneurysmal subarachnoid hemorrhage, a substantial number may experience depression or anxiety. All of these complications affect the quality of life, ability to resume responsibilities, and return to their former work.

Anoxic–Ischemic Encephalopathy

Conspicuously lacking in many outcome studies of any useful size is information about the neurologic examination, use of sedation, neuroimaging, and postresuscitation hemodynamic and organ function, all of which can be crucial in deciphering patients' prognoses [37]. One of the most important principles in postcardiopulmonary-arrest prognostication is that a good outcome remains very difficult to predict. Another observation over the many years of developing prognosticators is that only a few clinical findings stand out; no new findings have been introduced since the late 1980s [2, 38]. Most clinical studies have been done by non-neurologists, which may have led to generalities when the examination is concerned [39]. Poor prognostic neuroimaging (severe cortical involvement on MRI) and EEG findings (burst suppression and no reactivity) have more recently emerged as reliable indicators, but these studies have been available for many decades and may already have been used by practitioners in judging the severity of anoxic–ischemic injury. The more recent ERC/ESCIM guideline is also a modification of the prior first-reported guidelines and performs well.

Outcome after anoxic–ischemic encephalopathy may be determined by a number of factors including early decisions to forgo intensive care. Reasons for withdrawal of support by general intensivists are not exactly known and may depend on unknown factors. However, there may also be less critical scrutiny of the neurologic condition and associated tests [40]. Some patients with initially poor neurologic findings improve rapidly with a return of pupillary and corneal reflexes after blood pressure returns to consistently normal levels. Most studies focus on indicators of lack of improvement. The brainstem is typically spared. This indicates that absent pupillary or corneal reflexes are much less common and that motor response to noxious stimuli is the best indicator of the degree of injury in most patients. Absent pupillary responses in a midposition-size pupil remain particularly indicative of a poor outcome and reflect a far more severe anoxic–ischemic injury damaging the brainstem. Our findings confirm that status myoclonus, confirmed by a neurologist, is associated with poor outcomes in a large contemporary series of patients with postresuscitation encephalopathy. We also showed that not all forms of myoclonus portend the same prognosis. Status myoclonus is predictive of poor outcome and was associated with other neurologic predictors, including absent pupillary reflexes, a poor motor exam, and malignant EEG findings, compared to nonstatus myoclonus.

Very few studies have carefully looked into cognitive abnormalities of survivors following therapeutic hypothermia. In our experience, approximately 40% of patients had memory and attention difficulties, but the majority of patients returned to work. A recent study from Pittsburgh provided a number of specialists (a third neurologist or neurointensivist) with descriptive vignettes of clinical course and then asked them to offer a prognosis; the responses were significant for their misguided optimism [41]. The "prognosticators" were inaccurate in about a third of their assessments, and most of the errors were optimistic. Outcome was fortunately determined by neurological factors (89%) including exam (81%) and EEG (37%). Nonneurological factors were next commonly used in 73%.

Interviewing survivors, who felt their lives had been significantly disrupted, has identified a number of major concerns [42]. Symptoms included fatigue, pain (from ribs broken while being resuscitated), substantial weight loss, diminished muscle strength, and increased breathlessness. Cognitive functioning could also be affected leading to poor memory, decision-making difficulties, and speech problems [43]. Many expressed anxiety about their physical vulnerability and a possible repeat occurrence or worse.

Impaired cognition may improve. Two cognitive screening exams, the MoCA (or the CAMCI) are useful to identify intellectual impairment gaps in cardiac-arrest survivors [44]. Evaluation of neurocognitive function after cardiac arrest has been markedly hampered by missing data. Not infrequently, subjective statements or telephone interviews are put forward as evidence. When it comes to MoCA, only the most severely impaired may be unable to perform the tests, and those with milder cognitive impairment may be unidentified because of crude measures. This may bias the documentation of neurocognitive impairment and its consequences. A 2018 detailed neuropsychological assessment of cardiac-arrest survivors 6 months post arrest found substantial cognitive impairment (>3 abnormal cognitive domains) in 26% [45]. In clinical trials, there is no consensus concerning the time point for neurocognitive assessment mainly because of the lack of evidence regarding longitudinal change. In 2016, Ørbo et al. published results from a detailed neuropsychological examination at 3 months post arrest, showing impairment in verbal and visual episodic memory, executive functioning, and processing speed [46]. At 12 months, they found improvements in visual memory and executive function but not in other domains. They included a larger number of cardiac arrest survivors but administered fewer tests for neurocognitive assessment. At 2 weeks, 54% of survivors were impaired in one or more tests. Most cognitive recovery occurred in the first 3 months following cardiac arrest, with 43% of survivors showing impairment at both 3 and 12 months. Emotional problems are common and persistent. One study found a psychiatric diagnosis in a quarter of patients who survived, with depression being the most common and more frequent in women and younger patients [47].

CNS Infections

This category of injury to the brain is too heterogeneous to predict a robust clinical trajectory. First, what matters most is whether the patient has a competent immune system. CNS infections in immunocompromised patients not only have different etiologies but also a worse outcome; we cannot anticipate a trajectory to full functional recovery. Conversely, in an immunocompetent patient, we can expect a favorable clinical trajectory in any other type of brain infection, although the immediate clinical course is determined by the time it took to recognize the cause and initiate treatment with adequate and appropriate doses of antimicrobial medication. Another important determinant is whether the patient with bacterial meningitis has received early treatment with corticosteroids. In some instances, brain edema in fulminant meningitis responds well to osmotic diuretics and additional high-dose IV corticosteroids, and acute obstructive hydrocephalus should be treated with placement of a ventriculostomy.

Some infections are neurosurgical emergencies. This includes any cranial or spinal epidural empyema, cerebral abscesses, and ventriculitis. Each of these disorders can be treated effectively, washed out or drained, and treated with multiple antibiotics. With an aggressive approach, clinical recovery occurs within weeks.

Recovery from a recent CNS infection – regardless of the cause – takes time. We must resist the temptation to de-escalate care if the patient has "plateaued" for several weeks and, in particular, if there is no other reason that makes improvement unlikely [48].

The outcome of a single brain abscess is dependent on whether the patient is a good surgical candidate. Indeed, most patients with an isolated lesion are good surgical candidates. The threshold for a neurosurgical approach should also be low if the abscess is superficially placed or in the cerebellum. Antimicrobial therapy seems more appropriate in patients with deep-seated abscesses, multilocular abscesses, and in immunosuppressed patients with disseminated microabscesses. Enlargement of the abscess with gradual clinical deterioration is an absolute indication for stereotactic aspiration or extirpation. The chances of recovery may correlate with location of lesion (deep-seated), progression to coma before intervention, and particularly, ventricular rupture. When an abscess ruptures into the ventricular system and causes fulminant meningitis, the patient declines rapidly, and the outcome is often very guarded or even fatal [49].

The long-term clinical course of encephalitis remains hard to predict, but age, duration of disease, and level of consciousness matters. Patients younger than 30 years of age who remain largely alert have a much higher chance of returning to preinfection normal life than older patients with altered consciousness. Patients requiring mechanical ventilation often have a poor prognosis – but not all of them; indeed, some show substantial recovery over a number of years. However, predicting which patients will improve under these circumstances is impossible.

We have limited knowledge on later clinical trajectories and long-term outcomes, and this is a significant knowledge gap. There has been a low value placed on impairment and disability, and most data involve mortality from systemic complications. Systematic screening for impairments with validated tools has not been done in a number of CNS infections, which may lead to misclassifications and underreporting [50].

Acute Neuromuscular Disorders

Acute neuromuscular disorders may rapidly cause a major paralysis, and the patient and family often ask desperately for an estimation of outcome. Knowledge of prognosis in acute neuromuscular disorders serves an important purpose. It is not unreasonable to define a good outcome as: (1) the patient returns to walking with or without support; (2) the patient needs no ventilatory support and tracheostomy has been removed; (3) the patient has recovered the swallowing mechanism; (4) the patient is able to read with or without corrective measures; (5) the patient displays no signs of fatigue or depression; and (6) the patient has no major disturbances in micturition. Each of these criteria will, in some way, define the quality of life in a patient struck with this disorder.

Who is going to recover fully and why? What is the intensity level of rehabilitation? How do we measure disability in these disorders? For example, in Guillain–Barré syndrome (GBS), once the patient is mechanically ventilated, we can anticipate a prolonged ICU stay and a high likelihood of tracheostomy placement despite early administration of plasma exchange or IVIG treatment. Still, mechanically ventilated patients have the ability to walk independently; this occurs in about 75% of the patients and may extend up to 2 years post onset. Upper limb paralyses at peak disability may predict a lesser ability to ambulate. Nonetheless, patients may show clinically significant improvement beyond a year or two; therefore, supportive treatment and aggressive rehabilitation is warranted. Highly intensive rehabilitation (almost daily for 12 months), including bladder and bowel training as well as motility and transfers, improved disability in GBS 1 to 12 years after diagnosis. However, it remains impossible to predict who can eventually discard a wheelchair and who must remain in one permanently. Fortunately, most regain the ability to walk. Generally, in GBS, we expect to see the first upper-limb movements in 6 weeks and first speech in 2 months, after Passy-Muir valve placement. Weaning from the ventilator, movements in the lower limbs, and ability to transfer should all occur after 3 months. The ability to walk unassisted usually returns in 7 to 8 months.

There is much more variability with myasthenia gravis. Onset of myasthenia gravis in middle age has not only less severe manifestations but also lower probability of full remission and higher mortality when compared with early-onset myasthenia. A worse outcome can be expected in patients with malignant thymoma but only if the tumor has breached the capsule and caused metastasis. Ultimately, quality of life is largely determined by the severity of muscle weakness, which may include weakness of neck muscles and constant head drop, dysphagia and chewing problems, ptosis, diplopia, speech impediments, and the secondary effects of immunosuppressive therapy, which may include recurrent infections, osteoporosis, cataract, or malignancy in patients taking azathioprine for many years.

Acute Spinal Cord Injury

Mortality following acute spinal cord injury is comparatively high, and patients often succumb to their injuries before transport to hospital (~20%). Age greater than 20 years, male sex, severe systemic injuries (Injury Severity Scale ≥ 15), concurrent traumatic brain injury, one or more comorbidities, neurologic status, and level of care as judged by lack of admission to a level 1 trauma center are factors associated with death during initial hospitalization.

As expected, chances of recovery correlate inversely to the severity of the neurological injury. Only 10 to 15% of patients with a complete lesion on initial neurological examination will convert to incomplete status. Rates of recovery for incomplete lesions are better: one third of ASIA B patients will convert to ASIAC and another one third to ASIA C or D. Initial ASIA C status has an even better outcome; 70% will convert to ASIA D or E. However, very few patients improve from ASIA D to normal neurological status. The level of injury also influences potential for neurological recovery. Among patients with complete SCI, those with cervical-level injuries have significantly better neurological recovery when compared with those with injury at a thoracic level. The rates of neurological recovery for those with incomplete injuries appear to be similar for both cervical and thoracic SCIs.

In traumatic spinal cord injury at baseline, 73% of subjects were ASIA A, and among them, 15% converted to incomplete motor strength. In these patients, there are more variable outcomes of recovery below T9. Functionality declines with age without a clear cut-off [51].

Recovery after a spinal cord infarction is mixed. At their worst, about half of patients had severe impairment (ASIA grades A and B). At their nadir, over 75% of patients required a wheelchair for mobility, and nearly all were catheterized. A third of the patients complain of pain on their initial examination. Severe impairment (ASIA A or B) on initial examination, absence of Babinski sign, presence of sensory level, longitudinally extensive MRI lesions, and MRI lesions at the highest level in the thoracic region were associated with wheelchair and catheter use at 3 years follow-up. But half of the patients ambulated with or without a gait aid after a number of years passed. Meaningful recovery is possible in a substantial minority of these patients [52].

More Reflections

Prediction of outcome is expected of any physician and part of the practice of medicine.

At its worst, catastrophic brain injury often defines itself quickly. Outcome is often de-escalation of care and patients succumb. Some progress to brain death and may become organ donors. Mortality is generally a less common outcome in neurocritically ill patients with severe brain injury. In-hospital death may be due to additional medical complications. Later mortality is often subjected to "autopsy and retrospectives," even more these days because they affect hospital ratings. A physician of my acquaintance has categorically claimed that every bad outcome can be avoided and that retrospective review of any unsuccessful episode of care will reveal mismanagement or error. I would argue, however, that severe, persistent brain injury is always a setup for complications, many of which cannot be prevented or adequately treated. But mortality and morbidity conferences remain useful because they can identify a "system issue" such as potential lapses in communication (Chap. 12) or interpretation and delays in treatment.

Neurocritical care has just barely scratched the surface, and neurointensivists have to deal with shifting priorities of care or the choices we make, less than optimal assessments of outcome (cognitive vs. physical), and changing social support structures. A key (but mostly unanswered) question is whether short-term pessimism can change into qualified long-term optimism. (Also remember that pessimists do not see that outcome of certain disorders may have improved over the years. Optimists continue to ignore futility). Many very sick patients can eventually be rehabilitated into an acceptable functional outcome (as rated by patient or next of kin), but "acceptable" is a highly subjective judgment call. Preconceived notions of a poor outcome by family may affect outcome, but they are hard to objectify (and may not be as preconceived as we think). Overall, the premorbid state, clinical performance, and support systems (financial and social structures) will matter most. Who is there to take care of the affected person and how would that affect (i.e., ability to stop or change work) the caregiver? One crucial issue is that in countries in the Western Hemisphere, patients and families consider independence to be of utmost importance and do not want to go to a nursing home or skilled nursing facility. The burden is on the neurologic prognosticator to determine future independence (or lack thereof); this is far more important than awakening from coma or the presence of a residual neurologic handicap. Values differ in the Eastern Hemisphere, where independence as an outcome criterion is rarely a major determinant. One thing is for certain: there is a Goldilocks effect – everything has to go just right for many months for patients to reach a good functional outcome.

One final thought: for at least a century, there was a clear asymmetry in knowledge of outcome between patients, families, and responsible physicians. Physicians, mostly unwillingly, held some power over a patient's outcome. Claims of authority were common. We now share this responsibility and have a much better sense of what patients and families want. Occasionally, some family members seek to subvert the physician's moral obligation to give the patient bad news. Most of the time, however, communication with families is cordial and compassionate. Make no mistake – neurologists in their heart of hearts want patients to pull through.

Pointers and Takeaways

- Outcome trajectories based on clinical examination alone can be defined for most neurocritical illnesses.
- Outcome is indirectly determined by patients' prior wishes, extent of intervention, and preexisting conditions.
- Outcome generally is determined at onset, but another injury from a complication may jeopardize the chances of full recovery. How a patient fares in the hospital in the following week is a major determinant of prognosis.
- Outcome can be satisfactory in younger patients over a longer time but is far less likely in the elderly.
- Outcome (good or bad) remains hard to predict accurately. A good outcome may be offset by a subsequent major complication; a poor outcome may be offset by aggressive, attentive care, and unexpected (unmeasurable) innate resilience.
- Outcome is generally poor in patients who awaken late (months later).
- Outcome in acute brain injury is often more determined by nonphysical (strength and gait)

domains such as emotion, decision making, stress relief, mood changes, and personal interactions.

References

1. Bates D, Caronna JJ, Cartlidge NE, et al. A prospective study of nontraumatic coma: methods and results in 310 patients. Ann Neurol. 1977;2:211–20.
2. Levy DE, Bates D, Caronna JJ, et al. Prognosis in nontraumatic coma. Ann Intern Med. 1981;94:293–301.
3. Hocker S, Wijdicks EF. Recovery from locked-in syndrome. JAMA Neurol. 2015;72:832–3.
4. Twomey F, O'Leary N, O'Brien T. Prediction of patient survival by healthcare professionals in a specialist palliative care inpatient unit: a prospective study. Am J Hosp Palliat Care. 2008;25:139–45.
5. Finley Caulfield A, Gabler L, Lansberg MG, et al. Outcome prediction in mechanically ventilated neurologic patients by junior neurointensivists. Neurology. 2010;74:1096–101.
6. Pratt AK, Chang JJ, Sederstrom NO. A fate worse than death: prognostication of devastating brain injury. Crit Care Med. 2019;47:591–8.
7. Wijdicks EFM. Predicting the outcome of a comatose patient at the bedside. Pract Neurol. 2020;20:26–33.
8. Edlow JA, Rabinstein A, Traub SJ, Wijdicks EF. Diagnosis of reversible causes of coma. Lancet. 2014;384:2064–76.
9. Collaborators MCT, Perel P, Arango M, et al. Predicting outcome after traumatic brain injury: practical prognostic models based on large cohort of international patients. BMJ. 2008;336:425–9.
10. Lingsma HF, Roozenbeek B, Steyerberg EW, Murray GD, Maas AI. Early prognosis in traumatic brain injury: from prophecies to predictions. Lancet Neurol. 2010;9:543–54.
11. Maas AI, Marmarou A, Murray GD, Teasdale SG, Steyerberg EW. Prognosis and clinical trial design in traumatic brain injury: the IMPACT study. J Neurotrauma. 2007;24:232–8.
12. Broderick JP, Adeoye O, Elm J. Evolution of the modified Rankin scale and its use in future stroke trials. Stroke. 2017;48:2007–12.
13. Quinn TJ, Dawson J, Walters MR, Lees KR. Reliability of the modified Rankin scale: a systematic review. Stroke. 2009;40:3393–5.
14. Wilson JR, Grossman RG, Frankowski RF, et al. A clinical prediction model for long-term functional outcome after traumatic spinal cord injury based on acute clinical and imaging factors. J Neurotrauma. 2012;29:2263–71.
15. Osa Garcia A, Brambati SM, Brisebois A, et al. Predicting early post-stroke aphasia outcome from initial aphasia severity. Front Neurol. 2020;11:120.
16. Watila MM, Balarabe SA. Factors predicting post-stroke aphasia recovery. J Neurol Sci. 2015;352:12–8.
17. Kertesz A, McCabe P. Recovery patterns and prognosis in aphasia. Brain. 1977;100(Pt 1):1–18.
18. Forkel SJ, Thiebaut de Schotten M, Dell'Acqua F, et al. Anatomical predictors of aphasia recovery: a tractography study of bilateral perisylvian language networks. Brain. 2014;137:2027–39.
19. Nijland RH, van Wegen EE, Harmeling-van der Wel BC, Kwakkel G, Investigators E. Presence of finger extension and shoulder abduction within 72 hours after stroke predicts functional recovery: early prediction of functional outcome after stroke: the EPOS cohort study. Stroke. 2010;41:745–50.
20. Jenkinson PM, Moro V, Fotopoulou A. Definition: asomatognosia. Cortex. 2018;101:300–1.
21. Akkersdijk WL, Roukema JA, van der Werken C. Percutaneous endoscopic gastrostomy for patients with severe cerebral injury. Injury. 1998;29:11–4.
22. Young B, Ott L, Yingling B, McClain C. Nutrition and brain injury. J Neurotrauma. 1992;9 Suppl 1:S375–83.
23. Zarbock SD, Steinke D, Hatton J, Magnuson B, Smith KM, Cook AM. Successful enteral nutritional support in the neurocritical care unit. Neurocrit Care. 2008;9:210–6.
24. Crary MA, Mann GD, Groher ME. Initial psychometric assessment of a functional oral intake scale for dysphagia in stroke patients. Arch Phys Med Rehabil. 2005;86:1516–20.
25. Galovic M, Stauber AJ, Leisi N, et al. Development and validation of a prognostic model of swallowing recovery and enteral tube feeding after ischemic stroke. JAMA Neurol. 2019;76:561–70.
26. Weeden M, Bailey M, Gabbe B, Pilcher D, Bellomo R, Udy A. Functional outcomes in patients admitted to the intensive care unit with traumatic brain injury and exposed to hyperoxia: a retrospective multicentre cohort study. Neurocrit Care. 2021;34:441–8.
27. Williamson T, Ryser MD, Abdelgadir J, et al. Surgical decision making in the setting of severe traumatic brain injury: a survey of neurosurgeons. PLoS One. 2020;15:e0228947.
28. Delcourt C, Sato S, Zhang S, et al. Intracerebral hemorrhage location and outcome among INTERACT2 participants. Neurology. 2017;88:1408–14.
29. Eslami V, Tahsili-Fahadan P, Rivera-Lara L, et al. Influence of intracerebral hemorrhage location on outcomes in patients with severe intraventricular hemorrhage. Stroke. 2019;50:1688–95.
30. Arima H, Tzourio C, Butcher K, et al. Prior events predict cerebrovascular and coronary outcomes in the PROGRESS trial. Stroke. 2006;37:1497–502.
31. Biffi A, Anderson CD, Battey TW, et al. Association between blood pressure control and risk of recurrent intracerebral hemorrhage. JAMA. 2015;314:904–12.
32. Wermer MJH, Greenberg SM. The growing clinical spectrum of cerebral amyloid angiopathy. Curr Opin Neurol. 2018;31:28–35.
33. Arboix A, Vall-Llosera A, Garcia-Eroles L, Massons J, Oliveres M, Targa C. Clinical features and

functional outcome of intracerebral hemorrhage in patients aged 85 and older. J Am Geriatr Soc. 2002;50:449–54.
34. Carbajo-Garcia AM, Cortes J, Arboix A, et al. Predictive clinical features of cardioembolic infarction in patients aged 85 years and older. J Geriatr Cardiol. 2019;16:793–9.
35. Ariyada K, Ohida T, Shibahashi K, Hoda H, Hanakawa K, Murao M. Long-term functional outcomes for world federation of neurosurgical societies grade V aneurysmal subarachnoid hemorrhage after active treatment. Neurol Med Chir (Tokyo). 2020;60:390–6.
36. Risselada R, Lingsma HF, Bauer-Mehren A, et al. Prediction of 60 day case-fatality after aneurysmal subarachnoid haemorrhage: results from the International Subarachnoid Aneurysm Trial (ISAT). Eur J Epidemiol. 2010;25:261–6.
37. Steinbusch CVM, van Heugten CM, Rasquin SMC, Verbunt JA, Moulaert VRM. Cognitive impairments and subjective cognitive complaints after survival of cardiac arrest: a prospective longitudinal cohort study. Resuscitation. 2017;120:132–7.
38. Wijdicks EF, Hijdra A, Young GB, Bassetti CL, Wiebe S, Quality Standards Subcommittee of the American Academy of N. Practice parameter: prediction of outcome in comatose survivors after cardiopulmonary resuscitation (an evidence-based review): report of the Quality Standards Subcommittee of the American Academy of Neurology. Neurology. 2006;67:203–10.
39. Sandroni C, D'Arrigo S, Cacciola S, et al. Prediction of poor neurological outcome in comatose survivors of cardiac arrest: a systematic review. Intensive Care Med. 2020;46:1803–51.
40. Moseby-Knappe M, Westhall E, Backman S, et al. Performance of a guideline-recommended algorithm for prognostication of poor neurological outcome after cardiac arrest. Intensive Care Med. 2020;46:1852–62.
41. Steinberg A, Callaway C, Dezfulian C, Elmer J. Are providers overconfident in predicting outcome after cardiac arrest? Resuscitation. 2020;153:97–104.
42. Whitehead L, Tierney S, Biggerstaff D, Perkins GD, Haywood KL. Trapped in a disrupted normality: survivors' and partners' experiences of life after a sudden cardiac arrest. Resuscitation. 2020;147:81–7.
43. Blennow Nordstrom E, Lilja G. Assessment of neurocognitive function after cardiac arrest. Curr Opin Crit Care. 2019;25:234–9.
44. Koller AC, Rittenberger JC, Repine MJ, et al. Comparison of three cognitive exams in cardiac arrest survivors. Resuscitation. 2017;116:98–104.
45. Juan E, De Lucia M, Beaud V, et al. How do you feel? Subjective perception of recovery as a reliable surrogate of cognitive and functional outcome in cardiac arrest survivors. Crit Care Med. 2018;46:e286–93.
46. Orbo M, Aslaksen PM, Larsby K, Schafer C, Tande PM, Anke A. Alterations in cognitive outcome between 3 and 12 months in survivors of out-of-hospital cardiac arrest. Resuscitation. 2016;105:92–9.
47. Desai R, Singh S, Patel K, Fong HK, Kumar G, Sachdeva R. The prevalence of psychiatric disorders in sudden cardiac arrest survivors: a 5-year nationwide inpatient analysis. Resuscitation. 2019;136:131–5.
48. Stahl JP, Mailles A. Herpes simplex virus encephalitis update. Curr Opin Infect Dis. 2019;32:239–43.
49. Widdrington JD, Bond H, Schwab U, et al. Pyogenic brain abscess and subdural empyema: presentation, management, and factors predicting outcome. Infection. 2018;46:785–92.
50. Roos KL, Tunkel AR. Bacterial infections of the central nervous system. Preface. Handb Clin Neurol. 2010;96:ix.
51. Lee BA, Leiby BE, Marino RJ. Neurological and functional recovery after thoracic spinal cord injury. J Spinal Cord Med. 2016;39:67–76.
52. Robertson CE, Brown RD Jr, Wijdicks EF, Rabinstein AA. Recovery after spinal cord infarcts: long-term outcome in 115 patients. Neurology. 2012;78:114–21.

12 Communicating Clinical Findings

Neurocritical care is fundamentally observation, examination, and classification before intervention. There will be a clinical diagnosis and a trajectory—a part already known from history (Chap. 1) and a part that can be anticipated knowing the state the patient is in (Chap. 6). The emerging information must be reviewed to determine what one needs to know and what one can ignore. All this activity and bits of data are documented in electronic medical records and populated with often more information than practitioners request or want. It has been aptly named cognitive overload.

Tremendous volumes of clinical patient information—thousands of data points in 24-hour periods—consume vast amounts of physicians' time, and there is a growing need to develop tools to customize these records to something more straightforward. Electronic records also have created a larger digital divide between the younger and the older healthcare professionals. The younger generation truncates personal interactions between nursing staff and physicians to chats on the electronic medical record portal. This leaves a useful and verifiable trace. The senior physician, however, often feels that he or she is catching up with the changing technology.

All this streamlining is an improvement, but practice is potentially far more chaotic and in the moment. We need to make the right calls, set goals, and even pull the right strings to make things happen. And the information provided is only worthwhile if others understand what they are told. And we tell each other a lot. Amidst the intensive care information noise, we write and read notes, talk, call, text, chat, and email about our patients [1–4]. We can be reached instantaneously (and we want this contact), but we are often juggling several balls in the air. We now have accepted a medical environment with expectations for reachability.

And, how about handoffs or sign-outs?—a moniker for the communication when a patient transfers within the hospital or from another hospital. We know that night interns refer to the sign-out in order to answer questions that arise during the night [5]. At least one survey suggests that a "better" sign-out would reduce adverse events [6]. It is a complex problem. For example, another survey asked clinicians to identify whether they anticipated any nighttime events for each patient and what specific type of event they were most worried about after hearing the clinical history and course of events. In over 300 patient handoffs, these intensivists were interviewed immediately after the handoff but before caring for patients. Sadly, these nighttime clinicians could only identify 53% of the daytime clinicians' diagnoses [7]. These were clustered into five predefined domains (hemodynamic, respiratory, metabolic, neurologic, and hematologic). And not surprisingly, clinicians

E. F. M. Wijdicks, *Examining Neurocritical Patients*, https://doi.org/10.1007/978-3-030-69452-4_12

were less likely to diagnose patients with altered consciousness correctly—a condition hard to define and to communicate (Chaps. 3 and 7). If anything, this study showcased a fundamental misunderstanding of a patient's course. More seriously, communication errors may have contributed to underestimation of aspiration risk and cardiac arrhythmia [7].

A patient handoff loaded with irrelevancies, extraneous detail, and a less-than-succinct summary clearly increases the risk for missed vital information and misunderstanding [8]. The risk of not communicating potentially relevant information in a standardized handoff leaves the recipient healthcare provider poorly equipped to respond to medical emergencies [9–12]. In any case, improved provider perceptions of transfer workflow efficiency and patient safety may not be enough; communication must also include solutions for active medical problems and an outline of anticipatory guidance in the event of an acute change in a clinical condition ("what else–what if–what then" scenarios).

Checklists improve communication [13, 14]. Incorporating the checklist into the electronic health record allows data to auto-populate and eliminates reliance on the provider's memory for specific details of, for instance, medication dosages or administration times. The remaining problem is the sustainability of checklists and whether they may eventually disappear or remain unfiled. Checklists may not reduce ICU readmissions or rapid-response team (RRT) calls [13]. Rapid readmissions or so-called "bounce backs" remain difficult to predict, and communication failure is just one factor [15–17].

Generally, inadequate handoffs lead to diffused responsibility, for which we, and the patients, pay a price. Physicians tend to overestimate the effectiveness of their handoff communication, and moreover, we seldom agree on what is the most important piece of communicated information.

We can assume a connected life in the hospital, and we shape them all the time. Information ripples through these networks. Communication networks may involve a hierarchal flow (junior residents to other junior residents to seniors to fellows to staff) or directly (junior to staff and staff to staff communication). Circles of communication must be intact and people in it must be cooperative for information not to get lost (Fig. 12.1).

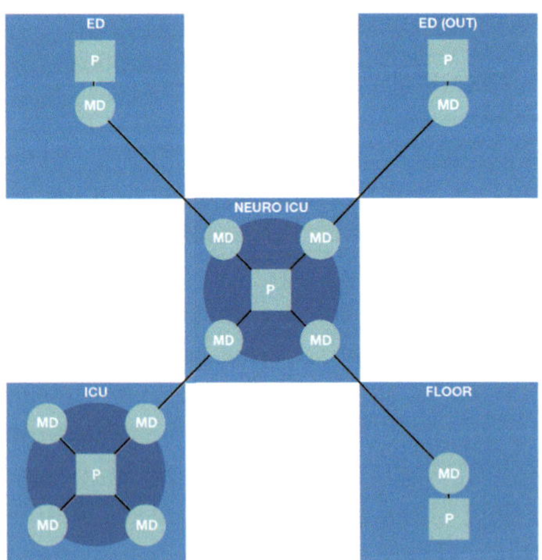

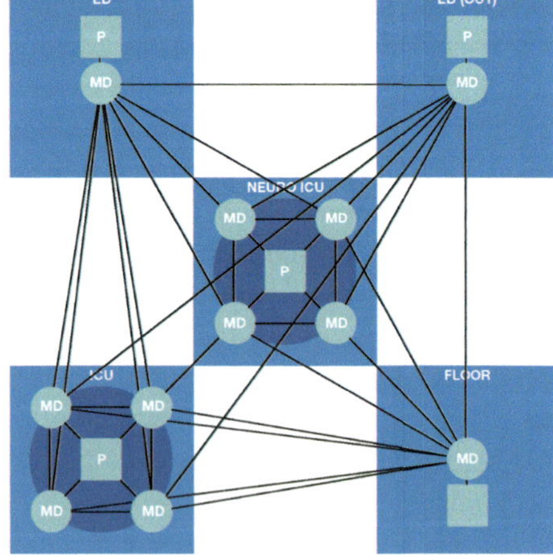

Fig. 12.1 A network of communication *Left*: Note that usually one MD from outside the neurosciences ICU communicates to a group of physicians taking care of the patient. These physicians ideally communicate with each other (text, email, rounds), and this may include a certain hierarchy (resident to fellow to staff). Who is informed first may vary. (P: patient; MD: attending physician.) *Right:* The networks can become quite crowded and we may not be able to reconstruct who talked to whom—a Babylonian confusion with the same language. Reciprocity is key

No currently existing tool, requirement, or system provides a standard system of communication. In-hospital communication protocols have used the "Situation-Background-Assessment-Recommendation" (SBAR) model, but it lacks specificity [18]. This tool typically covers the following questions: (1) what is the situation and what is happening with the patient that requires immediate attention? (2) What is the background information of the patient? (3) What is the most reasonable assessment of the clinical (neurologic) state? (4) How should the problem be corrected and what is recommended? The SBAR protocol was first introduced in 2003 to provide some structure in conversations between doctors and nurses about situations requiring immediate attention. Some experience has been reported in ICUs and acute pediatric settings [19–21]. Use by nursing staff is inconsistent, and use by ICU physicians is virtually nonexistent. However, major progress has been made with the Handoffs and Transitions in Critical Care (HATRICC) prospective study, which determined that handoff standardization effectively improved information exchange, teamwork, and professionalism, for mixed (cardiac and noncardiac) surgical populations but disappointingly increased the handoff time by 100% and had no effect on patient outcomes or ICU stay [22, 23].

This final chapter explores the commonly encountered scenarios where communication may falter and fracture or disintegrate care continuity, methods to improve communication skills, and how to acquire the necessary skills to resolve ambiguities. The argument I like to make is that our clinical neurologic examination may become meaningless without personal face-to-face communication. It often involves physician-to-neurointensivist interactions, but the principles outlined here must apply to any healthcare provider. The very ill patients in the neurosciences intensive care unit are at greater risk for communication errors, due to their higher acuity and consequent higher complexity. A high patient complexity introduces another source of error in the handoff process—the tendency to focus on recent details of patient care (e.g., ventilator settings, medication infusions) and ignore a broad overview of the patient's condition.

The Risks to Effective Communication

Communication means acquisition and dissemination of necessary information. So why can we not do it effectively? What are the barriers?

Challenges to efficient communication are manifold. First is ambiguous responsibility. Simply, who needs to know what and why? Providing crucial, actionable information to a person who is less informed (intern, float, or anyone who says "I am just covering") is a common mistake. It may lead to inadequately rendered acute care. Second is lack of structure. Just as a triage of patients requires structural and response models, information sharing has to "check a few boxes" before we can be confident of having the full picture. Even basic chronology may be missing. We should immediately suspect a critical flaw in communication if several follow-up questions from the person on the receiving end lead to more crucial information. Third is the tendency to ignore information when interrupted and worse when interrupted while overwhelmed with a high volume of patients. We all recognize the challenge of writing notes while also having to channel emergency calls, review neuroimaging ("can I run this case quickly by you?"), and respond to texts. It is obviously detrimental to care if these interruptions disrupt usual routines and competing activities disturb concentration. The pressure of time has only worsened, but we will have to adjust. The big question is if we can.

The Anatomy of Miscommunication

Misinformation is rampant, and many of us are discombobulated. In all sincerity, why do some messages fall on deaf ears? We may believe that

Table 12.1 Communication bumfuzzles

"Altered"
"Not talking and moving only on the left side"
"Going in and out of consciousness"
"Pupils are now sluggish"
"Babinski's are mute"
"Staring and not responding"
"No responses anywhere"
"It is bad, I mean really bad"
"Shakin' all over"
"Blood pressure is on the soft side"
"Heart rate is somewhat increased"
"Laboratory tests are stable"
"Laboratory values are creeping up"
"CT scan shows no bleed"
"Patient is borderline for the ward"

we are getting everything we need in a timely manner—but not necessarily. I have heard a few amusing bumfuzzles over the years (Table 12.1), and it is necessary to expose them in order to understand what we are not doing right. Often, we use lingo not understood by non-neurointensivists and vice versa. With no intended disparagement of colleagues (and humbly aware that none of us is infallible), I offer a few illustrative examples inspired by real events. Typically, the most error-prone situations are the triage and direct admissions to the neurosciences intensive care unit and the communications on management or transfer out of the intensive care unit. Some communications are clear-cut while others are inscrutable. I often find that continuing to ask (friendly) prodding questions will ultimately lead to more, extremely relevant information ("the plot thickens" phenomenon). Sometimes communications are premature ("CT scan of the brain has been ordered; I will call you back with the result") or do not make sense at least when heard for the first time ("the patient is not talking and there is no arm or leg movement on the left"). As noted in the opening chapter, taking histories from patients or family members entering emergency departments and moving later to the intensive care is a challenge. As such, it is a major determinant of successful communication.

Example 1: Communication With Physician Outside the Hospital

Physician A:	*I have a patient here who is pretty much unresponsive after a fall, and CT scan shows a hemorrhage. I want to send him over.*
Neurointensivist:	*What more can you tell me?*
Physician A:	*We are going to intubate him.*
Neurointensivist:	*What does his examination show?*
Physician A:	*GCS of 3*
Neurointensivist:	*Can you tell me more?*
Physician A:	*That is pretty much it.*

Take away: The CT scan shows an acute subdural hematoma with shift. The brainstem reflexes are intact, but the pupil on the same side of the hemorrhage is wider. INR returns markedly elevated. Best advice is to initiate osmotic diuretics and adequate reversal of anticoagulation with PCC and IV vitamin K and to call a neurosurgeon before he arrives.

Any neurointensivist will instinctively ask "what is the matter with this patient?" and look for patterns even when our colleagues fail to explain things clearly and with sufficient detail. We have trouble understanding those who do not seem to understand. Inarguably, communicating insufficient information about the patient could result in insufficient or inappropriate care before transport, such as in the example above. Perfect orchestration in collaborative communication may be too much to ask, but we should continue to consider these three most pertinent questions to calls from physicians asking for transfer: (1) Is the patient deteriorating and from what? (2) Is a neurosurgical (craniotomy) or neurocritical care intervention (osmotic agents) urgently needed? (3) Is an endovascular procedure needed?

Example 2: Communication with Physician Inside the Hospital (Coming to the ICU)

Physician A: *I have a patient on the ward who just RRTd for a seizure and needs to go to the neuro ICU.*
The neurointensivist: *Why?*
Physician A: *The nurses are uncomfortable.*
The neurointensivist: *How about you?*
Physician A: *Well, she does not seem to awaken after the seizure.*

Take-away: This patient has a glioma and has had focal seizures throughout the day and multiple doses of lorazepam. Secretions and airway protection are an issue, and we need a blood gas to exclude CO_2 retention. A new CT scan also will need to be done before the patient arrives.

Rapid-response team (RRT) calls are common reasons for intensive care admissions, and the rapid responder in question has the upper hand. Information is fragmented, and rapid transfer takes preference over detailed planning. Some arriving patients look just fine; others are very unstable. Regardless, their arrival often leads to some sort of surprise.

Example 3: Communication with Physician Inside the Hospital (Coming out of the ICU)

The neurointensivist: *We are going to transfer a patient after a traumatic brain injury to you because he does not need any more ICU level of care and we have a bed crunch.*
The floor consultant: *Anything else I need to know?*
The neurointensivist: *He has been doing fine. We kept him longer because there were some secretion and blood pressure issues.*
The floor consultant: *Do you think the nursing staff will be comfortable with taking care of him?*
The neurointensivist: *I think so.*

Take away: This patient just recovered from a ventilator-associated pneumonia. Moreover, there was an escalation of blood-pressure medication and only 24 hours of significant reduction of secretion burden. Six hours after transfer, the patient bounced back to the ICU with difficulty clearing secretions and a blood-pressure surge.

There are no well-established criteria for transfer when physicians and nursing staff feel the patient "does not need ICU level of care." Some have used 6- to 12-hour stability of blood pressure, heart rate, and breathing rate, but this is obviously an arbitrary cut-off. Often, patients are transferred with mild bradycardia or tachycardia, which may trigger a RRT when the shift changes and the new person (rightly or wrongly) "feels uncomfortable."

A Useful Mnemonic

How can healthcare providers effectively communicate urgent admissions to the neurosciences ICU? Ideally, correct information is provided to the receiving end at the outset, and no further questions arise. I recognize that many mnemonics have been used and a recent systematic review of published handoff mnemonics identified 46 articles describing 24 different handoff mnemonics [24]. There are no rigorous outcome studies demonstrating the effectiveness of handoff mnemonics or the accuracy of recall. Many are formulaic approaches that are highly unusual, (NUTS, GRRR), or unduly complex (I PASS the BATON). There is a decided lack of specificity in the whole thing, and many of these mnemonics are imitative. Moreover, while standardized handoffs for physicians and other healthcare providers may simplify data transfers, they do not necessarily lead to a focused consequential discussion about the patient's management. Effective mnemonics

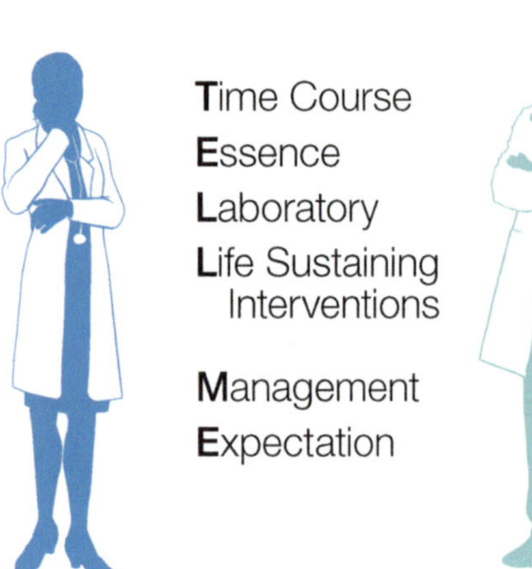

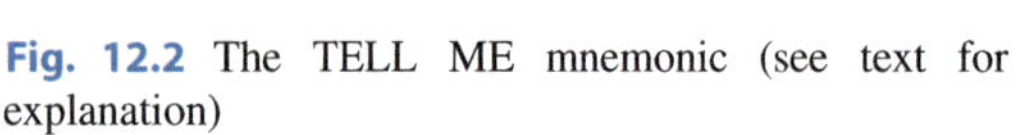
Fig. 12.2 The TELL ME mnemonic (see text for explanation)

are memory aids, catchy, symbolic, simple, serviceable, and invoke a visual image. Most have not been rigorously studied and accepted because they worked in practice.

I devised another mnemonic called TELL ME (Fig. 12.2), which helps physicians (or any other referring communicating healthcare professional) to standardize their communication. These are: knowing the time course (new and unexpected or ongoing for some time), extracting essential information from irrelevancies, communicating which tests are pending (CT and laboratory), relaying how much crucial support will be needed (e.g., secretion burden, intubation), knowing which emergency drugs have been administered (e.g., mannitol, antiepileptics, tranexamic acid), and planning for worst-case scenarios. I believe that with this information, the receiving staff better knows what to expect and is well prepared to intervene quickly. The mnemonic is not passive and invites a back-and-forth discussion, because it queries why crucial support is needed and how it has been managed. The mnemonic has been well received. (At the time of this writing, we are preparing a validation study.)

Rounds and Relaying Best Information

Rounds and the method of presenting patients have changed. Most ICUs no longer hold sit-down discussions with examination findings noted on blackboards with systematic precision and using decision trees. A high census in the ICU has slashed the time taken to discuss a single case and its pathophysiology comprehensively; now each case may be presented with a few explanatory remarks but without a step-by-step approach. Teaching or simply showing clinical findings at the bedside still takes place, but imaging has taken precedence over clinical interpretation. Our current methods in the ICU are not necessarily better; in fact, they may be worse [25, 26]. Senior consultants may express concern that having everyone standing outside the patient's room and glued to a screen undermines effective teaching during rounds. It is true that standing outside the room of the patient in question, offers every healthcare worker (and the family) the opportunity to hear a reasoned plan of care, but this method may not necessarily cover discrepancies in physical examination or provide an explanation for the findings. Do we provide explanations of changes in the clinical examination? If we do not set time apart, teams are more likely to report negative scores regarding teaching during rounds [27]. One study advocates using checklists to increase time for teaching, but most staff satisfaction studies have concentrated on communicating a plan rather than interpreting the examination [28].

With presentations of findings (and events of the night) usually proceeding through systems, dividing the body into a template of organ systems is an effective way of organizing thoughts. Most patients with an acute neurologic injury have other medical issues, which often flare up at night or soon after admission. Neurologists have all seen worsening congestive heart failure, new cardiac arrhythmias, poorly controlled diabetes mellitus, new gastrointestinal hemorrhage, and not infrequently, unstable blood pressures. It is essential to obtain information about past medical history, medication, allergies, and history of drug and alcohol use. In addition, it is important to obtain evidence of prior dementia (rarely admitted or volunteered by family members), which increases the risk of in-hospital delirium. It is even more important to obtain information about possible injuries that may have occurred as a result of a fall (due to hemiparesis).

The full medical physical examination that follows should include pulmonary evaluation with listening to breath sounds, cardiovascular status with listening for new murmurs and confirmation of the patency of the peripheral vasculature, and full abdominal evaluation for possible tenderness and expansion. The neurologic examination may reveal numerous findings, or there may be a true paucity of findings. The first major task is to reach some sort of working diagnosis, even if it seems to have escaped us completely. Patient concerns voiced by the nursing staff should be addressed and resolved if possible. The need for continuing intensive care management is addressed; if transfer to the floor (ward) is considered, the nursing staff plays a decisive role.

One of the more common difficulties in communicating is identifying changes in the level of consciousness that may be diagnostically informative. The term "unresponsiveness" is used freely in conversations and communication but is, of course, in no way a physical finding. Unresponsive may mean unreceptive, no verbal output, or simply unwilling to respond. Most comatose patients are actually quite responsive (to noxious stimuli).

Locked-in syndrome and catatonia are commonly missed by physicians. It is important to clarify that locked-in syndrome (unresponsive but receptive) must be considered in the appropriate clinical circumstances because it often is confused with coma. Acute locked-in syndrome can be seen with an embolus to the basilar artery causing ischemia in the dorsal pons and, unfortunately, may occur with neuromuscular-blocking agents if the patient is not adequately sedated (the patient is then completely unable to signal anything). Locked-in syndrome is due to a ventral pons lesion that makes the patient unable to move anything (no horizontal eye movements, no grimacing, no swallowing, and no head or limb movements). Comatose patients have their eyes closed, but the eyes of a patient with locked-in syndrome are open and capable of blinking and spontaneous vertical eye movement. Locked-in syndrome is horrific for the patient; therefore, recognition is critical (and the FOUR score (Chap. 2), when rigorously applied, will find it). The patient is fully aware of their surroundings but unable to communicate distress other than through vertical eye movements or blinking, hopefully picked up by an examiner.

Some unresponsive patients are aphasic and mute. An aphasic patient is often easy to diagnose despite minimal word output. Aphasia can be incorrectly interpreted as sudden unresponsiveness by the layperson or bystander. Acute muteness may occur in acute frontal syndrome, acute cerebellar lesions, or as the result of nonconvulsive status epilepticus or complex partial seizure.

Seizures can present with postictal agitation but also with unresponsiveness; they may even be perceived as unwillingness to cooperate. Nonconvulsive status epilepticus may have unusual presentations. Typically, the patient has rolling eye movements, and some have a fine nystagmus. There is often eyelid quivering and intermittent twitching of the extremities. The patient does not blink to threat and is not following or tracking. The diagnosis can be immediately confirmed with an EEG.

Psychogenic unresponsiveness is very uncommon and, therefore, often dismissed as a possibility—although, trust me, I have seen a few. Another important cause of unresponsiveness is catatonia. It usually presents with a posture, rigidity, mutism, and no response to commands. On examination, waxy flexibility and negativity is found. Some patients may suddenly speak, repeat a phrase continuously, or make repetitive movements and behaviors. There is often a glassy, fixed gaze; some may present echopraxia (mimicking examiner's movements), echolalia (mimicking examiner's speech), gegenhalten, or mitgehen (defined as a slight pressure causing the patient to move in their direction).

Patients with Parkinson's disease or neurodegenerative disease with parkinsonian features (e.g., progressive supranuclear palsy) may develop an akinetic crisis resulting in complete unresponsiveness. It may occur without any apparent change in antidopaminergic drug administration.

Communication of "altered" patients is perhaps the most difficult task in whole of Neurology (perhaps even Medicine), and throwing around nonsensical terms does not help ("the patient is so unconscious; there is nothing more to say"). All this vagueness is not helpful, and we have heard it too many times. Over many years, neurologists have developed labels for patients who do not respond normally: somnolent, encephalopathic, drowsy, disoriented, and when agitated, delirious. I approached these issues in Chap. 7, but it is good to reiterate the difficulties with describing "altered mental status" or "delirium" because they occur so often. Obviously, the terminology is far from clear (and attempted fixes may even exacerbate it [29]). Many patients with renal or hepatic encephalopathy show daytime sleepiness and nighttime agitation, also known as sundowning, Typically, delirium has several components of perturbation (arousal, language, perception, orientation, mood, sleep). Restlessness is associated with pallor, sweating, and tachycardia. Vivid visual and acoustic hallucinations are common but are not complex and simple (screams, bells ringing) and may last for a week. Neurologically, in alcohol withdrawal delirium markedly dilated pupils and hyperactive reflexes are found.

Delirium often has been diagnosed when confusion is accompanied by autonomic symptoms such as profuse sweating, muscle twitching, and fidgety movements, including more focused movements, such as pulling lines and catheters. *Hypoactive delirium* has been proposed to characterize patients with less attention and paucity of movement as opposed to *hyperactive delirium*, which is characterized by increased attention, agility, and exaggerated response to a simple stimulus. Mixed forms are defined as a combination of hypoactive and hyperactive delirium. Studies have found hyperactive delirium to be far less common than hypoactive delirium or a mixed form. The so-called "hypoactive" form has also been called "quiet delirium" and remains one of the most problematic designations for any neurologist. One can also easily imagine a patient with so-called "hypoactive delirium" harboring a CNS infection, nonconvulsive status epilepticus, a new metabolic derangement such as hyperammonemia, or the major side effect of an administered drug. Questions arise about patients in a quiet delirium with facial twitching, eye deviation, or, in less common situations, cortical blindness (a common presentation in patients with posterior reversible encephalopathy syndrome, PRES). And who wants to miss abulia in an acute frontal lobe infarct?

Presenting clinical neurology findings systematically may be ideal but hardly realistic. Diagnosis is based on a number of key findings; complexity will stymie clinical reasoning. Use simple clarity, not only in terms of the deficit, but what it means for the patient. So describe the following patient with a cerebral hemorrhage as follows:

Mr. Bed #9 has a hemorrhage in the putamen extending into the temporal lobe on the left, and accordingly, he cannot speak in full sentences, has lost all sense of syntax, has no way of naming objects such as a wedding ring, and cannot perform a simple task such as pointing to the ceiling. He is alert and frustrated by that. He has a pyramidal paresis with more involvement of the extensor muscles of both his arm and leg and cannot support his weight. We cannot reliably know his sensory deficit.

If new, even transient, findings have occurred at night, try to describe them and explain them in the usual context.

Mrs. Bed #10 with a subarachnoid hemorrhage had a brief period of poor awakening during routine neuro-checks but improved with a fluid bolus. The resident found no evidence of a new hemiparesis or speech impediment, but she was quiet and mute, suggesting abulia from vasospasm in the anterior cerebral artery.

One way to communicate during rounds in a patient with an acute neurologic illness is shown in Table 12.2. Finally, it is not uncommon that trainees fail to recognize the problem at hand, and it is fully understandable. For a trainee to speak up and ask for a detailed clarification is unusual (and may take courage), but it is needed to understand the reasoning behind the diagnosis and treatment. Defense mechanisms (Table 12.3) are a common human behavior, but they should be recognized and avoided.

Communicating to Families

Every ICU has families to support, and this is an exceedingly important job. Families are at the bedside and often stay overnight. Their physical presence should not be ignored or minimized. Because most illnesses we see come suddenly,

Table 12.2 Communicating findings with rounds

Link your findings to the nature of the disease and location in brain or spine
Ask yourself if it makes sense
Characterize a finding and recognize a pattern
Doubt hearsay (and even established facts)
With any conundrum, find out if it is new or known in the literature
Admit you do not know (there's a high likelihood that more experienced colleagues will not know either)
Imagine why an unfavorable outcome would be a major issue for the patient
Summarize expectations in clinical trajectory
Summarize why a test can be useful and why a test would not change anything
Curiosity is the best virtue a good physician can hope for

Table 12.3 Communication defense mechanisms

"That's what I am thinking"
"Fair enough"
"I am not sure if there is more to do"
"I think that has been done"
"I haven't heard"

we should expect families of neurocritically ill patients to be shocked and struggling to understand the acute situation. It is hard to imagine having breakfast with your loved one and then at night find yourself making decisions about the level of care, mechanical ventilation, or a craniotomy. It is very overwhelming; many are flinching from this uncomfortable situation and very few are ready for a detailed exposé of their loved one's condition. Families may become rapidly exhausted. When we see them at our first encounter, we rarely see a well-composed, well-adjusted group of family members; the healthcare worker needs to empathize and provide a calming environment. A separate, preferably spacious and well-lighted room is needed for an introduction and preliminary conversation about the condition of the patient at arrival. Communication requires compassion. By compassion, we mean accountability, honesty, authenticity, and genuineness. The ability to show compassion is a personality trait. We all hope to have it and, if not, will try to develop it. Maturity may increase compassion, but some of us never lose the inclination to patronize. Phrases such as "I am fighting for the patient," "we're doing all we can," or "we just pulled him back from the brink" are not only inaccurate but inappropriate and aggrandizing. Families recognize and despise jargon and pontification. Do not be deceived into thinking otherwise. Moreover, patients are not fighters. They have to face an unrelenting disease or overwhelming injury, and we help them overcome it by adjusting to a new reality and to optimize recovery if such exist. If we concur with this martial terminology, must we then think that did the patient not fight hard enough if there is a bad outcome?

Adequate compassion includes coaching, assisting, and supporting families while they are enduring a very tough period. It is also time. If you cannot find the time, find another person to do it or postpone it. Palliative care services are specifically focused on doing this work and have moved from support in chronic terminal conditions to more acute situations, and the field has matured significantly. But to "call in a new team," after weeks of prior discussions with family members, might be totally counterproductive and perhaps even inappropriate. Neuropalliative care (specializing in the conditions discussed in this book) is new and finding its way and welcomed by many.

Encouraging a reasonable sense of hope (more than the proverbial glimmer) is also part of compassion. However, we cannot let a family's strong belief in miracles consume and restrict us; indeed, the occurrence of a so-called "miracle" actually may signal a serious error in assessment.

In our first encounter, we must provide factual information [30, 31]. Rather than focusing on one medical diagnosis, the neurologist should almost immediately explain the big picture. During these conversations, a treatment plan is formulated with a careful explanation of options. Patients often present with significant comorbidity, and families often know that. Many already have decided previously that one more event would be too much to handle for their loved one. Families may quickly bring up advanced directives, which are important for decisions on resuscitation status and any type of escalation of care. When all resuscitative measures are continued and aggressive care is preferred with going the extra mile ("Please do everything you possibly can"), it may lead to further decline of the patient's functional state, dramatic changes in the person's appearance, and enduring weakness. Generally, patients before a major brain injury are not the same after the brain injury. Opportunities for complete cures are few and far between.

We must always strive to be kind to the patient and family; clarity of speech and undivided attention to families' concerns are necessary components [32, 33]. Mutual respect between physician and family members requires recognition of diversity and cultural sensitivity. Failures on this front can quickly turn a cordial relationship into an embarrassment. In an increasingly diverse world, different cultures may have a completely different view on how communication should proceed. Many ethnic groups display a strong sense of obligation to their family members and seek to do everything possible.

In Asian cultures, nondisclosure to the patient of a poor prognosis is common. Discussion of the details of serious illness may be perceived as confrontational, disrespectful, robbing hope, or creating too much anxiety for the patient. Directness is often seen as cruel to the patient and perceived as doing more harm (which, of course, it does not). The elderly are highly respected and revered. Speaking positively is expected. Physicians tend to inform patients' family members rather than the patient, which allows families to decide what to say. Not telling the truth is unethical in western medical societies, and this includes intentional nebulosity.

Religions emphasize that life always has value even if there is suffering. For example, the Catholic Church has always advocated supporting life in devastating illness but to also support suffering families. On balance, physicians should respect all faiths, and their only task is to say words of comfort [34]. Religious leaders of various denominations can often help clarify misbeliefs families might hold regarding their traditions. Chaplains are crucial in helping families bring closure by relieving guilt related to their religious beliefs, and we often invite them to be present.

Barriers to communication, including difficult family dynamics, should be identified early. Resolution of conflicts can only come with taking the necessary amount of time with the family to explain the plan and expectations. If the patient's condition appears frankly hopeless but families are determined to persevere with aggressive care, several steps may defuse the situation. First, physicians should temper their enthusiasm about eventual recovery to discourage false hope. Second, clear limits should be set once a major intervention is undertaken. When there is no benefit, there should be no continuation of aggressive care. Third, physicians should clearly explain why there are no good solutions and nothing can be done. In my view, offering family members vague, platitudinous descriptions of the medical condition may lead to higher expectations and, later, inability to cope with a disappointment. Nevertheless, we should (1) acknowledge their uncertainty, (2) acknowledge that our early prediction is not infallible, and (3) be willing to acknowledge that the situation may change.

Some patients with a catastrophic injury to the brain remain comatose and have been in this state for weeks without any hint of improvement or awareness. When sleep-and-wake cycles emerge and eyes are open for considerable periods of time, a persistent vegetative state is considered and can be provisionally made in some patients. As we noted in Chap. 7, it is very rarely permanent or even persistent. We can endlessly debate whether the adjective is justified.

Whether the patient will be persistently uncommunicative and not aware comes up regularly in a family conference. The term "vegetative state" has been associated with "vegetable" but typically by families and other laypersons. "Vegetable" should never be used in professional conversations by physicians or other healthcare providers, and the word is an erroneous conjecture. In many European countries, the word "plant" is used, which is equally dismissive and disrespectful.

Families understand the concept of a vegetative state, although their understanding is generally much broader and include any uncommunicative patient requiring full, extensive, and prolonged nursing home care. But the diagnosis—which requires a prolonged focus of clinical attention to be certain—is rarely cemented in acute intensive care settings. It proves difficult to be certain in the acute stages with too many confounders and fluctuations in examination. Neurointensivists must have many personal examples of patients thought to be in a persistent vegetative state but then get an email or phone call from the nursing home their patient is now smiling and nodding. Neurorehabilitation specialists have created the term "minimally conscious state," but that remains a vague category with no well-defined boundaries and certainly no neurologic characteristics.

For us as diagnosticians, this is all the more reason to be tentative and self-critical. Brain function studies (fMRI) may reveal clinical shortcomings with evidence of activity in these patients. As discussed in Chap. 7, studies suggest that some patients clinically in a vegetative state may respond to familiar voices or show cortical activity when asked to do a task. These studies on very limited number of patients (due to its very low prevalence) have been criticized (they may not be in a vegetative state), are not widely available, and, even if the technology is available, require carefully set and tested protocols. They do not aid clinical practice, and their findings may further confuse families. These tests should therefore be discouraged, and neurologists and (neuro)intensivists should clarify how these tests can be misinterpreted as evidence of cognition.

More Reflections

How can we improve communication in a teaching center where consultants are more or less dependent on residents and fellows to "tell the story" and "make a plan?"

Residents and fellows often ask me for help in improving their communication techniques. Moreover, it is not uncommon for them to ask, "what more would you like to know about this patient?" Or I may ask, "what else can you tell me about the patient that I should know?" I have several suggestions.

1. *Reverse the order.* Try starting with your final diagnosis and then discuss how you arrived there. The listener's mind is better attuned to hearing the key findings first, followed by an abbreviated time line. While a few options can be mentioned, a full differential diagnosis takes too long and should be considered later.
2. *Extract the essence.* Vague descriptions (e.g., patients are unresponsive, seizing, thrashing around) are not helpful. Avoid digressions as well as unnecessarily complex, repetitive, and long-winded exposition. Cut to the chase! Most physicians have very short attention spans. Some information, while significant, may also not be immediately essential; for example, the symmetry or the absence of reflexes is not essential information (unless the patient has Guillain–Barré syndrome) and same with tone (unless the patient has a serotonin syndrome or exhibits dysautonomic storming). Communicating the most essential findings is a learned skill and remains the crux of the problem. Physicians should lean toward simple diagnoses and simple solutions but remain able to recognize something really unusual. But even then, avoid too much truncation. Recipients have all been victims of communication "curve balls"; that is, essential information not initially volunteered but only elicited later in response to questioning (Table 12.4).
3. *Avoid numbers that lack specifics.* "This patient was found with a GCS of 5 and NIHSS of 10." Unsurprisingly, scales are too simplistic to be informative or have little clinical applicability (Chap. 2). Moreover, some scores may hide important information, and numbers are only useful with clinical trial statistics. Scales and sum scores have

Table 12.4 Communication curve balls

Failure to mention anticoagulation or thrombocytopenia
Failure to mention timing of CT scan
Failure to mention hypotension
Failure to mention fever
Failure to mention airway compromise

inflicted serious harm on how we communicate the results of the neurologic examination and neuroimaging. "The CT scan shows no bleed" is not an appropriate description. Instead, tell me what you looked for; for example, "the CT scan shows no hyperdense MCA or basilar artery sign, the basal cisterns are open, the ventricles are normal in size, and there are no fractures."

4. *Avoid language we cannot understand.* Some vagueness can easily slip in. Becoming articulate takes time. We cannot expect superior knowledge of the neurologic examination and expert grasp of the meaning of clinical patterns from those inexperienced in assessing acutely ill neurologic patients. Conversely, inexperienced physicians should not replace lack of knowledge with misfired guessing.
5. *What do I need to know for the night?* One of the major paradoxes of neurocritical care is that the talkative patient sitting in a chair after a ruptured cerebral aneurysm and ventriculostomy is, in actuality, critically ill. Critical illness is not defined by systemic criteria but by the high likelihood the patient may deteriorate quickly from consequences of the initial injury—the second wave of neurologic injury, so to speak. Summarize the examination and what it would look like when deterioration occurs.
6. *Which studies should be repeated and when?* Which pending studies need review, and when are they expected? Electronic devices can set up alerts from electronic medical records in some institutions, but we still need to know when to expect the study. What are the potential expectations? Watch for new blossoming contusions, enlarging subdural hematoma, worsening hydrocephalus, swelling of the cerebellum in the tight subtentorial fossa—to name a few.
7. *What is the threshold to call a consultant?* Very few consultants routinely remain in the ICU overnight; night is covered by residents and fellows. With standard staffing, a historical paradigm in the unit is an intensivist onsite during the daytime and taking calls from home at night, returning to ICU as deemed necessary. Shift work may not be the best solution. In fact, avoidance of shift work is the most common reason for retirement among senior emergency medicine physicians [35]. Easy remote access to patient data has reduced the need to be onsite. When asked by fellows about my preferences for being called, without being unnecessarily amusing, I share my criteria—call me 5 minutes before you get nervous and do not let me miss the excitement. Consultants taking call should be thought of as 911 operators who answer the phone immediately day and night. I cannot promise anything less while on call. I am on call for a reason: to help when I can and to have a two-way conversation. These directives have always been understood and worked very well.
8. *Text or email?* We should not email if we can text. E-mails should not be time-sensitive or require a response within a few hours. Texts, on the other hand, should be considered urgent; a breezy subject line such as "heads up" is inappropriate and misleading about the urgency.
9. *Is your "listener" actually listening?* We are not expecting gimlet-eyed listeners, but some behaviors clearly show that your listener is, indeed, not listening. These include looking at the phone, an unfocused gaze, or mentioning another, imminent commitment (e.g., "Sorry, I have to run to another meeting," "I have to take this call"). Learn to recognize perfunctory attention from someone who is preoccupied or focused on something else. (We all find ourselves doing exactly that.)
10. *Do you become easily annoyed?* Can you avoid being dismissive or overwrought? Do you recognize loss of resilience? Much may go unrecognized, but burnout manifests itself first in communication and is notable for abrasive, snarky remarks. Dark sarcasm and compassion fatigue follow. It is eventually destructive and adversely affects care and professionalism. Work accompanied by moral distress, incivility, and conflict among

colleagues are important drivers of burnout. One European multinational landmark ICU study found personal animosity, mistrust, and communication gaps to be the most common interprofessional conflict-causing behaviors [36].

The perfect note in the medical record may tell it all, but there is no substitute for a face-to-face conversation [37]. Once we have completed a detailed neurologic and physical examination, we should know what we are facing and what might happen later. Ropper famously asked two crucial questions in "How to determine if you have succeeded at neurology residency." He provocatively asked, "Can you present a case to an intelligent colleague in 2 minutes?" and "Can you tell who is sick?" [38] We all should continue to hone our interpersonal communication skills, and the aforementioned fundamentals can help to change a culture of fragmented communication. It is recognizing what is important (and what is not) and how to avoid wasting time. We must strive to present in a manner that will lead to asking the right questions. However, as a whole, better communication comes with better knowledge and knowing what your listener wants or needs to hear. Ideally, the communicator is logically succinct, and the listener asks answerable questions. Every complex problem can be easily summarized in a few sentences with training. Despite significant improvement over many years, recognizing (and reprimanding) microaggressions, racist implications, and mansplaining (Table 12.5), it is a sad indictment of our profession that we have learned to live with disinformation, delayed information, or no information at all (and even no patient identity). Some physicians create contradictions resulting in incoherence. Many spare words. Some give a cold shoulder. Some do not recognize communication is a reciprocal action and easily blame others for not informing them or overreact to the misinformation. In the business world, communication is an industry with numerous self-help books. It involves skills of persuasion and how to trade ideas and build an image. In medicine, we must communicate about the patient in the most effective way and to offer direction. We always need to prioritize process above personalities (even if we know too well the two are interrelated). Moreover, communication may break down to a resident level with staff rarely bothered except for triage matters. Arguably, there is some need for research into how communication affects management. Only with those data can communication be taught, learned, and practiced. Our greatest fallacy may be the belief that we always know what is going on. My mnemonic was not borne out of exasperation but inspiration to do a bit better. We must recognize that in certain cases (and with a large number of involved consultants and their teams), we may not see the forest for the trees. Those of us working in ICUs are always trying to find better means of communication through process and quality control. But, despite all these concerns, hospital practices work quite well in partnerships and glitches are far and between. It is not an "omnishambles." There is a lot to be proud of. When communication works, the responsible team is effective, and the patient does well, it is immensely gratifying.

Table 12.5 Good news about communication

Better available medical records (in fact everywhere where there is an Internet connection)
Standards and criteria through order sets
Improved language support through online services
The recognition of the importance of nonverbal communication
The recognition of unconscious bias
The recognition of microaggressions and mansplaining

Pointers and Takeaways

- Neurologic examination is meaningless if we cannot adequately communicate.
- Neurologic examination should include instability of vital signs, particularly if they are related.
- It is fallacious conjecture to assume we are generally well-informed and over-informed and need not worry too much about it.
- Communicating the neurologic picture must involve a clinical course and how this period was managed (or not).
- The mnemonic TELL ME can help the physician to focus on the bare essentials of the

patient's condition but also what will be likely expected in the coming hours and days.
- Communication with families is essential to create trust.
- Communication assumes reciprocity.
- Good communication of clinical signs should improve patient care (and universal patient and family satisfaction, if they are assured the attending team of healthcare professionals is well informed).

References

1. Firdouse M, Devon K, Kayssi A, Goldfarb J, Rossos P, Cil TD. Using texting for clinical communication in surgery: a survey of academic staff surgeons. Surg Innov. 2018;25:274–9.
2. Gellert GA, Conklin GS, Gibson LA. Secure clinical texting: patient risk in high-acuity care. Perspect Health Inf Manag. 2017;14:1d.
3. Goyder C, Atherton H, Car M, Heneghan CJ, Car J. Email for clinical communication between healthcare professionals. Cochrane Database Syst Rev. 2015;(2):CD007979.
4. Rokadiya S, McCaul JA, Mitchell DA, Brennan PA. Leading article: use of smartphones to pass on information about patients - what are the current issues? Br J Oral Maxillofac Surg. 2016;54:596–9.
5. Fogerty RL, Rizzo TM, Horwitz LI. Assessment of internal medicine trainee sign-out quality and utilization habits. Intern Emerg Med. 2014;9:529–35.
6. Sorokin R, Riggio JM, Hwang C. Attitudes about patient safety: a survey of physicians-in-training. Am J Med Qual. 2005;20:70–7.
7. Dutra M, Monteiro MV, Ribeiro KB, Schettino GP, Kajdacsy-Balla Amaral AC. Handovers among staff intensivists: a study of information loss and clinical accuracy to anticipate events. Crit Care Med. 2018;46:1717–21.
8. Cohen MD, Hilligoss PB. The published literature on handoffs in hospitals: deficiencies identified in an extensive review. Qual Saf Health Care. 2010;19:493–7.
9. Gandhi TK. Fumbled handoffs: one dropped ball after another. Ann Intern Med. 2005;142:352–8.
10. Riesenberg LA, Leitzsch J, Massucci JL, et al. Residents' and attending physicians' handoffs: a systematic review of the literature. Acad Med. 2009;84:1775–87.
11. Salerno SM, Arnett MV, Domanski JP. Standardized sign-out reduces intern perception of medical errors on the general internal medicine ward. Teach Learn Med. 2009;21:121–6.
12. Wayne JD, Tyagi R, Reinhardt G, et al. Simple standardized patient handoff system that increases accuracy and completeness. J Surg Educ. 2008;65:476–85.
13. Coon EA, Kramer NM, Fabris RR, et al. Structured handoff checklists improve clinical measures in patients discharged from the neurointensive care unit. Neurol Clin Pract. 2015;5:42–9.
14. Moseley BD, Smith JH, Diaz-Medina GE, et al. Standardized sign-out improves completeness and perceived accuracy of inpatient neurology handoffs. Neurology. 2012;79:1060–4.
15. Coughlin DG, Kumar MA, Patel NN, Hoffman RL, Kasner SE. Preventing early bouncebacks to the neurointensive care unit: a retrospective analysis and quality improvement pilot. Neurocrit Care. 2018;28:175–83.
16. Murray NM, Joshi AN, Kronfeld K, et al. A standardized checklist improves the transfer of stroke patients from the neurocritical care unit to hospital ward. Neurohospitalist. 2020;10:100–8.
17. Rojas JC, Lyons PG, Jiang T, et al. Accuracy of clinicians' ability to predict the need for intensive care unit readmission. Ann Am Thorac Soc. 2020;17:847–53.
18. Haig KM, Sutton S, Whittington J. SBAR: a shared mental model for improving communication between clinicians. Jt Comm J Qual Patient Saf. 2006;32:167–75.
19. Kotsakis A, Mercer K, Mohseni-Bod H, Gaiteiro R, Agbeko R. The development and implementation of an inter-professional simulation based pediatric acute care curriculum for ward health care providers. J Interprof Care. 2015;29:392–4.
20. Ozekcin LR, Tuite P, Willner K, Hravnak M. Simulation education: early identification of patient physiologic deterioration by acute care nurses. Clin Nurse Spec. 2015;29:166–73.
21. Panesar RS, Albert B, Messina C, Parker M. The effect of an electronic SBAR communication tool on documentation of acute events in the pediatric intensive care unit. Am J Med Qual. 2016;31:64–8.
22. Lane-Fall MB, Pascual JL, Massa S, et al. Developing a standard handoff process for operating room-to-ICU transitions: multidisciplinary clinician perspectives from the handoffs and transitions in critical care (HATRICC) study. Jt Comm J Qual Patient Saf. 2018;44:514–25.
23. Lane-Fall MB, Pascual JL, Peifer HG, et al. A partially structured postoperative handoff protocol improves communication in 2 mixed surgical intensive care units: findings from the handoffs and transitions in critical care (HATRICC) prospective cohort study. Ann Surg. 2020;271:484–93.
24. Riesenberg LA, Leitzsch J, Little BW. Systematic review of handoff mnemonics literature. Am J Med Qual. 2009;24:196–204.
25. Artis KA, Dyer E, Mohan V, Gold JA. Accuracy of laboratory data communication on ICU daily rounds using an electronic health record. Crit Care Med. 2017;45:179–86.
26. Segall N, Bennett-Guerrero E. ICU rounds: "what we've got here is failure to communicate". Crit Care Med. 2017;45:366–7.

27. Cao V, Tan LD, Horn F, et al. Patient-centered structured interdisciplinary bedside rounds in the medical ICU. Crit Care Med. 2018;46:85–92.
28. Ingram TC, Kamat P, Coopersmith CM, Vats A. Intensivist perceptions of family-centered rounds and its impact on physician comfort, staff involvement, teaching, and efficiency. J Crit Care. 2014;29:915–8.
29. Oldham MA, Flaherty JH, Maldonado JR. Refining delirium: a transtheoretical model of delirium disorder with preliminary neurophysiologic subtypes. Am J Geriatr Psychiatry. 2018;26:913–24.
30. Azoulay E, Sprung CL. Family-physician interactions in the intensive care unit. Crit Care Med. 2004;32:2323–8.
31. Fassier T, Azoulay E. Conflicts and communication gaps in the intensive care unit. Curr Opin Crit Care. 2010;16:654–65.
32. Truog RD. Patients and doctors – evolution of a relationship. N Engl J Med. 2012;366:581–5.
33. Wijdicks EFM, Rabinstein AA. The family conference: end-of-life guidelines at work for comatose patients. Neurology. 2007;68:1092–4.
34. Bulow HH, Sprung CL, Baras M, et al. Are religion and religiosity important to end-of-life decisions and patient autonomy in the ICU? The Ethicatt study. Intensive Care Med. 2012;38:1126–33.
35. Hall KN, Wakeman MA, Levy RC, Khoury J. Factors associated with career longevity in residency-trained emergency physicians. Ann Emerg Med. 1992;21:291–7.
36. Azoulay E, Timsit JF, Sprung CL, et al. Prevalence and factors of intensive care unit conflicts: the conflicus study. Am J Respir Crit Care Med. 2009;180:853–60.
37. Coleman C, Gotz D, Eaker S, et al. Analysing EHR navigation patterns and digital workflows among physicians during ICU pre-rounds. Health Inf Manag. 2020:1833358320920589.
38. Ropper AH. How to determine if you have succeeded at neurology residency. Ann Neurol. 2016;79:339–41.

Index

Italic page numbers refer to figures and **Bold** page numbers refer to tables.

E. F. M. Wijdicks, *Examining Neurocritical Patients*, https://doi.org/10.1007/978-3-030-69452-4

9783030694548